Urban Sustainability

The Urban Sustainability Book Series is a valuable resource for sustainability and urban-related education and research. It offers an inter-disciplinary platform covering all four areas of practice, policy, education, research, and their nexus. The publications in this series are related to critical areas of sustainability, urban studies, planning, and urban geography.

This book series aims to put together cutting-edge research findings linked to the overarching field of urban sustainability. The scope and nature of the topic are broad and interdisciplinary and bring together various associated disciplines from sustainable development, environmental sciences, urbanism, etc. With many advanced research findings in the field, there is a need to put together various discussions and contributions on specific sustainability fields, covering a good range of topics on sustainable development, sustainable urbanism, and urban sustainability. Despite the broad range of issues, we note the importance of practical and policy-oriented directions, extending the literature and directions and pathways towards achieving urban sustainability.

The series will appeal to urbanists, geographers, planners, engineers, architects, governmental authorities, policymakers, researchers of all levels, and to all of those interested in a wide-ranging overview of urban sustainability and its associated fields. The series includes monographs and edited volumes, covering a range of topics under the urban sustainability topic, which can also be used for teaching materials.

Vinay Kandpal · Tanupriya Choudhury
Editors

Blue Revolution: Sustainable Water Use and Circular Economy Innovations

Springer

Editors
Vinay Kandpal
Department of Management Studies
Graphic Era Deemed to be University
Dehradun, Uttarakhand, India

Tanupriya Choudhury
School of Computer Science
University of Petroleum and Energy Studies
Dehradun, Uttarakhand, India

ISSN 2731-6483 ISSN 2731-6491 (electronic)
Urban Sustainability
ISBN 978-981-95-1915-6 ISBN 978-981-95-1916-3 (eBook)
https://doi.org/10.1007/978-981-95-1916-3

This Springer imprint is published by the registered company Springer Nature Singapore Pte Ltd.
The registered company address is: 152 Beach Road, #21-01/04 Gateway East, Singapore 189721, Singapore

Preface

Water is much more than a resource—it is the basis of life, an impetus for development, a symbol of purity, and increasingly a driver of innovation. Today, with water-linked challenges becoming both more urgent and more complex—from climate-induced scarcity to urban consumption, from governance to pollution—the publication represents a timely and interdisciplinary addition to the global conversation about sustainable futures for water.

This edited volume has contributed to the multi-dimensional growth of water-related scholarship by inviting scholars, professionals and research participants across disciplines and geographic locations to engage with the evolving arena of water management across three integrative themes: Circular Economy and Community Water Solutions, Technological Innovations and Digital Transformation, Finance, Policy, and Socio-Spiritual Dimensions of Water. Each theme has integrated a call for embracing holistic and systems-leveraging management of water solutions that blend ground-up initiatives, innovative technologies, financial capital, and culture.

Each chapter of this volume presents a wide-ranging array of themes with implications for water decision-making and management, including smart cities and blockchain water tracking, microalgae as a waste-to-value solution, and the spiritual aspect as it pertains to cultural traditions of water. Policy mechanisms have emerged, including blue bonds, and global case studies can be found in communities spanning from India to the Democratic Republic of Congo. New tools, such as AI and bibliometric analyses, are being brought to bear on understanding the paths forward for research and policy endeavors.

This volume is unique in its attempts to wade into the tension between sound research, application in the "real world", and cultural aspects undergoing water; this in itself reflects the collaborative spirit of the authors and editors. We hope to advance knowledge and understanding while also pushing the limits for practitioners, students, and policymakers to utilize in innovative thinking and praxis.

A very special thank you to all of our authors—we have nothing without your contributions. Thank you to all evaluators who provided their ordering suggestions, and a thank you to our academic zones for unwavering support and encouragement.

Lastly, thank you to readers—students, scholars, architects of policy, or stewards of water—and it is our intention (in part) to find knowledge and inspiration in our pages.

As the world stands at the threshold of environmental responsibility, we trust that this volume will contribute to rethinking water as our common inheritance, public trust, and collective responsibility.

Thanks,

Dr. Vinay Kandpal
Professor, Department of Management Studies
Graphic Era (Deemed to be University)
Dehradun, India

Dr. Tanupriya Choudhury
Professor, School of Computer Science
University of Petroleum and Energy Studies (UPES)
Dehradun, India

Acknowledgements

We, the editors, thank everyone who contributed to the conception, execution, and publication of this edited volume. First, we would like to express our gratitude to the co-authors for their quality research and collaborative team-working attitudes. The thoughtful and diverse contributions have created the heart of this book, enriching the volume across disciplines, geographies, and sectors. We would like to acknowledge our institutions, Graphic Era (Deemed to be University), Dehradun & University of Petroleum and Energy Studies (UPES), Dehradun, for their unwavering support, encouragement, and research-friendly environments, which allowed us to complete this work.

In addition, we would like to express our appreciation to the editorial and production team at Springer Nature for guidance, professionalism, and support during the publication process. With a commitment to scholarly excellence and innovative publishing, the journey of development was peerless due to the nature of collaboration.

We also would like to thank the anonymous reviewers and academic peers who provided feedback, constructive criticism, and support to ensure all chapters achieved the best level of scholarly integrity and thematic coherence. Finally, we would like to acknowledge our families, colleagues, and mentors for their constant positivity and our readers: researchers, students, practitioners, and policymakers who will adopt this work into their own. We look forward to conversations and actions in sustainable water management that emerge from this book.

Yours sincerely,

Dr. Vinay Kandpal
Professor, Department of Management Studies, Graphic Era (Deemed to be University), Dehradun, India

Dr. Tanupriya Choudhury
Professor, School of Computer Science, University of Petroleum and Energy Studies (UPES), Dehradun, India

Contents

About the Editors

Dr. Vinay Kandpal, a distinguished Professor at the Department of Management Studies, Graphic Era (Deemed to be University), Dehradun, Uttarakhand, India, has over 19 years of experience in education and business. Holding a D.Litt. and a Ph.D. in Finance from Kumaun University, India, along with a Postdoctoral Fellowship from PG Administracao-Paulista University (UNIP), Brazil, Dr. Kandpal has made significant contributions to the field of finance and business management. He has authored more than 81 scholarly articles and 15 books covering topics like sustainable finance, financial technology, smart urban development, risk management, circular economy, and financial inclusion.

Dr. Tanupriya Choudhury is a highly accomplished individual with an impressive academic and professional background. He holds a bachelor's degree in CSE from West Bengal University of Technology, Kolkata, India; a master's Degree in CSE from Dr. MGR University, Chennai, India; and a Ph.D. degree, which he received in 2016. In addition to his educational qualifications, Dr. Choudhury has fourteen years of experience teaching and conducting research related to Human Computing, Soft Computing, Cloud Computing, Data Mining, etc., making him an ideal candidate for the Professor position at UPES Dehradun's School of CS Cluster of Informatics. Recently, Dr. Tanupriya was honored with the Global Outreach Education Award for Excellence, the best young researcher award 2018 (GOECA). This prestigious award recognizes his contribution towards advancing knowledge within this field through various domestic and international conferences, resulting in over 250 published papers since then! Furthermore, he has filed 25 patents to date and received 26 copyrights from the MHRD due to the development of his own software programs. His achievements speak volumes about his dedication to furthering technology-related research while inspiring others around him by demonstrating that hard work leads to success! He has been associated with many conferences throughout India as a TPC member, session chair, etc. He is a lifetime member of IETA, a senior member of IEEE, and a member of IET(UK) and other renowned technical societies. He is associated with Corporate and a Technical Adviser of Deetya Soft Pvt. Ltd., Noida,

IVRGURU, Mydigital360, etc. He holds the post of Honorary Secretary in IETA (Indian Engineering Teachers' Association-India). He is also holding the Senior Advisor Position in INDO-UK Confederation of Science, Technology and Research Ltd., London, UK, and the International Association of Professional and Fellow Engineers-Delaware-USA.

Integrating Circular Economy Principles in Higher Education: Pathways to Sustainable Transformation

Chahat Sahani, Navneet Rawat, Navya Bhat, and Vinay Kandpal

Abstract The circular economy transition stands as a vital method to reach sustainable development goals because of worsening climate issues and diminishing natural resources. Higher education institutions (HEIs) serve as key drivers for this transition because they create knowledge while promoting innovation and behavioral change. This research investigates how higher education institutions implement circular economy principles through recent global developments and institutional strategies that advance circular thinking in curriculum design, campus operations, research activities, and community partnerships. The chapter draws from current research to show how HEIs function as living laboratories for circular innovation through interdisciplinary teaching, green infrastructure, sustainable procurement, and student-led initiatives. The chapter also examines how digital technologies, including artificial intelligence, Internet of Things (IoT), and data analytics, serve as drivers for integrating circular practices into university systems. The chapter identifies obstacles to implementation, such as policy misalignment, funding constraints, and institutional inertia, and provides actionable recommendations for stakeholders.

Keywords Circular economy · Higher Education · Sustainability · Education for Sustainable Development (ESD) · Green campus · Curriculum innovation · Digital technologies · Artificial Intelligence · Sustainable Development Goals (SDGs)

1 Introduction

Circular economy (CE) has progressed from a niche sustainability activity to a structural strategy in the world climate action and resource efficiency action plans. The CE is identified as the priority strategy to reduce waste, manage to maximize resource

C. Sahani (✉) · N. Rawat · N. Bhat · V. Kandpal
Department of Management Studies, Graphic Era Deemed to Be University, Dehradun, India
e-mail: chahatsahani@geu.ac.in

N. Bhat
e-mail: navya.bhat@geu.ac.in

V. Kandpal and T. Choudhury (eds.), *Blue Revolution: Sustainable Water Use and Circular Economy Innovations*, Urban Sustainability,
https://doi.org/10.1007/978-981-95-1916-3_1

productivity and support continuous material use by the United Nations Sustainable Development Goals (SDGs) and Goal 12: Responsible Consumption and Production, in particular [48]. The role of higher education institutions (HEIs) in implementing standards and creating mainstreaming of CE principles should not be overlooked because higher education institutions have a key role in controlling knowledge and becoming sites of innovation acceleration [33]. This chapter examines how HEIs integrate CE thinking in the development of curriculum, university campus management, research practices, and involvement with the communities, and how the HEIs in this way bring the education sector to regenerative education models.

Institutions of higher learning are shifting in terms of curriculum development to incorporate CE or thematic-based curricula rather than disciplinary learning. This instructional change is increasing the systems-thinking abilities of students so they can solve challenging sustainability issues. Prieto-Sandoval et al. [37] note that adequate CE education should unify the disciplines of environmental science, economics, design thinking, and systems thinking to train graduates who will be able to create circular solutions. Engaging students in experiential types of learning, such as lifecycle assessment, circular product design, and sustainability audit, is being employed to actively engage students in the pedagogical process and develop environmental citizenship [41]. The CE paradigm has been increasingly disseminated across all academic areas, and it currently applies to classes outside the fields of engineering and environmental sciences [1, 17], including business, law, and social sciences. Outside of the classroom, numerous universities operate in circular ways, involving zero-waste programmes, sustainable procurement, composting, and reuse programmes, and hence the CE can be observed and learned through on-campus experiences [12, 32].

One of the contributors to this transformation is digital technology. Due to the introduction of AI, big data analytics, and the Internet of Things (IoT), university institutions are using the resources efficiently, reducing energy and water consumption, and implementing smart waste management processes. With the help of such tools, it is possible to create intelligent campuses that will not only be environmentally friendly but will also allow student-driven research and CE experiments to have a full source of data [28]. In addition, digital environments also allow scholars and researchers all over the world to collaborate globally, improve knowledge exchanges, and construct transdisciplinary CE learning ecosystems [3, 43]. Along with these innovations, there remain some systematic inhibitions. The inertial impact, lack of funding, obsolete policies, and countering organizational change may be the factors that never allow the more profound integration of CE principles in HEIs [15]. The absence of industry standards in measuring CE also makes it hard to evaluate satisfactorily incorporated progress across universities [50, 49]. This chapter matches the larger dream of the Blue Revolution that stresses the sustainable management of water and resources via advancement. As a way of promoting this vision, the implementation of CE in higher education is developing students who are sustainability-minded, water-literate, and empowered to adopt circular solutions within and across the sectors [47]. The transformation can be instigated by HEIs that adopt the Education for Sustainable Development (ESD) framework and make sustainability a technical goal and a cultural norm. The chapter presents practical recommendations on what curriculum developers, academic administrators, policy development agents, and facility managers can do to address the issue of barriers to implementation. These are enhancing interdepartmental cooperation, incorporating CE into the institutional ratings and accreditation systems, awarding student-led engagement, and establishing partnerships with industry and local governments to co-design solutions. HEIs must become agents of change in the post-linear future by raising a new generation of leaders who will not only adapt to sustainable practices but also build them. Higher education can lead the way to achieving a new, regenerative, and truly circular economy through coordinated educational, operational, and digital initiatives.

1.1 The Evolving Role of Higher Education in Circular Economy Transitions

The circular economy (CE) represents a transformative approach that moves away from traditional linear "take-make-dispose" operations to establish a regenerative system that minimizes waste while maximizing resource efficiency and material recycling. Higher education institutions (HEIs) now play a vital role in implementing circularity throughout society, although the circular economy has already gained substantial momentum in policy and industry sectors. Higher education institutions function as innovation centers, research hubs, and critical thinking platforms to facilitate the transition toward sustainable circular economic systems [22, 44]. Universities must lead the transformation of sustainable futures because they function as centers for information generation and skill development. The global academic discourse and institutional initiatives and sustainability frameworks of universities have experienced significant growth in incorporating CE principles during the last few years. Higher education institutions function as active social and environmental change agents through their implementation of CE principles in their strategic development plans and institutional goals. Higher education institutions function as "living laboratories" because they directly apply theoretical knowledge through immediate experimental and innovative practices during CE transitions. Higher education institutions are transforming their instructional paradigms by adopting cross-disciplinary circular thinking as part of their educational approach. Educational institutions now support interdisciplinary learning through CE programs, which unite engineering with environmental science and design, and economics and social studies under one general framework while breaking down traditional subject boundaries.

The curriculum changes include project-based learning, sustainability hackathons, and partnerships with local businesses and government agencies, which provide students with practical problem-solving abilities. Higher education institutions gain a stronger influence in promoting CE adoption through collaborative efforts and partnerships. Various institutions form partnerships with governmental organizations, nonprofits, and private businesses to evaluate circular innovations at the regional and national levels. A review of 77 articles considered the role of the HEIs in promoting the circular economy. A study investigates the way universities can function as catalysts with the industry and government, through education, research, and linkages [8]. Higher education institutions shape policy development and community conduct by working with outside partners to establish circular practices that extend past their campus limits. The transition to circular economy requires a learning process that demands new attitudes, values, and skills beyond technological and financial aspects. Higher education institutions transform from information repositories into active change agents to establish a circular future.

The global sustainability landscape depends on higher education institutions because they connect education with research, innovation, and societal engagement. The responsibility to promote resilient, regenerative, inclusive systems for future generations directly influences their capacity to advance such systems. A transition at a systemic level is needed to achieve greater circularity in plastics production and consumption, and advance progress toward SDG12. Key barriers linked to data-information-knowledge maps interact with and are embedded in policies and standards, market and technology, socio-cultural norms, networks, and business models. Signals are more visible in the packaging sector, and institutions supporting processes of information sharing, such as labelling and certification, are more advanced [23].

1.2 Integrating the Circular Economy in Curriculum, Campus Operations, and Research

Higher education institutions' (HEIs') application of circular economy (CE) principles extends beyond philosophy to include actual implementations in institutional research, campus operations, and curriculum design. Higher education institutions are becoming more aware of their responsibility to educate future professionals who can understand and implement circularity in various fields. Changes to the curriculum are being made to incorporate CE frameworks, lifecycle thinking, and sustainable innovation into undergraduate and graduate programs. These programs prepare students for new green employment opportunities and circular industries, aligning with the global movement toward sustainability competencies [42].

The circular economy teaching at TU Delft, University of Exeter, and Arizona State University demonstrates global momentum for curricular reform through innovation studios and CE-focused degree tracks. The integration of CE concepts in the curriculum uses interdisciplinary approaches that link environmental science with economics, engineering, design, and policy. The adoption of project-based learning and transdisciplinary modules enables students to gain practical experience with real-world sustainability problems while proposing circular solutions for product design, supply chain optimization, and waste minimization [19]. Higher education institutions have started to implement circular approaches throughout their campus operations, which extend beyond traditional classroom activities. The controlled environment of university campuses enables the testing and practical implementation of CE models, including closed-loop water systems, energy-positive buildings, sustainable food procurement, and zero-waste systems. The operational strategy transformation through campus repair cafés, materials reuse centers, and plastic-free dining halls serves as the best example of circular change [45]. The institutionalization of green procurement practices leads colleges to select products that are either recyclable, biodegradable, or made from renewable resources. Research centers dedicated to circularity have emerged as vital entities for knowledge development and sharing. Various institutions have established living labs and circular hubs that collaborate with external stakeholders, including local governments and businesses, to develop solutions through experimentation with waste valorization, bio-based materials, and digital tools for circularity [24].

Urbanization is increasing, creating pressure on waste management infrastructure. As a result, Living Labs have been identified as a promising mechanism to support building collaborations between citizens, local authorities, private operators, and associations. We still, however, do not have adequate infrastructure for waste sorting and are unable to coordinate all initiatives to realize fully circular cities [10]. These initiatives not only improve research outputs but also provide pathways for commercializing sustainable innovations. Research initiatives further enhance the CE integration in HEIs by promoting innovation, policy development, and industry engagement. Furthermore, HEIs are increasingly involved in CE research with industry partners and communities. Students and scholars are actively involved in circular practices outside of school through service-learning initiatives, internships, and knowledge-transfer partnerships. The research shows that this increased participation ensures that HEI contributions to circularity are not in isolation but are part of larger societal processes. The concept of universities as civic organizations that drive regional sustainability agendas supports this type of extended CE education, which promotes systemic change. The successful implementation of CE throughout operations, research, and curriculum demonstrates complete institutional dedication to sustainability while transforming higher education from knowledge transmission to knowledge transformation to develop future leaders who will direct regenerative transitions.

1.3 Digital Technologies as Enablers of Circular Innovation in Higher Education

Higher education institutions (HEIs) face modifications in their strategic environment because of digital transformation, which includes sustainability integration and circular economy (CE) implementation. The new technologies of artificial intelligence (AI), the Internet of Things (IoT), blockchain, and data analytics offer new opportunities to simplify circular practices in campus operations, academic curriculum, and research activities. The technologies function as speedup tools that enable HEIs to implement and expand circular methods throughout their systems. Digital innovation establishes itself as a core foundation that supports circular efforts, planning, execution, and oversight during college transformations into smart, sustainable campuses. The Internet of Things (IoT) enables the connection between digital intelligence and physical infrastructure. Real-time monitoring of waste production, water use, and energy consumption is made possible via sensors and smart meters placed throughout campus buildings. The data-driven insights from these systems enable preventive maintenance and sustainable asset management, which helps infrastructure operations follow circular principles [31]. Some HEIs are currently researching digital twins as virtual representations of real systems to create circular scenarios for buildings, labs, and logistical systems for experiential learning and planning purposes. The implementation of blockchain technology now promotes transparency and accountability in university sustainability reporting. Blockchain technology enables circular procurement and recycling procedure monitoring through secure decentralized tracking of waste management and supply chain transactions [27]. The verification of sustainable sourcing and end-of-life processing plays a vital role in campus supply chain ecosystems because it establishes legitimacy and ensures compliance. The monitoring of circularity indicators together with institutional coordination of efforts through international frameworks such as the Sustainable Development Goals (SDGs) requires data analytics and visualization technologies. Higher education institutions use dashboards to monitor their carbon footprint metrics, together with circular material usage rates and environmental benefits generated by the reuse initiative.

Circular economy (CE) is a term that has actually existed since the 1970s, and has been of increasing relevance in recent years, partly due to natural resources being mined from the environment at a rate exceeding that which can be replaced, and changes in consumer behavior. Emerging technologies such as the Internet of Things (IoT) and big data have the ability to enable organizations and society to take advantage of CE concepts, as they become more prominent in everyday life [34]. Digital tools enable students to become more involved and creative in CE. Through online incubators, MOOCs, and virtual labs, students from any location can access circular economy education. Student-led hackathons and open innovation competitions conducted on digital platforms drive the development of circular solutions while providing practical learning experiences. Technological empowerment through democratized knowledge creates a new generation of circular economy leaders who possess digital technology skills [46]. Digital technologies serve as effective tools to embed CE into the core operations of institutions of higher learning. The appropriate use of these tools enables HEIs to transform their operational methods, teaching practices, and innovation processes, which lead to the development of circular campuses that combine resilience and efficiency with inclusivity and future readiness.

1.4 Barriers and Enablers to Circular Economy Implementation in Higher Education

The development of effective interventions and enabling strategies requires knowledge about systemic, institutional, and operational factors that support or block higher education institutions (HEIs) from adopting circular economy principles. The practice of sustainability implementation faces challenges despite university declarations of commitment because of multiple obstacles. The implementation of CE principles faces obstacles because national and institutional policies lack sufficient circularity value and innovation incentives for teaching and campus management. The sustainability goals of HEIs remain general without detailed strategies to implement circular economy principles throughout their operational and research activities and educational programs. The primary barrier to innovation stems from institutional inertia, which perpetuates established practices and hierarchical structures, thereby avoiding change. Traditional academic silos work against the essential need for cross-disciplinary cooperation which circular thinking requires. The lack of proper training and information about CE concepts prevents administrative and faculty personnel from implementing these concepts in their departments and courses [21].

Sustainable Development (SD) is the universal imperative of our time, and universities must play an important role in operationalizing this through policies and practices. In addressing Sustainable Development goals, universities should adopt a global view while still remaining locally grounded in their approach. In the past decade, many theories and recommendations about how to integrate sustainability have been developed and adopted globally for higher education systems. In India, it is compulsory for environmental education to be included at all formal levels of the education system. However, it remains a strong argument that it is much more than simply a matter of changing syllabi and curricula. Rather, universities should be at the forefront of incorporating SD goals into their research and operations [6].

Barriers in Behavior and Culture

The implementation of the circular economy (CE) faces resistance from cultural traditions about teaching methods and university administrative structures. Teachers believe they lack both time and resources to transform their curricula, and they consider CE to be beyond their academic domain. The lack of circular concepts in orientation programs, elective courses, and extracurricular activities makes it difficult for students to become aware of them [2]. The main obstacle to CE value dissemination exists in different academic fields. The applications of circular economy extend across business, humanities, arts, and social sciences, even though it is commonly linked to engineering and environmental sciences. The narrow definition of CE creates a risk of excluding stakeholders who could contribute significantly to circular transitions through ethical, design, or behavioral perspectives.

Key Enablers of Circular Economy in Higher Education

Multiple strategic enablers exist to speed up the implementation of CE in HEIs despite these obstacles. The university administrators develop a clear circularity vision, which becomes part of institutional strategic plans and financial goals to show their strong leadership commitment. The creation of specialized sustainability offices and CE task forces promotes cross-functional cooperation and execution. The planning and direction of circular projects receive stakeholder engagement, particularly from students who participate in this process. Student-led organizations have launched campus recycling initiatives together with swap shops, upcycling workshops, and digital campaigns that promote circular habits [7]. The grassroots initiatives promote sustainability as a core campus value, which drives top-down decision-making processes. The implementation of CE benefits from financial backing, technical expertise, and practical learning opportunities through partnerships between NGOs, businesses, and municipal governments. Colleges can develop educational materials and test new projects through joint efforts with sustainable manufacturers and municipal waste management agencies. The previous section explained how digital technologies function as a cross-cutting enabler that helps HEIs track and enhance their CE initiatives through continuous assessment. The implementation of digital twins and AI-powered waste monitoring systems and dashboards serves to enhance transparency while helping organizations scale their successful solutions. National accrediting systems together with policy coherence should motivate institutions to enhance their circular economy commitment through funding plans, rankings, and curriculum standards [26].

2 Literature Review

2.1 The Role of Higher Education in Circular Economy Transitions

Universities can become 'living laboratories' through student-faculty collaboration, which develops practical solutions for circular economy challenges in actual environments. The dual function of HEIs stems from this concept, which shows their role as knowledge centers and practical sustainability implementers. Higher Education Institutions (HEIs) have gained recognition as essential entities that drive the transition to the circular economy (CE). The institution fulfills various functions through its educational mission, research activities, and community-based solution development. The knowledge creation and dissemination capabilities of HEIs enable them to develop essential critical thinking and innovative capabilities needed for sustainability system transformations [11]. Different policy frameworks worldwide provide support to higher education institutions (HEIs) in their leadership roles for sustainability efforts. The European Union's Green Deal and India's National

Education Policy (NEP) prioritize sustainability and a circular economy as their strategic objectives. The policies support universities in implementing circular practices through their institutional strategies, research activities, and curriculum development. The most important demonstration of the transition to circularity is sustainable campus operations, which numerous universities worldwide are currently developing [5]. Higher education institutions serve as essential catalysts for global sustainability transformation because they produce innovative solutions needed to execute circular economy principles. Higher education institutions encounter various obstacles to implementing circular economy principles because of institutional resistance, insufficient multidisciplinary courses, and limited funding opportunities. Leadership strength combined with ongoing institutional backing for sustainability helps eliminate these barriers. Higher education institutions have the ability to drive systemic change because they influence social, economic, and environmental systems, even though they face obstacles [4]. Universities function as knowledge centers to develop interdisciplinary methods that integrate economic principles with engineering and social scientific knowledge for developing future circular economies.

2.2 Integrating Circular Economy in Curriculum, Campus Operations, and Research

Higher education institutions apply circular economy principles through three main areas, which include curriculum design, campus operations, and research and partnerships. The implementation of circular economy thinking receives significant support through curricular design. Students gain a deeper understanding of CE principles through interdisciplinary courses that unite business management with engineering and design and environmental sciences. The delivery of theoretical CE knowledge by HEIs requires experiential learning to actively engage students in practical applications [29]. The education system needs to integrate project-based learning together with industry partnerships and real-world sustainability problems to prepare students for circular economy work after graduation. Higher education institutions have started to integrate CE as a core operational practice. The adoption of green infrastructure by universities worldwide includes water recycling systems and waste-to-energy technology, renewable energy sources, and energy-efficient buildings. The operational strategies achieve two essential goals by minimizing environmental effects from campus operations, offering students hands-on learning experiences, and establishing sustainability standards. The implementation of sustainable procurement methods by universities worldwide enables them to align their purchasing activities with the principles of the circular economy [36, 52].

2.3 Digital Technologies as Enablers of Circular Innovation in Higher Education

Higher education institutions are accelerating the adoption of circular economy principles through digital technology advancements. Higher education institutions can improve operational efficiency and reduce waste while optimizing resource management through the implementation of blockchain technology, artificial intelligence (AI), Internet of Things (IoT), and data analytics. AI algorithms develop predictive models to forecast recycling rates, waste production, and energy consumption patterns. AI examines large datasets to forecast energy peak usage, which helps optimize resource allocation across university facilities [20].

The implementation of these technologies enables universities to enhance their environmental efficiency and better integrate circular economy principles into their campus operations. The Internet of Things (IoT) functions as an essential tool that enables HEIs to achieve circular innovation. Smart meters and sensors, along with IoT-enabled real-time monitoring systems, allow for the collection of data about consumption patterns on campus resources. The university can analyze the data and optimize resource management through IoT systems by recognizing waste generation, water use, and energy use. IoT connects sensor-based digitized technologies to allow for smart campus operational practices through the monitoring and control of all infrastructure elements within the university system. Accessing real time data can enable the university to realize greater decision-making and resource management capabilities that enable better campus operational optimization, and ultimately, lower environmental footprints [18, 35]. Higher education institutions should adopt blockchain technology as a key innovation is needed to support their circular economy implementation. Organizations can have full visibility of their waste management and supply chain management operations through traceability and transparency with blockchain technology. Blockchain technology facilitates higher education institutions in tracking and managing the recycling process to verify materials are being recycled and not simply in a waste bin. Once colleges implement blockchain, they can verify sustainability credentials in supply chains, which builds stakeholder trust and establishes a secure system for tracking circular practices [39].

Through blockchain technology implementation, HEIs can achieve transparency in sustainability performance by securing data protection within sustainability reporting systems. The ability to generate useful information for HEIs that want to adopt circular economy principles depends on data analytics. Data analytics functions as a management tool for campus waste control and for measuring energy usage and evaluating sustainability policy success.

2.4 Barriers and Challenges to CE Implementation in HEIs

Higher education holds great potential to adopt circular economy principles, yet various barriers stop their general adoption. The main challenge to adopting circular economy principles is institutional inertia, which describes the resistance to change within established university structures. The role of political intervention is crucial in the initial phases of CE for consumers and firms, including disseminating a positive and sustainable image of the CE concept and model, but its costs for consumers and organizations are prohibitively expensive. In this regard, the support of CE research for that intervention is vitally important to create a CE pathway consistent with all three principles of sustainable development, that is, the society's aspirations and pertinence for facing present weather patterns and environmental upheaval in an unprecedented fashion [25]. The academic staff, along with administrative personnel, lack sufficient knowledge and resources to execute circular economy initiatives successfully. The main barrier to progress is the shortage of available funds. The implementation of circular economy initiatives requires substantial initial investments, which become particularly expensive when universities need to improve their infrastructure through energy-efficient buildings and sustainable waste management systems. The financial constraints of many Higher Education Institutions (HEIs) become more severe because they face reduced public funding alongside increasing research fund competition. The problem becomes more severe because sustainability measures need definition to calculate the financial benefits of circular activities and validate required investments. Universities need more specific guidelines and policy alignment with circular economy goals to develop successful sustainability strategies. Higher Education Institutions need policy coherence to successfully implement circular economy principles. The main challenge to implementing circular economy principles arises from policy misalignment between university objectives and national or international circular economy frameworks. The sustainability initiatives at many universities exist independently from national circular economy strategies.

3 Research Method

The study relies on secondary sources to acquire recent information and meaningful findings from existing research literature. The research study implements an exploratory and descriptive approach to investigate both the benefits and drawbacks of adopting circular economy principles within higher education institutions. Through qualitative research methods, the study delivers a complete understanding of the dynamic nature of sustainability practices in higher education. This study used a systematic review approach to gather comprehensive data from the literature. The research used academic sources to evaluate journal articles, institutional reports, and conference proceedings in order to discover appropriate frameworks, along with strategies and sustainability policies. Research sources were retrieved from Scopus, Web of Science, and Google Scholar databases. These databases were selected because they offered extensive peer-reviewed scholarly indexing.

The research focused on publications from the past ten to fifteen years because sustainability and circular economy practices in higher education underwent significant changes during this period. The research analyzed 50 to 60 publications that covered various subjects about circular economy and sustainability frameworks, together with implementation strategies and higher education challenges. The selected literature provides empirical evidence about sustainability practices in higher education by identifying success factors and obstacles as well as practical solutions to problems. The papers underwent thorough analysis to identify essential themes and patterns relevant to the study of circular economy implementation in academic institutions. The research applied thematic analysis to examine all literature review data. The research employed thematic analysis to organize and understand the findings through systematic categorization, revealing recurring themes and trends related to the implementation of higher education circular economy. The research uses thematic analysis to examine fundamental topics, which include curriculum integration, campus operations, research, community partnerships, and digital technology implementation for advancing circular economy practices. The research uses thematic analysis to develop practical advice for higher education stakeholders and supports continuous sustainability discussions in academic institutions. Research uses secondary data from peer-reviewed publications to conduct an exploratory and descriptive qualitative analysis of circular economy integration opportunities and challenges in higher education. The thematic analysis approach offers comprehensive literature review methods that enable researchers to identify significant patterns and relevant findings.

4 Conclusion

The research aimed to identify the essential functions that higher education institutions (HEIs) perform in teaching sustainable practices through their educational programs, operational management, research activities, and community partnerships. Higher education institutions stand as ideal institutions to drive sustainability transitions, stimulate innovation, and develop professionals who will lead the global circular economy transition because of increasing environmental challenges. The research demonstrates that higher education institutions possess strategic capabilities to function as circular innovation centers. Project-based experiential learning techniques have become more prominent because they increase student participation while ensuring students develop skills to address real-world sustainability challenges [30].

Students will engage in sustainable practice transformation throughout their college years and professional careers because academic curricula will incorporate circular economy tenets. Collaboration for the advancement of circular economy innovation requires research activities in combination with institutional partnerships, and Higher Education Institutions should support interdisciplinary research projects that explore new business models, technological innovations, and policy frameworks that better facilitate circularity. The innovation hubs in higher educational institutions act as an important conduit for collaborative work between local communities and business and academic partners [38]. The sustainable development (SD) agenda has been adopted by numerous higher education institutions (HEIs). In addition, multiple sustainability assessment tools (SATs) have been made available to HEIs to assist them in a systematic manner to measure, audit, benchmark, and provide evidence of sustainable development. Recently, many stakeholders have asked HEIs to demonstrate their impacts (what they say, do, and evidence) on sustainable development—impacts that extend beyond their organizational boundaries and take the form of the direct and indirect relationship an HEI has to society, the natural environment, and the economy [14].

The innovation hubs demonstrate the ways in which universities function as innovation accelerators through operationalizing solutions to sustainability challenges. The HEI external partnerships collaborating on co-operative research develop knowledge about circular economy practices that impact local and international policy and decision-making [16].

Technology continues to change the way we operate on campuses and develop new ways of learning about circular economy principles in research and education. This research developed an understanding of many barriers to the implementation of circular economy practices in higher education, despite many opportunities for diffusion. Full implementation of circular economy principles continues to face barriers (i.e., institutional inertia, financial barriers, and misalignment of policy with practice). Higher education institutions continue to experience barriers to implementing circularity because they find it difficult to develop pathways to their sustainability objectives within regulations that are international, national, and even campus-specific. The challenge of the circular economy is the barriers they face because institutions often have to take a leap in upfront funding to either renew their building by transferring to green technologies or sustainable procurement processes, and even so, this is complicated when many of these decisions are being taken under budget crises management with competing priorities. For these barriers to be successfully implemented, they require strong institutional leadership with coherent policy frameworks and continued investment in sustainability actions [38].

The main challenge stems from policy alignment difficulties. The pursuit of sustainability by many colleges remains independent, while they lack a unified policy framework that supports global circular economy objectives. The evaluation of sustainability progress in higher education becomes more complex because there are no established benchmarks available. HEIs need to work closely with legislators, business executives, and community members to create regulations that promote institutional circularity and enable effective practice scalability throughout the industry [13]. Governments should establish financial support systems, legislative frameworks, and operational frameworks to help HEIs successfully implement circular economy principles [9]. To implement circular economy principles successfully, institutions of higher education must partner with public organizations, private organizations, and local communities despite several barriers that exist. Universities can enhance their sustainability ambitions through partnerships that bring environmental and social change. Partnerships remove technological and financial barriers to circularity and foster an innovative environment. To successfully implement circular economy principles, universities will need to engage with the external stakeholders, who will help to develop solutions that have social benefits and implications for the higher education sector [40, 51].

5 Scope of Future Study

Future research on Circular Economy (CE) implementation in higher education institutions (HEIs) presents numerous opportunities for investigation. Future research should conduct comparative regional studies to identify multiple opportunities and barriers, while performing longitudinal studies to assess the long-term effects of CE practice and evaluate student-led initiatives for their ability to drive institutional change. Research should continue to explore how AI, IoT, and data analytics technologies enhance Circular Economy practices on college campuses, particularly through smart campus applications and resource management predictive analytics. Additional research is needed to develop standardized frameworks for HEIs to follow during their CE transition and to address implementation challenges such as institutional inertia and policy misalignment. Research should investigate how interdisciplinary teaching methods combined with experiential learning opportunities can enhance student engagement in CE ideas. Future research through these channels will help HEIs become essential sustainability facilitators that support the broader circular economy goals of society.

References

1. Abo-Khalil AG (2024) Integrating sustainability into higher education: challenges and opportunities for universities worldwide. Heliyon 10(9):e29946. https://doi.org/10.1016/j.heliyon.2024.e29946
2. Alves MN, Seixas C, Castro A, Leitão A (2024) Promoting the transition to a circular economy: a study about behaviour, attitudes, and knowledge by university students in Portugal. In: Sustainability, vol 16, Issue 1. https://doi.org/10.3390/su16010343
3. Antonopoulou K, Begkos C, Zhu Z (2023) Staying afloat amidst extreme uncertainty: a case study of digital transformation in Higher Education. Technol Forecast Soc Chang 192:122603. https://doi.org/10.1016/j.techfore.2023.122603
4. Bocken NMP, de Pauw I, Bakker C, van der Grinten B (2016) Product design and business model strategies for a circular economy. J Ind Prod Eng 33(5):308–320. https://doi.org/10.1080/21681015.2016.1172124

5. Brundiers K, Wiek A (2011) Educating students in real-world sustainability research: vision and implementation. Innov High Educ 36(2):107–124. https://doi.org/10.1007/s10755-010-9161-9
6. Chakraborty A, Kumar S, Shashidhara LS, Taneja A (2021) Building sustainable societies through purpose-driven universities: A case study from Ashoka University (India). In: Sustainability, vol. 13, Issue 13. https://doi.org/10.3390/su13137423
7. Corsini F, Gusmerotti NM, Bartoletti E, Testa F, Appolloni A, Iraldo F (2024) Addressing plastic concern: behavioral insights into recycled plastic products and packaging in a circular economy. Circ Econ Sustain 4(3):1961–1981. https://doi.org/10.1007/s43615-024-00367-5
8. Domenech T, Bahn-Walkowiak B (2019) Transition towards a resource efficient circular economy in europe: policy lessons from the EU and the member states. Ecol Econ 155:7–19. https://doi.org/10.1016/j.ecolecon.2017.11.001
9. Doran J, Ryan G (2016) The importance of the diverse drivers and types of environmental innovation for firm performance. Bus Strat Environ 25(2):102–119. https://doi.org/10.1002/bse.1860
10. Elbroumi S, Idrissi MA, Eddahmouny H, Chaaouan M (2025) Living labs supporting circular cities in Morocco: towards collaborative waste management in Casablanca. Sustain Futur 10:100868. https://doi.org/10.1016/j.sftr.2025.100868
11. Evans J, Jones R, Karvonen A, Millard L, Wendler J (2015) Living labs and co-production: university campuses as platforms for sustainability science. Curr Opin Environ Sustain 16:1–6. https://doi.org/10.1016/j.cosust.2015.06.005
12. Faitani NG, Lopez de Munain DN, Bielsa RO (2025) Circular economy framework as a tool for the construction of prospective scenarios for the energy management of a Higher Education institution. In: Brandli L, Rosa FD, Petrorius R, Veiga Avila L, Filho WL (eds) The contribution of life cycle analyses and circular economy to the sustainable development goals. Springer Nature Switzerland, pp 25–44. https://doi.org/10.1007/978-3-031-85300-5_2
13. Filho WL, Will M, Salvia AL, Adomßent M, Grahl A, Spira F (2019) The role of green and Sustainability Offices in fostering sustainability efforts at higher education institutions. J Clean Prod 232:1394–1401. https://doi.org/10.1016/j.jclepro.2019.05.273
14. Findler F, Schönherr N, Lozano R, Stacherl B (2019) Assessing the impacts of Higher Education institutions on sustainable development—an analysis of tools and indicators. In: Sustainability, vol 11, Issue 1. https://doi.org/10.3390/su11010059
15. Franco IB, Tracey J (2019) Community capacity-building for sustainable development. Int J Sustain High Educ 20(4):691–725. https://doi.org/10.1108/IJSHE-02-2019-0052
16. Geissdoerfer M, Savaget P, Bocken NMP, Hultink EJ (2017) The circular economy—a new sustainability paradigm? J Clean Prod 143:757–768. https://doi.org/10.1016/j.jclepro.2016.12.048
17. Giannoccaro I, Ceccarelli G, Fraccascia L (2021) Features of the Higher Education for the circular economy: the case of Italy. In: Sustainability, vol 13, Issue 20. https://doi.org/10.3390/su132011338
18. Gilman E, Tamminen S, Yasmin R, Ristimella E, Peltonen E, Harju M, Lovén L, Riekki J, Pirttikangas S (2020) Internet of Things for smart spaces: a university campus case study. In: Sensors, vol 20, Issue 13. https://doi.org/10.3390/s20133716
19. González-Domínguez J, Sánchez-Barroso G, Zamora-Polo F, García-Sanz-Calcedo J (2020) Application of circular economy techniques for design and development of products through collaborative project-based learning for industrial engineer teaching. In: Sustainability, vol 12, Issue 11. https://doi.org/10.3390/su12114368
20. Hariyani D, Hariyani P, Mishra S, Kumar Sharma M (2024) Leveraging digital technologies for advancing circular economy practices and enhancing life cycle analysis: a systematic literature review. Waste Manag Bull 2(3):69–83. https://doi.org/10.1016/j.wmb.2024.06.007
21. Henrysson M, Nuur C (2021) The role of institutions in creating circular economy pathways for regional development. J Environ Dev 30(2):149–171. https://doi.org/10.1177/1070496521991876

22. Herth A, Verburg R, Blok K (2025) How can campus living labs thrive to reach sustainable solutions? Clean Prod Lett 8:100078. https://doi.org/10.1016/j.clpl.2024.100078
23. Hsu W-T, Domenech T, McDowall W (2022) Closing the loop on plastics in Europe: the role of data, information and knowledge. Sustain Prod Consum 33:942–951. https://doi.org/10.1016/j.spc.2022.08.019
24. Innella C, Barberio G, Brunori C, Cappellaro F, Ceddia AR, Civita R, Dimatteo S, Ferraris M, Pentassuglia R, Sciubba L (2024) Experimenting urban living lab methodology on circular economy co-design activities in some Italian urban territories. Front Sustain Cities 6(July) https://doi.org/10.3389/frsc.2024.1406834
25. Kevin van Langen S, Vassillo C, Ghisellini P, Restaino D, Passaro R, Ulgiati S (2021) Promoting circular economy transition: A study about perceptions and awareness by different stakeholders groups. J Clean Prod 316(June):128166. https://doi.org/10.1016/j.jclepro.2021.128166
26. Kosta AD, Keramitsoglou KM, Tsagarakis KP (2025) Circular economy and sustainable development in primary education. Front Sustain 6(February):1–11. https://doi.org/10.3389/frsus.2025.1414055
27. Kouhizadeh M, Zhu Q, Sarkis J (2020) Blockchain and the circular economy: potential tensions and critical reflections from practice. Prod Plan Control 31(11–12):950–966. https://doi.org/10.1080/09537287.2019.1695925
28. Liu Q, Trevisan AH, Yang M, Mascarenhas J (2022) A framework of digital technologies for the circular economy: digital functions and mechanisms. Bus Strat Environ 31(5):2171–2192. https://doi.org/10.1002/bse.3015
29. Lozano R, Garcia I (2020) Scrutinizing sustainability change and its institutionalization in organizations. Front Sustain 1(May):1–16. https://doi.org/10.3389/frsus.2020.00001
30. Luederitz C, Schäpke N, Wiek A, Lang DJ, Bergmann M, Bos JJ, Burch S, Davies A, Evans J, König A, Farrelly MA, Forrest N, Frantzeskaki N, Gibson RB, Kay B, Loorbach D, McCormick K, Parodi O, Rauschmayer F, … Westley FR (2017) Learning through evaluation—a tentative evaluative scheme for sustainability transition experiments. J Clean Prod 169:61–76. https://doi.org/10.1016/j.jclepro.2016.09.005
31. Martins P, Lopes SI, Rosado da Cruz AM, Curado A (2021) Towards a smart & sustainable campus: an application-oriented architecture to streamline digitization and strengthen sustainability in academia. In: Sustainability, vol 13, Issue 6. https://doi.org/10.3390/su13063189
32. Mendoza JMF, Gallego-Schmid A, Azapagic A (2019) A methodological framework for the implementation of circular economy thinking in higher education institutions: towards sustainable campus management. J Clean Prod 226:831–844. https://doi.org/10.1016/j.jclepro.2019.04.060
33. Moreno M, De los Rios C, Rowe Z, Charnley F (2016) A conceptual framework for circular design. In: Sustainability, vol 8, Issue 9. https://doi.org/10.3390/su8090937
34. Nobre GC, Tavares E (2017) Scientific literature analysis on big data and internet of things applications on circular economy: a bibliometric study. Scientometrics 111(1):463–492. https://doi.org/10.1007/s11192-017-2281-6
35. Polin K, Yigitcanlar T, Limb M, Washington T (2023) The making of smart campus: a review and conceptual framework. In: Buildings, vol 13, Issue 4. https://doi.org/10.3390/buildings13040891
36. Popa I, Corboş R-A, Ştefan SC, Bunea O-I (2025) Circular economy practices, procurement strategies, and organizational competitiveness: insights from future business professionals. In: Systems, vol 13, Issue 5. https://doi.org/10.3390/systems13050344
37. Prieto-Sandoval V, Ormazabal M, Jaca C, Viles E (2018) Key elements in assessing circular economy implementation in small and medium-sized enterprises. Bus Strat Environ 27(8):1525–1534. https://doi.org/10.1002/bse.2210
38. Reike D, Vermeulen WJV, Witjes S (2018) The circular economy: new or refurbished as CE 3.0?—Exploring controversies in the conceptualization of the circular economy through a focus on history and resource value retention options. Resour Conserv Recycl 135:246–264. https://doi.org/10.1016/j.resconrec.2017.08.027

39. Rejeb A, Appolloni A, Rejeb K, Treiblmaier H, Iranmanesh M, Keogh JG (2023) The role of blockchain technology in the transition toward the circular economy: Findings from a systematic literature review. Resour Conserv Recycl Adv 17:200126. https://doi.org/10.1016/j.rcradv.2022.200126
40. Renfors SM (2024a) Education for the circular economy in higher education: an overview of the current state. Int J Sustain High Educ 25(9):111–127. https://doi.org/10.1108/IJSHE-07-2023-0270
41. Renfors SM (2024b) Supporting green transition in the Finnish tourism sector by identifying green skills. Eur J Tour Res 36(2024):1–16. https://doi.org/10.54055/ejtr.v36i.3223
42. Sachs JD, Schmidt-Traub G, Mazzucato M, Messner D, Nakicenovic N, Rockström J (2019) Six transformations to achieve the sustainable development goals. Nat Sustain 2(9):805–814. https://doi.org/10.1038/s41893-019-0352-9
43. Saeedikiya M, Salunke S, Kowalkiewicz M (2025) The nexus of digital transformation and innovation: a multilevel framework and research agenda. J Innov Knowl 10(1):100640. https://doi.org/10.1016/j.jik.2024.100640
44. Serrano-Bedia A-M, Perez-Perez M (2022) Transition towards a circular economy: a review of the role of higher education as a key supporting stakeholder in Web of Science. Sustain Prod Consum 31:82–96. https://doi.org/10.1016/j.spc.2022.02.001
45. Singh AB, Meena HK, Khandelwal C, Dangayach GS (2023) Sustainability assessment of Higher Education institutions: a systematic literature review. In: Engineering Proceedings, vol 37, Issue 1 https://doi.org/10.3390/ECP2023-14728
46. Strapchuk O, Koniordos M, Strapchuk S, Nitsenko V (2025) Sustainable business models in the digital transformation of higher education: evidence from Ukraine. Transform Sustain 1(2):87–99. https://doi.org/10.63775/hsfgga88
47. UNESCO (2020) Education for sustainable development: A Roadmap, vol 2009, Issue 18. https://doi.org/10.1111/j.2048-416x.2009.tb00140.x
48. United Nation (2016) Ensure sustainable consumption and production patterns. https://sustainabledevelopment.un.org/content/documents/19847SDG12_Highlights.updated_format.pdf
49. Valls-Val K, Ibáñez-Forés V, Bovea MD (2024) Measuring and monitoring the transition to the circular economy of universities: CExUNV. J Environ Manag 356:120492. https://doi.org/10.1016/j.jenvman.2024.120492
50. Valls-Val K, Ibáñez-Forés V, Lo-Iacono-Ferreira VG, Capuz-Rizo SF, Bovea MD (2023) Adequacy of existing circular economy assessment tools for higher education institutions. Sustain Prod Consum 39:399–413. https://doi.org/10.1016/j.spc.2023.05.011
51. Vergani, F (2024) Higher education institutions as a microcosm of the circular economy. J Clean Prod 435:140592. https://doi.org/10.1016/j.jclepro.2024.140592
52. Xu L, Jia F, Yan F, Chen L (2022) Circular procurement: a systematic literature review. J Clean Prod 365:132845. https://doi.org/10.1016/j.jclepro.2022.132845

Innovations in Microalgae-Based Wastewater Treatment and Biomass Valorization: An Integrated Circular Economy Approach

Lazarus Vijune Lawrence, Swagata Pal, and Dhanya Vishnu

Abstract One of the main critical environmental concerns is water pollution. This is caused by various contaminants, including heavy metals, organic pollutants such as dyes, and excess nutrients, leading to adverse ecological as well as health effects. Conventional treatment strategies, such as chemical precipitation, membrane filtration, and other techniques employed are often expensive and have limitations and thus necessitating sustainable and cost-effective alternatives. Microalgal-based biosorption has gained significant attention as an eco-friendly approach for water bioremediation owing to its high efficiency in adsorbing pollutants. Moreover, in addition to wastewater treatment, the harvested microalgal biomass holds extensive potential for valorization in various industrial applications. This chapter covers the importance of microalgal-based treatment strategies to mitigate pollutants in wastewater and highlighting its biomass utilization in biofuel production, biofertilizers, and other value added products. With perspectives of cost effective strategy, this chapter provides the key role of cultivating microalgae in wastewater as a synthetic medium. This integration of microalgal-based biosorption into wastewater treatment systems aligns with sustainable development goals, reducing environmental pollution while generating bio-based products. It also covers the challenges in optimizing biosorption efficiency and large-scale implementation. Thus, microalgal biosorbents present a promising, sustainable, and circular approach for water treatment and biomass valorization, bridging the gap between pollution control and sustainability.

Keywords Circular economy · Microalgae · Sustainability · Valorization · Wastewater treatment

---

L. V. Lawrence · S. Pal · D. Vishnu (✉)
Department of Biotechnology, School of Bio Sciences and Technology, Vellore Institute of Technology, Vellore, Tamil Nadu, India
e-mail: dhanya.v@vit.ac.in

V. Kandpal and T. Choudhury (eds.), *Blue Revolution: Sustainable Water Use and Circular Economy Innovations*, Urban Sustainability,
https://doi.org/10.1007/978-981-95-1916-3_2

1 Introduction

Water pollution remains a major environmental threat since the utility of water in various industrial sectors such as food, agricultural, petrochemical, pharmaceutical, tannery, textile, electronics and others mainly pollute water. In addition to pollution, another severe environmental risk is disposing such contaminated water directly to the nearby lakes, ponds, rivers and other water bodies could develop toxicity to the aquatic habitats [38]. The physicochemical characteristics of various wastewater revealed the presence of high amount of both organic and inorganic nutrient, biological oxygen demand (BOD), chemical oxygen demand (COD), and other pollutants [15, 46, 47]. These compounds when accumulated in the water bodies, it cause severe toxicity effects to aquatic organisms. The wastewater treatment remains challenging since conventional methods for treating heavy metals and removal of nutrients from such effluents is achieved to be not effective. The complexity and heterogeneity of the pollutants provide challenges for this wastewater, necessitating the use of highly effective methods to achieve the necessary water quality criteria [38].

The traditional wastewater treatment methods includes adsorption, ionic exchange, precipitation, membrane separation, coagulation-floculation, photodegradation and others offers steps for removing pollutants and nutrients in the wastewater. However, since the sustainable developmental goals (SDGs) were adopted in 2015, a number of studies have attempted to analyze the wastewater treatment facility's contribution to the goals [51]. A study by Obaideen et al. [37] found that the majority of these research concentrated on the contribution to SDG 6 (clean water and sanitation) or on performing a multi-criteria decision analysis to assess how emerging wastewater technologies could boost the contribution. Furthermore, the majority of earlier research has not focused on the study of wastewater treatment [37]. Moreover, Spherical Insights and Consulting updated, "The Global wastewater treatment services market is expected to reach USD 97.54 Billion by 2032, at a CAGR of 6.2% during the forecast period 2022 to 2032" [18].

Microalgae have emerged as a promising biological resource for wastewater treatment due to their ability to simultaneously remediate pollutants and produce valuable products from its biomass. Their efficiency in removing excessive nutrients, heavy metals, and organic pollutants from wastewater, coupled with their rapid growth rate and minimal resource requirements, makes microalgae a sustainable alternative to conventional wastewater treatment methods [49]. The use of microalgae not only addresses water contamination but also aligns with the principles of circular economy by converting waste into valuable by-products [35, 60]. The harvested microalgal biomass has extensive potential for valorization in various industrial applications, such as biofuel production, biofertilizers, bioplastics, and other value-added products [20, 21]. By utilizing wastewater as a medium for culturing microalgae, it is possible to create a system where wastewater is treated while simultaneously generating biomass for various sectors. This integration of microalgal biosorption into wastewater treatment systems offers a cost-effective and environmentally friendly alternative to traditional methods [60].

This chapter focuses on the effective wastewater treatment strategy by using microalgae in the perspectives of circular economy integration. It provides the comprehensive update of utilising wastewater as a cost effective media for harvesting biomass and it gives the data on physiochemical characteristics changes observed after treating with microalgae in various industrial real time wastewater. Moreover, it also focused on the heavy metal removal mechanism in both synthetic as well as real wastewater. Importantly, this chapter covers the integrating cultivation and remediation aspects of microalgae and valorizing its biomass for various applications.

2 Wastewater as a Cost Effective Media for Microalgae

The cultivation of microalgae from various sources offers effective wastewater treatment such as tannery effluent, textile wastewater, piggery wastewater, and other wastewater treatment. Moreover, it is also possible and reported that microalgae has remediation potential for removing various metal and other contaminants from such wastewater. In order to remediate the pollutants in the wastewater by using microalgae, acclimatization is crucial for its growth and survival. Rawat et al. [44] demonstrated microalgae cultivation in wastewater using a custom-built open pond under controlled LED lighting (300 μmol m^{-2} s^{-1}, 18:6 light–dark cycle). The inoculum, pre-cultivated in BBM to the log phase, was introduced at an OD of 0.2 (680_{nm}). Growth was monitored every two days over a 14-day period using spectrophotometry. This study highlights wastewater's potential as a cost-effective and nutrient-rich medium for microalgal biomass production. Zapata et al. [60] evaluated *Spirulina platensis* UTEX growth in Zarrouk medium, dairy wastewater and cheese whey wastewater. They reported that both wastewaters were filtered, supplemented with nutrients (K_2HPO_4 and urea), and adjusted pH before cultivation. Results demonstrated the potential of dairy effluents as low-cost alternatives to synthetic media for *Spirulina* cultivation.

Moreover, Morillas-España et al. [36] carried out wastewater-based microalgae cultivation using two 80 m^2 raceway reactors operated insides a greenhouse. One reactor was fed with municipal wastewater, while the other used freshwater with commercial fertilizers. The reactors were inoculated with a culture dominated by *Scenedesmus / Tetradesmus* and operated semi-continuously at a dilution rate of 0.2 day^{-1} for 14 days. Daily, 2.56 m^3 of biomass was harvested and concentrated using membrane filtration followed by centrifugation. This study demonstrated the potential of untreated municipal wastewater for large-scale algal biomass production under semi-real conditions. Liu et al. [29] reported that swine wastewater was effectively used as a culture medium for microalgae cultivation, enabling nutrient recovery and biomass generation. All these studies supported that wastewater proves to be an effective and low-cost medium for microalgae cultivation, offering essential nutrients for algal growth. It serves as a sustainable alternative to synthetic media, reducing operational costs while addressing environmental concerns. The use of wastewater in microalgae systems also provides an efficient method for wastewater treatment, promoting nutrient recovery and biomass production.

3 Simultaneous Cultivation and Remediation of Real-Time Wastewater by Microalgae

Microalgal such as *Chlorella vulgaris*, *Chlorella sorokiniana*, *Chlamydomonas reinhardtii*, *Tetradesmus* sp., *Neochloris* sp., *Micractinium* sp., *Desmodesmus* sp., *Tetradesmus dimorphus*, *Spirulina platensis*, *Radiococcus polycoccus*, *Chlorolobion braunii*, *Desmodesmus multivariabilis*, and many others have been reported for successful growth in various wastewater [6, 15, 19, 25]. These microalgae have demonstrated excellent bioremediation of pollutants present in the wastewater, nutrient removal and reduction in COD and BOD. Interestingly, in addition to bioremediation, these microalgae also reported for its increased biomass after the treatment of wastewater and thus, they produce value added products [15]. Figure 2.1 represents the process outline for treating wastewater and key characteristic changes observed in the wastewater after treatment of microalgae.

3.1 Dairy Wastewater

Dairy industry is one of the main sectors of the food industry and it has been reported that approximately 2 to 10 L of wastewater is generated for every 1 L of milk [22].

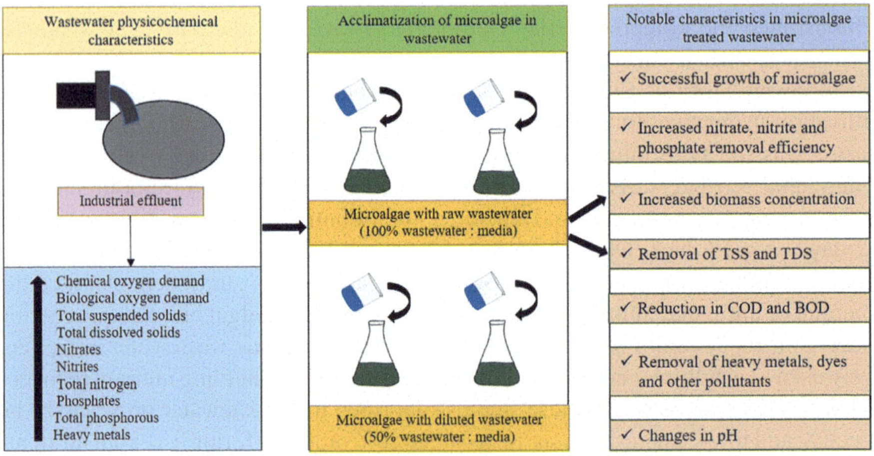

Fig. 2.1 Physiochemical characteristic changes observed in the wastewater after treating with microalgae

Numerous studies surveyed the dairy wastewater characteristics in terms of pH, dissolved oxygen (DO), COD, BOD, total dissolved solids (TDS) and total suspended solids (TSS). As conducted by Idusuyi et al. [22] reported that dairy wastewater contained pH of 4.35, BOD of 420 mg L^{-1}, COD of 1032 mg L^{-1} and TDS of 402 ppm. Dairy manure wastewater from a dairy farm was reported for pH of 5.34, COD of 2600–3000 mg L^{-1} and total phosphate (TP) of 0.3–1.0 mg L^{-1} [6]. Azari et al. [5] evaluated the physicochemical characteristics of raw dairy wastewater and reported pH of 7.0, BOD of 1166 mg L^{-1}, dissolved COD of 2655 mg L^{-1}, TSS of 198 mg L^{-1}, and TDS of 5164 mg L^{-1}. Custodio et al. [7] analysed dairy wastewater from dairy processing plants pH of 7.0, BOD of 584.8 mg L^{-1}, COD of 833.2 mg L^{-1}, and TSS of 85.8 mg L^{-1}. All these physiochemical characteristics from real dairy wastewater revealed excessive concentrations of BOD, COD and other water quality parameters, severely impacted environment.

Isolation and cultivation of microalgae from dairy wastewater offers effective remediation strategy for treating such effluents. For instance, Kusmayadi et al. [25] cultivated *Chlorella sorokiniana* SU-1 with unsterilized anaerobic digestate of dairy wastewater along with BG-11 medium at various concentration (25, 50, 75 and 100% of wastewater) for 10 days. They have observed COD percentage removal efficiency of 75.5% for 25% wastewater and 59.1% removal efficiency for 100% wastewater. However, the increase in inoculum size influence the remediation performances. Co-culturing microalgae in dairy wastewater also exhibited for enhanced biomass productivity, remediate wastewater, and accumulate biochemical components. For example, co-culturing high-value microalgae sp. with filamentous microalgae sp. had demonstrated for effective remediation of dairy wastewater with removal of 86.7% COD, more than 88.7% and 98.5% of nitrate and phosphate content. The reported study also recovered 78.5% biomass than the monocultures (32.4%). This co-cultivation of microalgae and filamentous algae using dairy wastewater offers excellent prospects for the production of sustainable biomass because the wastewater is a rich source of nutrients. In order to be effective, numerous of challenges must be carefully addressed when scaling up this strategy for industrial applications [20]. Interestingly, a study found that microalgae cultivation in dairy effluents offers enhanced biomass and production of more value added products [19]. Moreover, genetic recombinant microalgal strains also reported for recycling dairy wastewater and improved the bioremediation of wastewater. For instance, Gramegna et al. [19] cultivated recombinant *Chlamydomonas reinhardtii* microalgae in dairy wastewater; where lactose concentration was found to be ~4% w/v and has been noted for producing thermostable β-glucosidase.

3.2 *Tannery Wastewater*

The leather industry contributes significantly to environmental contamination due to the vast volumes of solid and liquid waste established throughout various stages of manufacturing. As similar to dairy wastewater, tannery effluents also highly reported for the presence of increased COD, BOD, TSS, TDS and other water parameters. For example Reda [46] conducted the Modjo tannery effluent wastewater to determine the physicochemical parameters levels of BOD and COD of 960.34 mgL^{-1} and 2011.00 mgL^{-1}, respectively. The pH level was 9.33 and concentration of TDS and TSS were observed as 15670.00 mg L^{-1} and 9370.00 mg L^{-1} respectively. Similarly, water quality characteristics were revealed by physicochemical characterization of the effluent to the Ethiopian tannery examined by Abate et al. [1] and reported that the effluent pH of 7.5, COD of 1,070 mg L^{-1}, TDS of 222 mg L^{-1}. Recently, Monzón Martínez et al. [33] found nitrate, nitrite, ammonia, total nitrogen, phosphate and total phosphorous of 1641 mg L^{-1}, 0.15 mg L^{-1}, 180 mg L^{-1}, 495.36 mg L^{-1}, 31.05 mg L^{-1}, and 46.32 mg L^{-1} respectively. The BOD, COD and TSS also been found to be 4368 mg L^{-1}, 6720 mg L^{-1}, and 4960.56 mg L^{-1} respectively. Singh et al. [53] reported that the tannery effluent pH and COD range was found that 6.5–8.5 and 250 mg L^{-1} respectively. The presence of heavy metals, particularly chromium have been reported in the tannery effluent. For instance, Tarahomi et al. [55] identified 3286 mg L^{-1} of chromium in leather factory effluent, which is found to be beyond permissible limits set by World Health Organization (WHO).

Acclimatization of microalgae in tannery effluents are one of the promising approaches to remediate such wastewater and this process enable microalgae to produce and accumulate biomolecule components. Ranjan et al. [43] acclimated *Tetradesmus* sp. and *Neochloris* sp. isolates to 50% raw tannery effluent. Interestingly, the study found that the acclimated microalgal strains remediate tannery effluent with complete ammonia removal of 12 mg L^{-1}. Moreover, 12.7–19.8% and 24.5–37.2% of carbohydrate and protein content observed with increased biomass concentrations. *Tetradesmus dimorphus* microalgal strain showed excellent treatment of tannery wastewater with ~ 96.59% removal of COD. Microalgae-cyanobacteria consortia also noted for being a promising approach for wastewater treatment. Since cyanobacteria also have capacity to remediate tannery wastewater like microalgae, mixed consortia of both microalgae-cyanobacteria could also render to be effective strategy for low cost wastewater treatment process. Monzón Martínez et al. [33] evaluated the tannery wastewater physiochemical characteristics after treating with microalgae-bacteria consortia and observed highest COD consumption, along with nitrogen and phosphorous removal. They also obtained increased biomass concentration after the treatment process. It is noteworthy to observe the treatment results of introducing microalgae and cyanobacterium sp. upon light/dark cycle. It was reported that those microalgae sp. have ability to remove high percentage of inorganic nitrate and phosphate in addition to other water quality parameters. Thus, the microalgae have ability to metabolize the nutrients that present in the wastewater and have role in the generation of value added products or metabolites increased microalgal biomass concentration. The integration of advanced oxidation process with microalgae cultivation exhibited for effective tannery wastewater treatment. To illustrate, Urbina-Suarez et al. [57] developed the integrated approach of advanced oxidation process with *Chlorella* sp. cultivation and observed this system achieved

BOD, COD, and TOC removal of 55–75%. Similar to other approaches, this integrated system functionality also significantly demonstrated both tannery effluent purification as well as microalgal biomass valorization.

3.3 Piggery Wastewater

Large-scale farming methods produce a lot of livestock wastewater, which must be appropriately disposed of to reduce the environmental impact of meat production. Regarding pig farming, a recent report projects a general upward tendency that will reach an annual production volume of 21.7 million tons by 2030 [48]. The discharge of excessive untreated wastewater raises the cost of water treatment in addition to generating pollutants in rivers and ditches. Additionally, it causes heavy metal accumulation in shallow groundwater layers, organic debris and ammonia infiltration into groundwater, and health hazards from pathogen exposure. The COD, ammonia nitrogen, phosphate, and TSS of raw piggery wastewater was identified to be 3759 mg $O_2 \cdot L^{-1}$, 1500 mg L^{-1}, 97.5 mg L^{-1}, and 2575 mg L^{-1} respectively [13]. Raw swine wastewater collected from swine farm at Vietnam had pH, BOD, TN, ammonia, nitrate, TP of 6.9, 4576.52 mg L^{-1}, 975.45 mg L^{-1}, 623.86 mg L^{-1}, 377.43 mg L^{-1}, and 415. 34 mg L^{-1} respectively. Piggery wastewater from pig dung from open pig farming system evaluated for its physicochemical characterization and revealed average of 7.51 of pH, 45 mg L^{-1} of TSS, and 1992 mg L^{-1} of soluble chemical oxygen demand (SCOD) [24].

Cultivation of microalgae in piggery wastewater had documented for enhanced nutrient removal. For example, *Spirulina platensis* cultivated in the four-fold dilution of piggery wastewater and observed removal efficiency of 76%, 72%, and 93.1% for TN, TP and COD respectively. They also reported for increased biomass concentration of 0.56 g L^{-1} [27]. Certain microalgae such as *Desmodesmus* sp. CHX1, have isolated from piggery wastewater. Recently, Flores-Zambrano et al. [15] isolated four microalgal strains such as *Radiococcus polycoccus, Micractinium* sp., *Chlorolobion braunii,* and *Desmodesmus multivariabilis* from oxidation pool for piggery effluents and piggery wastewater and shown effective wastewater treatment approach. Nitrogen compounds such as ammonium and nitrate removal using *Chlorolobion braunii* from piggery wastewater have reported to remove such compounds effectively after 12 days. The recorded results were demonstrated removal of 67.73% for ammonium and 30.59% nitrate under acidic conditions [15]. Microalgae-bacteria consortia for piggery wastewater treatment had been revealed for the removal of ammonia, phosphate and COD of 90, 90 and 59% respectively with increased biomass concentration (10.7 g m^2 d^{-1}). This research suggest that the spread of nitrogen content is reduced to the land by 77% with a perspectives of developed nitrogen valorised biofertilizers [48]. Pre-treatment process such as electrocoagulation, Photo-Fenton, ammonia stripping, and constructed wetlands methods which would enhance the performance of microalgae based treatment. For example Ferreira et al. [13] studied various pre-treatment strategy as mentioned above, and concluded that Photo-Fenton pre-treatment demonstrated more higher removal efficiency nearly 90% for water parameters except for ammonia. Similarly, recent investigation also found that solar photo-Fenton pre-treatment contribute sustainable and cost-effective strategy for allowing microalgal growth and achieved removal of 59.2–77.1% and 83.8–94.7% of TCC and COD [12]. However, pre-treatment methods for wastewater should be also achieve the performance in both lab as well as large scale reactors.

3.4 Textile Wastewater

Typical dye concentrations in real textile industry wastewaters range from 25 to 2500 mg L^{-1}. If such wastewater is released into water bodies without being properly treated, it poses a major risk to all living things, including humans, animals, and plants. The raw textile wastewater collected from handloom city of Pilkhuwa-Hapur, Uttar Pradesh, India revealed the properties of 7.91, 7,479 mg L^{-1}, and 4,252 mg L^{-1} for pH, TDS and TSS. Additionally, COD and BOD were found to be 3209 mg L^{-1} and 1214 mg L^{-1} [39]. Digital printing technology, which has been used by numerous textile sectors over the past 15 years, allows for exceptional colour definition, requires minimal manufacturing time, and requires cheap material and equipment costs. The average values of wastewater from textile digital printing were 0.16 g L^{-1}, 0.13 g L^{-1}, 1.00 g L^{-1}, 198 mgN L^{-1}, and 764 mg L^{-1} for TSS, volatile suspended solids (VSS), nitrate, ammonium, and COD respectively [30]. It is well-known that textile industrial effluents predominantly contaminated with hazardous acidic and basic dyes [10]. Textile wastewater from industrial washing process of jeans had been reported mainly for the presence of indigo dye [47]. Moreover, textile wastewater from a dyeing industry at northeast Brazil was characterized and revealed the pH, TOC, BOD, COD, and TS of 7.8, 734 mg L^{-1}, 289 mg L^{-1}, 1130 mg O_2 L^{-1}, and 3570 mg L^{-1} respectively [56].

Cultivation of microalgae in untreated textile wastewater demonstrated excellent removal of COD, increased biomass concentration and lipid accumulation. Javed et al. [23] cultivated *Chlorella vulgaris* without and with diluted (25, 50, and 75%) real textile wastewater. The best results obtained from those concentrations is identified for 50% due to its high removal of COD (82%), and lipid content. However, when cultivated with mixotrophic condition found to be 94% of COD removal than the cultivation. Similarly, textile effluents from a Portuguese textile company revealed the successful growth rate microalgae *Chlorella vulgaris* ranging from 0.234 to 0.290 d^{-1} and noted for more than 98, 95 and 96% for phosphate, ammonia nitrogen and nitrate contents [31]. As discussed on microalgal-bacterial consortia remediation for other wastewater, textile wastewater can also be remediate using such consortium for environmental sustainability. Co-culturing both microalgae and bacteria under various conditions indicated the maximum removal efficiency of 58.57, 86.42, and 91.5% for nitrate, phosphate and COD [45]. Recently, Simsek et al. [52] cultivated *Neglectella solitaria* in the real textile wastewater and reported the higher removal efficiencies of 90.3–98.3%, 91–97%, 60.1–96.5% and 27.7–84.8% for COD, phosphate, nitrite and nitrate contents. This microalgal species achieved excellent textile untreated water even without acclimatizing the culture and noted increased biomass concentration. Cultivation of microalgae consortia in wastewater from textile digital printing had removed textile pollutants and increased biomass as well as pigment concentration [30]. Integrated cultivation of microalgae *Chlorella vulgaris* with textile wastewater in both diluted and undiluted water have reported for excellent COD removal. The removal efficiency achieved higher in diluted textile wastewater (99.7%) than the undiluted textile wastewater (76.3%) [11].

4 Removal of Heavy Metals in Wastewater by Microalgae

The widespread heavy metals in the environment is one of the major threats to both human as well as environment. Tanning industry, petrochemical products, printing, textile, fertilizer manufacturing, and battery manufacturing are some of the industries that release heavy metals that leads to accumulation in the environment. These metals are found to be beyond permissible limits set by WHO [38]. Moreover, it has lack of biodegradability, stays persistent in the environment and cause toxic effects to the ecosystem. The heavy metal removal process can be achieved by employing physical, chemical or biological methods. However, in order to look up into sustainability, biological sources using various microbes aids the biosorption of heavy metals in the wastewater [42]. Particularly, microalgae has been noted for the effective removal of heavy metal that present in the wastewater and could possibly convert toxic form of metal into less toxic form. Figure 2.2 depicts the mechanism of heavy metal removal by microalgae.

Biosorption of heavy metals such as hexavalent chromium (Cr (VI)), copper (Cu (II)), arsenic (As (III)), zinc (Zn (II)), lead (Pb (II)) and other industrial processing metals in both synthetic as well as real time wastewater. *Scenedesmus quadricauda* microalgae achieved with 67.03% removal of Cr (VI) at lag phase within 12 days of introducing Cr metals in the medium [9]. *Craticula subminuscula*, a freshwater microalgae also been reported for the removal of Cr (VI) in the aqueous solution. The maximum removal % of Cr was obtained for this species is 95.32% where the biosorption isotherm and kinetic models follows the Langmuir and pseudo-second order respectively; which represents the monolayer adsorption process [50]. *Chlorella thermophilia* demonstrated removal of both Cr (VI) as well as As (III) with removal efficiency of 50–65% and 70–91% at 0.2–0.6 mg L^{-1} and 0.5–10 mg L^{-1} metal ion initial concentrations respectively [16]. Similarly, *Scenedesmus almeriensis* microalgae also achieved 93% and 76% of Zn and Cu metal removal at a initial concentration of 100 mg L^{-1} respectively [3]. *Chlamydomonas* sp. also been reported for both Cu and Cr (VI) removal under simulated wastewater conditions. The recorded results suggest that this microalgae had removal efficiency of 41.4% and 58.11% for both Cu and Cr (VI) metal ions [17]. These findings suggests that microalgae can effectively remove more than one metal with higher removal efficiency simultaneously. However, *Chlamydomonas* sp. introduced for the monometallic removal of

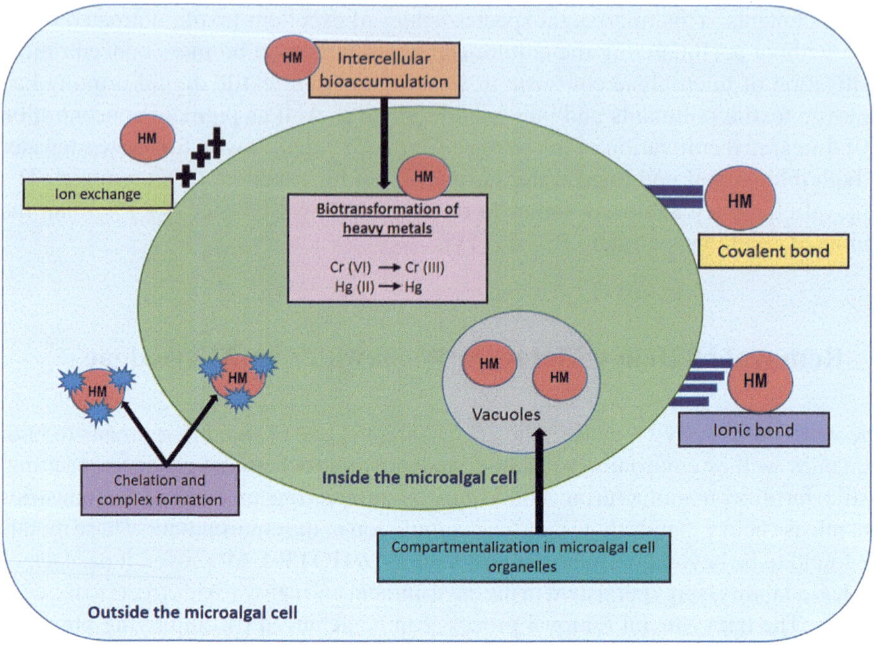

Fig. 2.2 Mechanism of heavy metal biosorption and biotransformation by microalgae

Cr (VI), it effectively removes Cr with 94.17% removal efficiency under aqueous solutions. Interestingly, Liu et al. [28] found maximum metal removal efficiency for Cu (II) achieved by *Desmodesmus* sp. CHX1 was 88.35% under real time piggery digestate, suggesting that these microalgae effectively can remove the heavy metals under wastewater conditions.

5 Integrated Approach for Treating Wastewater and Biomass Valorization for Value Added Products

It is interesting to note that various microalgal species have ability to utilize the wastewater and reported for the increased biomass production [34, 42, 43]. Figure 2.3 depicts the integrated microalgae cultivation as well as the biomass valorization.

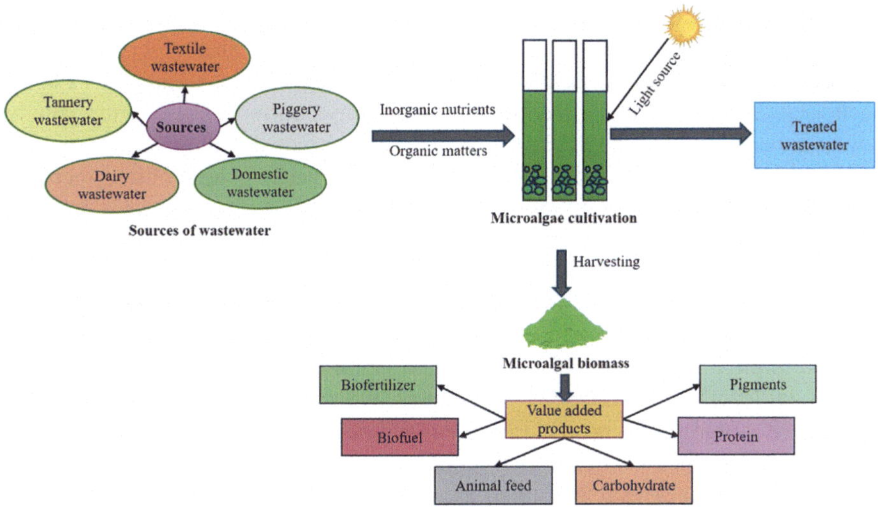

Fig. 2.3 Simultaneous wastewater treatment and microalgal biomass harvesting for value added products

5.1 Biofertilizers

Generally, microalgae biomass and its extracts are extensively studied for enhancing plant growth and development; which effectively reported as biofertilizer and biostimulant activity [26]. Figure 2.4 outlines the role and advantages of microalgal biomass and extracts for plant productivity.

The harvested microalgae after the treatment of wastewater, the biomass has been used for enhanced plant growth and development *via* both soil and foliar applications.

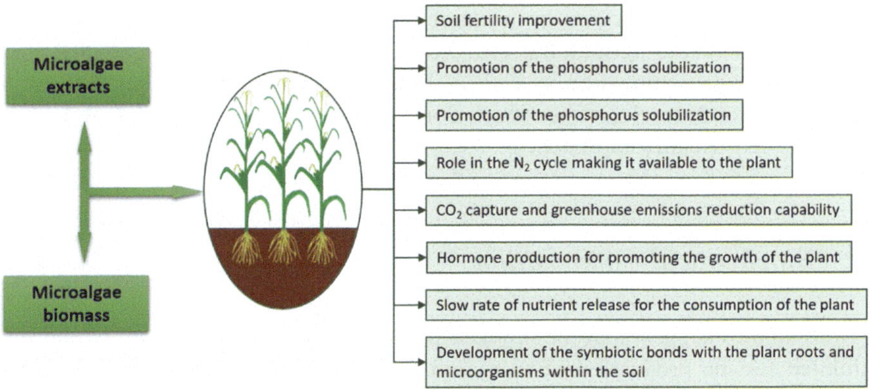

Fig. 2.4 Role and advantages of microalgal biomass and extracts for plant productivity

Zapata et al. [60] cultivated *Spirulina platensis* in dairy wastewater and the harvested biomass had developed for biofertilizer application. They also observed notable phytohormones such as Indole-3-acetic acid, Salicylic acid, Jasmonic acid, Phenylacetic acid, Abscisic acid, Gibberellin, and other phytohormones also. Interestingly, they have been reported that those hormones were extracted in the microalgae cultivated in the dairy wastewater than the microalgae cultivated in the synthetic medium. This study suggest that the wastewater can be utilized as the effective alternative media than the synthetic wastewater, where wastewater can also be remediated as well as the harvested biomass also can be used as plant productivity agents [60]. Another study by Rawat et al. [44] studied the potential of microalgae cultivated from wastewater for biofertilizer applications. They demonstrated that the algal biomass, obtained through wastewater treatment, could be utilized to prepare bio-stimulants, which significantly enhanced seed germination and promoted shoot and root growth in maize (Zea mays) and beans (*Phaseolus vulgaris*). GC-MS and NMR analyses identified key compounds in the microalgal extract that contribute to plant growth.

Ferreira et al. [14] studied microalgae for treating piggery wastewater and harvested the biomass for the application of biostimulants and biopesticides. Four microalgae strains such as *Tetradesmus obliquus*, *Chlorella vulgaris*, *Chlorella protothecoides*, and *Synechocystis sp.* were tested, and showed significant nutrient removal from piggery wastewater. The biomass from *Tetradesmus obliquus* and *Chlorella protothecoides* promoted seed germination and root growth for cucumber seeds. Moreover, they also studied that these microalgae particularly, *Tetradesmus obliquus* and *Chlorella protothecoides* inhibited *Fusarium oxysporum* growth, demonstrating biopesticides activity. Moreover, Alling et al. [2] cultivated *Chlorella vulgaris* and *Scenedesmus obliquus* in untreated municipal wastewater using a 16 L photobioreactor. The resulting biomass and supernatant were tested on tomato and barley seeds to assess biostimulant effects. *Chlorella vulgaris*, especially intact cells and supernatant, significantly improved germination percentage (up to 25% higher) and reduced germination time by 0.5–1 day compared to control. The germination index was also higher in *Chlorella vulgaris* treatments for both crops. This suggests that *C. vulgaris* grown in municipal wastewater is a promising biostimulant for sustainable agriculture. All these studies demonstrated that dual integrated microalgae cultivation as well as biomass utilization in the application of plant growth and development, which contribute environmental sustainability via practising eco-friendly approaches.

5.2 Biochemical Composition

Various studies suggest that the biomass harvested from microalgae in wastewater, exhibited for the production of increased biochemical composition like carbohydrate, lipid and protein. To illustrate, *Chlorella sorokiniana* SU-1 cultivated in a dairy wastewater was studied for the lipid content of cultivated microalgae in various concentrations. The reported results showed that lipid production of $10.3 \pm 5\%$ in

100% wastewater, 10.4 ± 9% in 25% wastewater, and 10.7 ± 10% in 0% wastewater (BG-11 medium) [25]. They also investigated the carbohydrate content and the results showed that carbohydrate accumulation decreased with increasing digestate concentration. The highest carbohydrate content, 34.4 ± 4%, was observed in 0% wastewater (BG-11 medium), while lower values were recorded at 25%, 50%, 75%, and 100% wastewater with 23.9 ± 2%, 26.8 ± 5%, 21.7 ± 3%, and 19.1 ± 11% respectively. However, they observed 31.8–268.3% increase in carbohydrate concentrations from day 7 to day 15, with the highest productivity (24.31 mg L^{-1} d^{-1}) recorded for the *Scenedesmus–Tetradesmus* (S:T) consortia on day 9 [20]. Moreover, for the lipid production, they have reported that the lipid content increased in response to prolonged nutrient limitation, with *Chlorella* showing a rise in lipid concentration from day 7 to day 15, reaching a maximum of 70.8 mg L^{-1}. Although the lipid concentration peaked on day 15, lipid productivity declined after day 9. The highest LP, 4.94 mg L^{-1} d^{-1}, was observed on day 9 in the *Chlorella–Tetradesmus* (C:T) consortia, which was significantly higher—by 126.03% and 58.63%;—than in the *Chlorella* and *Tetradesmus* monocultures, respectively [20].

Furthermore, Gupta et al. [21] found that co-cultivating *Scenedesmus acutus* and *Tetradesmus* in a 3:1 mix of anaerobically digested and untreated dairy wastewater significantly enhanced carbohydrate content, increasing by 30–264%. During fed-batch cultivation, the carbohydrate content further improved by 44–100% over five cycles, indicating a shift towards carbohydrate accumulation due to nitrogen limitation. The study also revealed that lipid content increased by 30–264% under optimal conditions in the mixed dairy wastewater, with a 44–100% improvement observed during fed-batch cultivation over five cycles. This increase was accompanied by a decrease in the protein-to-lipid ratio, further suggesting moderate nitrogen deficiency in the system, which promotes lipid accumulation.

6 Challenges and Future Prospects

Microalgae-based wastewater treatment presents numerous benefits. The majority of the study on treating wastewater focused on lab scale level experimental setups. However, several challenges must be addressed to optimize this approach for large-scale applications. One of the primary challenges is the efficient harvesting of microalgal biomass. Advanced harvesting methods, such as flotation, centrifugation, or filtration, need further refinement to reduce operational costs and increase the overall efficiency of the process. Another significant challenge is the variability in the composition of wastewater from one to another. Different types of wastewater, such as those from industrial, agricultural, or municipal sources, contain varying concentrations of pollutants as disused earlier, which can affect the growth and efficiency of microalgae. This variation requires the acclimatization of microalgae strains or treatment systems that can perform effectively under diverse conditions.

Future works should focus on improving the scalability and cost-effectiveness of microalgae-based systems. Optimizing algal cultivation conditions, such as light intensity, pH, and nutrient levels, will enhance biomass productivity and pollutant removal efficiency. Also, scaling up the batch scale level biosorption process to pilot scale level for effective wastewater treatment with real time wastewater is suggesting for enhanced real time industrial scale application. Genetic engineering and synthetic biology could also be employed to develop microalgae strains with improved resistance to heavy metals and other toxic substances. Furthermore, there is a need for more extensive studies on the valorization of microalgal biomass, particularly in the development of high-value products such as bioplastics, pharmaceuticals, and other valuable compounds. Studies also should focus on real time wastewater treatment plants using microalgae for further opportunities in terms of cost reduction and increased efficiency.

7 Conclusion

In conclusion, microalgal-based biosorption offers a sustainable and eco-friendly alternative for mitigating water pollution caused by heavy metals, dyes, and nutrients. Unlike conventional treatment methods, microalgae provide a cost-effective strategy with the added advantage of biomass valorization. Cultivating microalgae in wastewater not only aids in pollutant removal but also reduces the need for synthetic media. The harvested biomass holds immense potential for conversion into biofuels, biofertilizers, and other value-added products, thus supporting a circular economy. Ultimately, this integrated approach of using wastewater as a medium for microalgal growth offers promising strategy in mitigating environmental pollution.

8 Competing Interests

The authors declare no competing interests (Tables 2.1 and 2.2).

Table 2.1 Cultivation and treatment of microalgae in various wastewater

Wastewater	Cultivated microalgae	Mode of cultivation	Concentration of wastewater	pH	COD (mg L^{-1})	BOD (mg L^{-1})	Biomass concentration (g L^{-1})	Reference
Textile effluent	*Diplosphaera mucosa* VSPA, and *Scenedesmus obliquus*	Mixotrophic co-cultivation	-	8.2	135	87.2	6.14	[54]
Domestic effluent	*Diplosphaera mucosa* VSPA, and *Scenedesmus obliquus*	Mixotrophic co-cultivation	-	7.3	520	372	8.87	[54]
Anaerobic digestate of dairy wastewater	*Chlorella sorokiniana* SU-1	Mono-cultivation	100%	-	4800	-	4.8	[25]
Tannery wastewater	*Chlorella* sp.	Mono-cultivation	50%	-	6080	2830	1.86	[57]
Piggery wastewater	*Radiococcus polycoccus*	Mono-cultivation	-	7.09	43.5	220	1.08 g dw L^{-1}	[15]
Oxidation Pond of piggery effluents	*Micractinium* sp.	Mono-cultivation	-	8.52	274.1	1050	0.56 g dw L^{-1}	[15]
Sewage wastewater	*Chlorella vulgaris*	Mono-cultivation	-	8.41	920	198	0.67	[40]
Anaerobically digested abattoir effluent	*Chlorella* sp.	Mono-cultivation	Raw wastewater	7.1–7.4	302–514	-	Increased 28%	[58]
Anaerobic digestate abattoir effluent	*Arthrospira platensis*	Mono-cultivation	Untreated wastewater	-	302	-	1.55	[32]
Domestic wastewater	*Scenedesmus* sp. LX1	-	Sterilized domestic wastewater	8.16	548	-	5.26	[59]

Table 2.2 Removal of heavy metals using microalgae

Microalgae	Nature of wastewater	Heavy metals	Initial concentration (mg L^{-1})	Removal efficiency (%)	Reference
Craticula subminuscula	Aqueous solution	Cr (VI)	0 – 30	95.32	[50]
Chlorella vulgaris and bacteria	Synthetic tannery wastewater	Cr (VI)	50	93	[8]
Chlorella thermophilia	Synthetic wastewater	Cr (VI)	0.2–0.6	50–65	[16]
Chlorella thermophilia	Synthetic wastewater	As (III)	0.5–10	70–91	[16]
Scenedesmus almeriensis	Synthetic wastewater	Cu	100	76	[3]
Scenedesmus almeriensis	Synthetic wastewater	Zn	100	93	[3]
Chlamydomonas sp.	Aqueous solution	Cu	57.61	58.11	[17]
Chlamydomonas sp.	Aqueous solution	Cr (VI)	57.61	41.4	[17]
Desmodesmus sp. CHX1	Piggery digestate	Cu (II)	1.00	88.35	(L. [28])
Chlamydomonas sp.	Aqueous solution	Cr (VI)	250	94.17	[4]

Acknowledgements The authors would like to express their gratitude to the management of Vellore Institute of Technology, Vellore for the constant encouragement, help and support for extending necessary facilities.

References

1. Abate TA, Desta AF, Love NG (2021) Evaluating tannery wastewater treatment performance based on physicochemical and microbiological characteristics: an Ethiopian case study. Water Environ Res 93(5):658–669
2. Alling T, Funk C, Gentili FG (2023) Nordic microalgae produce biostimulant for the germination of tomato and barley seeds. Sci Rep 13(1):3509. https://doi.org/10.1038/s41598-023-30707-8
3. Antolín B, Torres A, García PA, Bolado S, Vega M (2024) Mechanisms of copper and zinc bioremoval by microalgae and bacteria grown in nutrient rich wastewaters. Chemosphere 355, 141803. https://doi.org/10.1016/j.chemosphere.2024.141803
4. Ayele A, Suresh A, Benor S, Konwarh R (2021) Optimization of chromium (VI) removal by indigenous microalga (Chlamydomonas sp.)-based biosorbent using response surface methodology. Water Environ Res 93(8):1276–1288

5. Azari P, Hasheminejad H, Zarean Mousaabadi K (2025) Synthesis and characterization of MIL-88 A(Fe)/C composite for treatment of dairy factory's wastewater by enhanced electro-Fenton method. Sci Rep 15(1):8503. https://doi.org/10.1038/s41598-025-92765-4
6. Chu Y, Jiao X, Li D, Yu J, Wang C, He M (2025) Integration of dairy manure wastewater treatment and biofuel production by microalgae-bacteria consortia in a biofilm reactor. Algal Res 86:103912. https://doi.org/10.1016/j.algal.2025.103912
7. Custodio M, Peñaloza R, Espinoza C, Espinoza W, Mezarina J (2022) Treatment of dairy industry wastewater using bacterial biomass isolated from eutrophic lake sediments for the production of agricultural water. Bioresour Technol Rep 17:100891. https://doi.org/10.1016/j.biteb.2021.100891
8. Dabai AI, Mohammed K (2024) Biosorption and desorption of chromium using hybrid microalgae-activated sludge treatment system. Appl Water Sci 14(2):16. https://doi.org/10.1007/s13201-023-02068-0
9. Daneshvar E, Zarrinmehr MJ, Kousha M, Hashtjin AM, Saratale GD, Maiti A, Vithanage M, Bhatnagar A (2019) Hexavalent chromium removal from water by microalgal-based materials: adsorption, desorption and recovery studies. Bioresour Technol 293:122064. https://doi.org/10.1016/j.biortech.2019.122064
10. Dutta S, Adhikary S, Bhattacharya S, Roy D, Chatterjee S, Chakraborty A, Banerjee D, Ganguly A, Nanda S, Rajak P (2024) Contamination of textile dyes in aquatic environment: adverse impacts on aquatic ecosystem and human health, and its management using bioremediation. J Environ Manag 353:120103. https://doi.org/10.1016/j.jenvman.2024.120103
11. Fazal T, Rehman MSU, Javed F, Akhtar M, Mushtaq A, Hafeez A, Alaud Din A, Iqbal J, Rashid N, Rehman F (2021) Integrating bioremediation of textile wastewater with biodiesel production using microalgae (Chlorella vulgaris). Chemosphere 281:130758. https://doi.org/10.1016/j.chemosphere.2021.130758
12. Ferreira A, Belachqer-El Attar S, Villaró-Cos S, Ciardi M, Soriano-Molina P, López JLC, Lafarga T, Marques-dos-Santos C, Acién G, Gouveia L (2025) Piggery wastewater treatment by solar photo-Fenton coupled with microalgae production. Water Res 271:122869. https://doi.org/10.1016/j.watres.2024.122869
13. Ferreira A, Figueiredo D, Cardeiras R, Nabais R, Ferreira F, Ribeiro B, Cordovil CMdS, Acién FG, Gouveia L (2022) Exploring different pretreatment methodologies for allowing microalgae growth in undiluted piggery wastewater. In: Agronomy, vol. 12, Issue 3. https://doi.org/10.3390/agronomy12030580
14. Ferreira A, Melkonyan L, Carapinha S, Ribeiro B, Figueiredo D, Avetisova G, Gouveia L (2021) Biostimulant and biopesticide potential of microalgae growing in piggery wastewater. Environ Adv 4:100062. https://doi.org/10.1016/J.ENVADV.2021.100062
15. Flores-Zambrano K, Tapia W, Castillejo P (2025) Microalgae strains isolated from piggery wastewater in Ecuador: effective nitrogen compound removal and growth potential in extremophile conditions. Biotechnol Rep 45:e00883. https://doi.org/10.1016/j.btre.2025.e00883
16. Ganguly A, Nag S, Bhowmick TK, Gayen K (2024) Bioremediation of chromium(VI) and arsenic(III) using isolated microalgal (Chlorella thermophilia): analysis of growth, biomolecular compositions (carbohydrate, protein, chlorophyll) and biosorption kinetics. Algal Res 82:103635. https://doi.org/10.1016/j.algal.2024.103635
17. Ghribi I, Elleuch J, Dubessay P, Michaud P, Abdelkafi S, Fendri I (2025) Biotechnological potential of newly isolated microalga strain in Cu and Cr biosorption from single and bimetallic systems. In: Water, vol 17, issue 7. https://doi.org/10.3390/w17070999
18. Global Wastewater Treatment Services Market (nd) https://www.sphericalinsights.com/reports/wastewater-treatment-services-market
19. Gramegna G, Scortica A, Scafati V, Ferella F, Gurrieri L, Giovannoni M, Bassi R, Sparla F, Mattei B, Benedetti M (2020) Exploring the potential of microalgae in the recycling of dairy wastes. Bioresour Technol Rep 12:100604. https://doi.org/10.1016/j.biteb.2020.100604
20. Gupta S, Marchetti JM (2024) Co-cultivation of high-value microalgae species with filamentous microalgae for dairy wastewater treatment. Npj Clean Water 7(1):119. https://doi.org/10.1038/s41545-024-00421-7

21. Gupta S, Marchetti JM, Wasewar KL (2024) Enhancing nutrient removal, biomass production, and biochemical production by optimizing microalgae cultivation in a mixture of untreated and anaerobically digested dairy wastewater. J Water Process Eng 63:105413. https://doi.org/10.1016/j.jwpe.2024.105413
22. Idusuyi N, Adebayo MA, Igwegbe CA, Aghogho OT, James A, Kazeem RA (2025) A sustainable approach to dairy wastewater treatment through electrocoagulation: from beverage cans to clean water. Waste Manag Bull 3(1):96–106. https://doi.org/10.1016/j.wmb.2024.12.012
23. Javed F, Rehman F, Khan AU, Fazal T, Hafeez A, Rashid N (2022) Real textile industrial wastewater treatment and biodiesel production using microalgae. Biomass Bioenergy 165:106559. https://doi.org/10.1016/j.biombioe.2022.106559
24. Jirasansawat K, Chiemchaisri W, Chiemchaisri C (2024) Enhancement of sulfide removal and sulfur recovery in piggery wastewater via lighting-anaerobic digestion with bioaugmentation of phototrophic green sulfur bacteria. Environ Sci Pollut Res 31(9):13414–13425. https://doi.org/10.1007/s11356-024-31920-2
25. Kusmayadi A, Leong YK, Lu P-H, Huang C-Y, Yen H-W, Chang J-S (2022) Simultaneous nutrients removal and bio-compounds production by cultivating Chlorella sorokiniana SU-1 with unsterilized anaerobic digestate of dairy wastewater. Algal Res 68:102896. https://doi.org/10.1016/j.algal.2022.102896
26. Lawrence LV, Venkat Kumar S, Vishnu D (2025) Emerging trends on biologically synthesized nanofertilizers from microalgal extracts and their enhanced productivity in foliar and soil applications. Energy Ecol Environ. https://doi.org/10.1007/s40974-025-00363-x
27. Liang C, Zhang N, Pang Y, Li S, Shang J, Zhang Y, Kuang Z, Liu J, Fei H (2023) Cultivation of Spirulina platensis for nutrient removal from piggery wastewater. Environ Sci Pollut Res 30(36):85733–85745. https://doi.org/10.1007/s11356-023-28334-x
28. Liu L, Lin X, Luo L, Yang J, Luo J, Liao X, Cheng H (2021) Biosorption of copper ions through microalgae from piggery digestate: optimization, kinetic, isotherm and mechanism. J Clean Prod 319:128724. https://doi.org/10.1016/j.jclepro.2021.128724
29. Liu X-Y, Hong Y, Zhang Y-W, Li L-H (2024) Valorization of treated swine wastewater and generated biomass by microalgae: their effects and salt tolerance mechanisms on wheat seedling growth. Environ Res 251:118664. https://doi.org/10.1016/J.ENVRES.2024.118664
30. Marazzi F, Fornaroli R, Clagnan E, Brusetti L, Ficara E, Bellucci M, Mezzanotte V (2023) Wastewater from textile digital printing as a substrate for microalgal growth and valorization. Bioresour Technol 375:128828. https://doi.org/10.1016/j.biortech.2023.128828
31. Martins RA, Salgado EM, Gonçalves AL, Esteves AF, Pires JCM (2024) Microalgae-based remediation of real textile wastewater: assessing pollutant removal and biomass valorisation. Bioengineering 11(1)
32. Matos ÂP, Vadiveloo A, Bahri PA, Moheimani NR (2021) Anaerobic digestate abattoir effluent (ADAE), a suitable source of nutrients for Arthrospira platensis cultivation. Algal Res 54:102216. https://doi.org/10.1016/j.algal.2021.102216
33. Monzón Martínez L, Atayupanqui Dueñas R, González Sánchez A, Montalvo Andia J (2025) Study on the growth of a native microalgae-bacteria mixed culture from the Peruvian Andes exposed to different tannery effluent dilutions and photoperiod. Environ Tech 1–16
34. Moreno-Cruz CF, Tzintzun-Camacho O, Gonzalez-Joaquin MC, Aguilar-Martinez XE, González-Mendoza D, Martinez-Quiroz M (2025) Livestock wastewater as a microalgae growth medium for potential production of biodiesel in arid areas of Mexico. Algal Res 86:103957. https://doi.org/10.1016/j.algal.2025.103957
35. Morillas-España A, Lafarga T, Sánchez-Zurano A, Acién-Fernández FG, González-López C (2022) Microalgae based wastewater treatment coupled to the production of high value agricultural products: Current needs and challenges. Chemosphere 291:132968. https://doi.org/10.1016/J.CHEMOSPHERE.2021.132968
36. Morillas-España A, Pérez-Crespo R, Villaró-Cos S, Rodríguez-Chikri L, Lafarga T (2024) Integrating microalgae-based wastewater treatment, biostimulant production, and hydroponic cultivation: a sustainable approach to water management and crop production. Front Bioeng Biotech 12:1364490. https://doi.org/10.3389/fbioe.2024.1364490

37. Obaideen K, Shehata N, Sayed ET, Abdelkareem MA, Mahmoud MS, Olabi AG (2022) The role of wastewater treatment in achieving sustainable development goals (SDGs) and sustainability guideline. Energy Nexus 7:100112. https://doi.org/10.1016/j.nexus.2022.100112
38. Oladimeji TE, Oyedemi M, Emetere ME, Agboola O, Adeoye JB, Odunlami OA (2024) Review on the impact of heavy metals from industrial wastewater effluent and removal technologies. Heliyon 10(23):e40370. https://doi.org/10.1016/j.heliyon.2024.e40370
39. Pandey A, Kant G, Chaudhary A, Amesho KTT, Reddy K, Bux F (2024) Axenic green microalgae for the treatment of textile effluent and the production of biofuel: a promising sustainable approach. World J Microbiol Biotechnol 40(3):81. https://doi.org/10.1007/s11274-023-03863-2
40. Pooja K, Priyanka V, Rao BCS, Raghavender V (2022) Cost-effective treatment of sewage wastewater using microalgae Chlorella vulgaris and its application as bio-fertilizer. Energy Nexus 7:100122. https://doi.org/10.1016/j.nexus.2022.100122
41. Priya AK, Gnanasekaran L, Dutta K, Rajendran S, Balakrishnan D, Soto-Moscoso M (2022) Biosorption of heavy metals by microorganisms: evaluation of different underlying mechanisms. Chemosphere 307:135957. https://doi.org/10.1016/j.chemosphere.2022.135957
42. Procopio LSD, Escudero A, Hunter C (2025) Chlamydomonas acidophila for phosphorus recovery from municipal wastewater treatment plants: effect of light intensity, temperature, different wastewaters and long-term semi-continuous feeding. Algal Res 86:103941. https://doi.org/10.1016/j.algal.2025.103941
43. Ranjan A, Lilly M, Deepnarain N, Welz PJ (2025) Valorisation of tannery wastewater with indigenous microalgal strains: acclimation, identification, biomass characterization and remediation. Algal Res 86:103953. https://doi.org/10.1016/j.algal.2025.103953
44. Rawat J, Nanda M, Kumar S, Sharma N, Sharma R, Joshi HC, Vlaskin MS, Hussain A, Kumar V (2024) Integrating wastewater treatment to bio-stimulant & biochar generation for plant growth promotion using microalgae. Process Biochem 145:187–194. https://doi.org/10.1016/j.procbio.2024.06.031
45. Raza N, Rizwan M, Mujtaba G (2024) Bioremediation of real textile wastewater with a microalgal-bacterial consortium: an eco-friendly strategy. Biomass Convers Biorefinery 14(6):7359–7371. https://doi.org/10.1007/s13399-022-03214-5
46. Reda AH (2016) Physico-chemical characterization of tannery effluent and its impact on the nearby river. J Environ Chem Ecotoxicol 8(6):44–50
47. Rendón-Castrillón, L., Ramírez-Carmona, M., Ocampo-López, C., González-López, F., Cuartas-Uribe, B., & Mendoza-Roca, J. A. (2024). Efficient bioremediation of indigo-dye contaminated textile wastewater using native microorganisms and combined bioaugmentation-biostimulation techniques. *Chemosphere*, *353*, 141538. https://doi.org/10.1016/j.chemosphere.2024.141538
48. Rossi, S., Pizzera, A., Bellucci, M., Marazzi, F., Mezzanotte, V., Parati, K., & Ficara, E. (2022). Piggery wastewater treatment with algae-bacteria consortia: Pilot-scale validation and techno-economic evaluation at farm level. *Bioresource Technology*, *351*, 127051. https://doi.org/10.1016/j.biortech.2022.127051
49. Sarma U, Hoque ME, Thekkangil A, Venkatarayappa N, Rajagopal S (2024) Microalgae in removing heavy metals from wastewater—an advanced green technology for urban wastewater treatment. J Hazard Mater Adv 15:100444. https://doi.org/10.1016/j.hazadv.2024.100444
50. Sbihi K, Elhamji S, Lghoul S, Aziz K, El Maallem A, Mabrouki J, El-Sheekh M, Aziz F (2024) Biosorption of Hexavalent Chromium by Freshwater Microalgae Craticula subminuscula from Aqueous Solutions. In: Sustainability, vol 16, issue 2. https://doi.org/10.3390/su16020918
51. Shamshad J, Rehman RU (2025) Innovative approaches to sustainable wastewater treatment: a comprehensive exploration of conventional and emerging technologies. Environ Sci Adv
52. Simsek K, Sisman-Aydin G (2025) The potential of growth and pollutant removal of green microalgae Neglectella solitaria for real textile wastewater. Environ Technol Innov 37:104014. https://doi.org/10.1016/j.eti.2024.104014
53. Singh K, Dave H, Prasad B, Kumari M, Dubey D, Rai AK, Ravi R, Manjhi J, Sillanpää M, Prasad KS (2024) NFeO decorated wood biochar as an adsorbent for aqueous Cr (VI) ions: hyphenated, coagulation-column treatment of tannery effluent. J Water Process Eng 59:105084

54. Singh V, Mishra A, Srivastava P (2023) Textile and domestic effluent treatment via co-cultivation of Diplosphaera mucosa VSPA and Scenedesmus obliquus. Biomass Bioenergy 172:106756. https://doi.org/10.1016/j.biombioe.2023.106756
55. Tarahomi DS, Hosseini SP, Abbaspour A, Jafari A, Mousavi SM (2025) Characterization and application of polysaccharides produced by Pseudomonas atacamensis M7D1 for chromium (III) removal from tannery effluent. Int J Biol Macromol 298:139944. https://doi.org/10.1016/j.ijbiomac.2025.139944
56. Torres NH, Souza BS, Ferreira LFR, Lima ÁS, dos Santos GN, Cavalcanti EB (2019) Real textile effluents treatment using coagulation/flocculation followed by electrochemical oxidation process and ecotoxicological assessment. Chemosphere 236:124309. https://doi.org/10.1016/j.chemosphere.2019.07.040
57. Urbina-Suarez NA, Salcedo-Pabón CJ, Contreras-Ropero JE, López-Barrera GL, García-Martínez JB, Barajas-Solano AF, Machuca-Martínez F (2025) Biotechnological strategy for tannery wastewater treatment: Bicarbonate/H2O2 oxidation integrated with microalgae cultivation. Case Stud Chem Environ Eng 11:101060. https://doi.org/10.1016/j.cscee.2024.101060
58. Vadiveloo A, Matos AP, Chaudry S, Bahri PA, Moheimani NR (2020) Effect of CO2 addition on treating anaerobically digested abattoir effluent (ADAE) using Chlorella sp. (Trebouxiophyceae). J CO2 Util 38:273–281. https://doi.org/10.1016/j.jcou.2020.02.006
59. Wang Q, Wang X, Hong Y, Liu X, Zhao G, Zhang H, Zhai Q (2022) Microalgae cultivation in domestic wastewater for wastewater treatment and high value-added production: species selection and comparison. Biochem Eng J 185:108493. https://doi.org/10.1016/j.bej.2022.108493
60. Zapata D, Arroyave C, Cardona L, Aristizábal A, Poschenrieder C, Llugany M (2021) Phytohormone production and morphology of Spirulina platensis grown in dairy wastewaters. Algal Res 59:102469. https://doi.org/10.1016/J.ALGAL.2021.102469

Desideratum of the Life Mission in India: Exploring the Need, Requisites, and Implications of Water Green Credits

Juhi Kamra, Manu Sharma, Ambica Prakash Mani, and Sudhanshu Joshi

Abstract Although water enwraps three-fourths of the earth's surface, only 3% of this water can be categorized as drinkable, non-saline, fresh water. Of this fresh water, 69% is stored in the form of glaciers and ice caps, and only 31% of it is accessible. In the context of India, though the region is a peninsula but it still has to deal with scarcity of water and generation of wastewater. India constitutes 17% of the global population, i.e., about 146 crores, and has 2.4% of the global landmass that consumes around 4% of the world's water resources. As per the last census, the per capita water availability is 1544 cubic meters, which is declining every year, signaling a threat to water sustainability. Over and above the scarcity of water, there lies concerns for wastewater disposal that is around 38,000 million litres per day generated from urban centres alone as per the last census. To vouch responsible water stewardship in the long run and hold individuals accountable for their wastewater generation and water usage, the Government of India introduced the Mission LIFE in 2021 at the UN Climate Change Conference (COP26) that has water conservation through green credits as one of the focus areas among other environmental sustainability domains. It focuses on transitioning from the "use-and-dispose" economy to a "circular economy" whereby individuals and organizations would be able to save, trade, and earn water green credits with a threshold being standardized for water consumption. The water green credits would work in a similar mechanism to carbon trading. To dive deep into the understanding of the water green credits mechanism, this research would assess the

J. Kamra (✉) · A. P. Mani
Department of Commerce, Graphic Era (Deemed to Be) University, Dehradun, Uttarakhand, India
e-mail: juhi.phd@geu.ac.in

A. P. Mani
e-mail: ambicamani9@gmail.com

M. Sharma
Department of Management, Graphic Era (Deemed to Be) University, Dehradun, Uttarakhand, India
e-mail: manu.sharma@geu.ac.in

S. Joshi
School of Management, Doon University, Dehradun, Uttrakhand, India
e-mail: sudhanshujoshi@doonuniversity.ac.in

V. Kandpal and T. Choudhury (eds.), *Blue Revolution: Sustainable Water Use and Circular Economy Innovations*, Urban Sustainability,
https://doi.org/10.1007/978-981-95-1916-3_3

need, the requisites, and its implications for India by analyzing it from the perspective of the carbon trading mechanism.

1 Introduction

Water is an irreplaceable and fundamental commodity for the subsistence of the world. But due to the increasing population, rapid urbanization, industrialization, and inconsiderate use of water, its sustainability is threatened [16]. Since only 3% of the water on the earth's surface can be deemed as drinkable, fresh water, which is swiftly declining [3], it becomes imperative to initiate steps towards saving, storing, and promoting efficient usage of water. Moreover, it is projected that nearly 3 billion people will experience severe water scarcity due to droughts resulting from global warming [21]. In the context of India, the region has the highest population in the world of around 1.4 billion, i.e., about 17% of the world's population [21], with 2.4% of global landmass [29] that expends 4% of global water resources, has around 1047 billion Cubic Meters per year of water footprints [16], and groundwater depletion rate of 23% in a decade from 2000–2010 [15]. As per a report of the Ministry of Environment, Forest and Climate Change (MoEFCC), Government of India, the agricultural sector in India accounts for 91% of water consumption, followed by the domestic sector with 7% consumption and the Industrial sector with 2% consumption of water as shown in Fig. 1 [24]. The report also asserts that the usage of water is inefficient in India as compared to other countries wherein India records a low irrigation efficiency of just 38% across its agricultural projects in comparison to 50–60% efficiency in developed nations, the industrial plants in India consumes 2–3.5 times of water than comparable plants of other nations, and the loss of water in domestic sector due to leakages in pipes and valves accounts for 30–40% [24]. This makes it indispensable for India to march towards water preservation and make plans to achieve efficient water usage. Moreover, India is a member country of the United Nations Sustainable Development Goals (SDGs) under which the 6th SDG pertains to ensuring accessibility and sustainable water stewardship and sanitation inclusivity [30]. Thus, international pressure also mandates India to take steps towards water conservation, resilient water stewardship, and the reduction of its water footprint. Angelo George, the CEO of a well-known beverage company in India, Bisleri collaborated with the TERI School of Advanced Studies to assess the water footprints of companies across their supply chain and proposed a water green credit system similar to the carbon credit certificate system to the government for efficient and sustainable usage and management of water [5]. This way, the companies can be held accountable for their water footprints and also be rewarded for their water conservation practices.

India took a pivotal move towards sustainable management of water through the Mission LiFE [18]. The Mission LiFE was introduced by the Hon'ble Prime Minister of India, Shri Narendra Modi, at the 26th Conference of the Parties (COP26) to the United Nations Framework Convention on Climate Change (UNFCCC) in Glasgow,

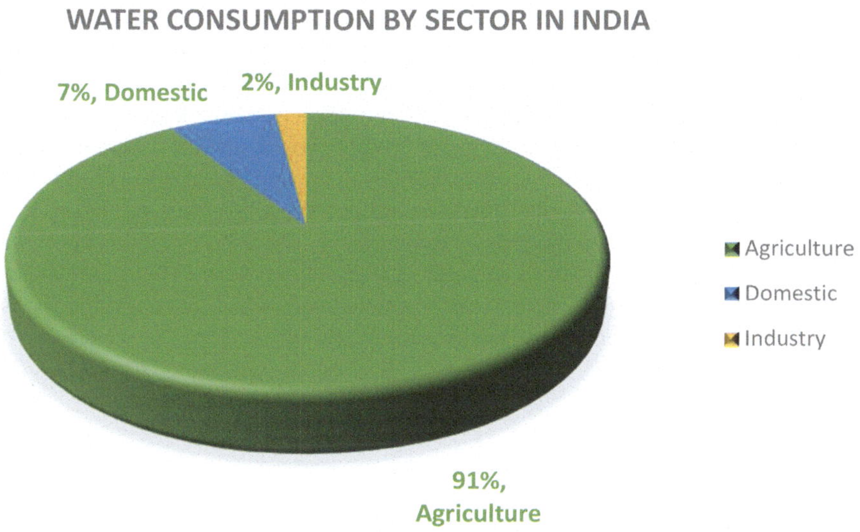

Fig. 1 Consumption of water in different sectors in India

UK. India incorporated the Mission LiFE into its Nationally Determined Contributions as part of its commitments to the Paris Agreement [19]. The Mission LiFE stands for Lifestyle for Environment, which aims to channel Indian and transnational citizens to engage cohesively for conserving the environment in the period between 2022–2028 [25]. This Mission will encourage individuals, institutions, and communities to take small strides for environmental conservation, thus leading the gentle steps to greener paths. The Mission prescribes a list of 75 actions in 7 environmental realms of energy, the embracement of sustainable food systems, water, waste reduction, adoption of a healthy lifestyle, single-use plastic, and reduction of E-waste. This Mission envisions translating the lifestyle of global citizens from a use-and-throw approach to a circular-economy approach. To further the achievement of this Mission, the MoEFCC, Government of India, launched the Green credit Program (GCP) on 13th October, 2023, supported and administered by inter-ministerial Steering Committee and the Indian Council of Forestry Research and Education (ICFRE), that will stimulate voluntary environment-friendly actions by individuals, private sector industries, communities, and companies [20]. Water conservation is one of the focuses of GCP in its initial phase, which will steer India towards fulfilling its commitment to SDG 6 of the Paris Agreement.

The Green credit program will work on the same grounds as the carbon market and trading mechanism, whereby carbon credits are bought and sold by companies [8]. The carbon trading mechanism works in two ways: through a compliance mechanism or through voluntary mechanism. In the compliance mechanism, the government sets a cap on the emissions of the companies and then enables them to trade the emission allowances with each other to maintain the caps. The voluntary mechanism works through independent standards or non-governmental organizations that allow companies to purchase carbon credits through green projects that trades off their emissions to keep their emissions within the limits [2]. In the Green credit program, methodologies are designed to grant green credits to the companies and are shared with stakeholders. These methodologies will provide a framework to assess activities for their environmental impacts, the green credits earned, and their exchanges. The government registers the green projects, verifies them, and grants green credits to the companies through a digital platform (www.moefcc-gcp.in), thus providing green credit certification upon verification of the projects. These green credits could then be traded among the companies to achieve their targets [20]. This study aims to explore the need for a water green credits mechanism, which is one of the focuses of GCP, its requisites, and its prospective implications.

2 Literature Review

There is meagre literature that talks about green water credits. Most of the research on green water credits relates to China, Kenya, and Morocco [6]. [4] examined carbon credit generation through irrigation water management by studying 6 different approaches in 3 different dimensions, viz. methodology and accounting that relate to the calculation, and assumptions involved in generating credits, impact beyond carbon that inculcates the impact of irrigation beyond environmental impacts like gender inequality, and Incentives for farmers that focuses on the direct and indirect benefits of the carbon credit schemes to the farmers. [27] identified the strengths and challenges of Green Water Credits in the agricultural sector to build resilience against drought conditions, whose success hinges on conserving soil and water, hydro-policies, agro-technologies, and payment for environmental services. [14], along with exploring the role of soil conservation measures in curbing soil erosion through quantitative modelling techniques in Kenya, also developed a business case model to explore the impact of the green water credit programme in implementing the soil conservation measures, showing the costs and benefits associated with it. [10] developed a methodological framework using public domain tools and datasets like the Green water and Blue water Assessment Toolkit (GBAT) to explore the effects of soil conservation measures in supporting the implementation of green water credit schemes that positively impact the upstream and downstream water conservation and availability in Upper-Tana basin in Kenya. [26] assessed the feasibility of implementing the Green Water Credit scheme in the Piedade watershed, in the Minas Gerais state of Brazil. Moreover, the study also explored the role of different stakeholders in aiding the rollout of the Green Water Credit mechanism using literature review, questionnaires, and interviews.

Other research in this context is in the form of official reports. A report by the International Soil Reference and Information Centre (ISRIC) designed a financing model for the Green Water Credit Investment Fund. For this purpose, the report identified 1. the choices farmers make in selecting investment mechanisms for soil and water preservation, 2. Models and potential financial frameworks and institutional arrangements for green water management, 3. Prospective sources for funding of soil and water conservation activities [23]. Another report by ISRIC undertook a small demonstration project through a Proof-of-Concept assessment, which assessed the feasibility of Green Water Credits in the Sebou River Basin in Morocco at the initial stage using the Soil and Water Assessment Tool (SWAT) [11]. Yet another report of ISRIC assessed the political, financial, and institutional framework for Green water credits in the Tana basin of Kenya by analyzing the land and water rights in the region, the management institutions, feasibility of operation of Green Water Credits, capacity and authority to incorporate changes in the system, the institutional environment, primary organizations and policies impacting land and water management in the region [17].

It could be inferred from the above discussion that the research related to the Green Water Credit hitherto prominently pertains to Africa, and there is an absence of studies in this field for India. Therefore, through this review, this study endeavours to conduct the foundational study for water green credits in India, specifying the need for water green credits in India, its requisites, and its implications by assessing and aligning the system with the carbon trading mechanism.

3 Methodology

Data has been searched using the terms like "water green credits" OR "green water credits" OR "carbon trading mechanism" OR "green credit mechanism" in Scopus, Google and Google Scholar. The search led us to the identification of official reports and a few articles related to the green credit system. The results have been analyzed through reviewing articles and grey literature in the form of government reports, news reports, etc.

4 Results

4.1 Need for Water Green Credit System in India

There exist many reasons for the implementation of the water green credit system in India, like rapid urbanization in India, India's commitment to the Paris Agreement, depletion of groundwater, increasing water usage for industrialization and agriculture, worsening natural disasters, etc [16]. The prominent reasons behind the implementation of the Water Green Credit System in India are depicted through Fig. 2 and are explained through the following points assessed through the review of literature:

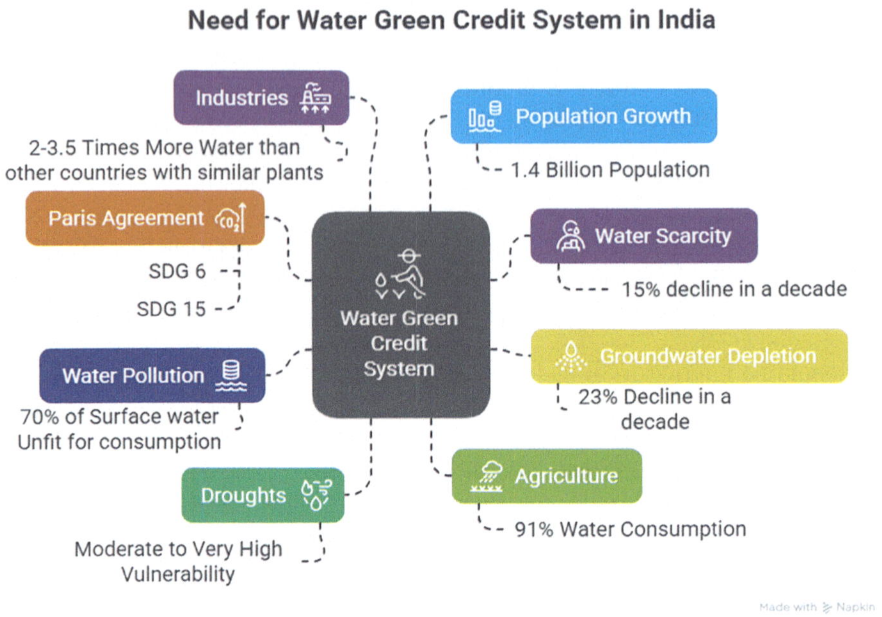

Fig. 2 Need for Water Green Credit mechanism in India

1. *India's commitment to the Paris Agreement:* India is a member country to the Paris Agreement that aims to curb climate change. Moreover, due to India's commitment to the United Nations Sustainable Development Goals, with SDG 6 that focuses on clean water and sanitation, and SDG 15 for Life on Land that emphasizes the sustainable use of ecosystems, it binds India to take initiatives for the sustainable use of water [30]. The Water Green Credit system may serve as an effective way to achieve this.
2. *Water Scarcity in India:* As per a report of the Ministry of Jal Shakti, Department of Water Resources, RD & GR, the per capita water availability for 2001 was 1816 cubic meter that declined to 1544 cubic meter in 2011 i.e., a decline of around 15% in a decade. Owing to this heavy decline of per capita water availability in India, it is indispensable for India to take a strong move for dealing with the water scarcity issue through Water Credit Mechanism [22].
3. *Depleting groundwater level:* The groundwater in India is being depleted at the rate of 23% in a decade from 2000–2010 [15]. This urges India to take strong action to preserve this drastic loss.
4. *Increasing water pollution and degrading water quality:* As per the report of the Ministry of Jal Shakti, Department of Water Resources, RD & GR, urban centres with population more than 50,000 in India generate around 38,000 million litres of wastewater per day (mld) which is expected to cross over 1,00,000 mld by 2050 with another 50,000 mld projected to be generated by rural areas [22]. As per the World Economic Forum, around 70% of the surface water in India is not

fit for consumption, and around 40 million litres of wastewater emanate into the rivers and water bodies in India. Moreover, it also asserts that this could not only lead to water pollution but could also decrease the GDP of India by half [9].

5. *Increased frequency of droughts:* As per the Ministry of Earth Sciences, Meteorological Department of India, most of the districts in India are prone to drought with a normalized vulnerability index between 0.25 to 1, i.e., between moderate to very high [12]. A mechanism like Water Green Credit could promote sustainable water management.
6. *Overuse of water in Agriculture:* 91% of water consumption in India accounts for agricultural usage [24]. A scheme like Water Green Credit will incentivize the farmers to use water-efficient and sustainable practices in agriculture and thus earn green credits through their green projects that could later be traded to companies who want to make up for their water consumption [1].
7. *Inefficient usage of water in industries:* The report by the National Water Mission asserts that water is used inefficiently in industries in India, wherein the industrial plants in India consume 2–3.5 times of water than the comparable plants of other nations [24]. To alleviate inefficiencies in water use and counterbalance the overconsumption of water by companies, the Water Green Credit system could serve as a solution wherein these companies can purchase water green credits from those small companies and farmers who earn and save the credits [7].
8. *Increasing population of India:* The population of India is the highest in the world, i.e., around 1.4 billion, which accounts for 17% of the global population. Moreover, it is expected to peak and reach about 1.7 billion in the early 2060s [28], which alarms for increasing water demand as well. In order to sustainably manage the water resources, it is an urgent need to launch systems like Water Green Credits, which will promote the preservation and efficient usage of water bodies.

4.2 Requisites of the Water Green Credit System in India

The requisites of the Water Green Credit program in India are assessed by analyzing the requisites of the carbon credit system and probing the key aspects of the carbon credit and trading system. Table 1 shows the different parameters that are required for the implementation of Carbon Credit Trading System which are also deemed necessary for the implementation of Water Green Credit System or the Green Credit Program launched by the Ministry of Environment, Forest and Climate Change on 13th October, 2023 with afforestation and water conservation as the key focus areas [20]. Moreover, Fig. 3 highlights the requisites that are yet to be adopted by India for the successful implementation of the Water Green Credit system.

Table 1 Requisites of the Water Green Credit System aligned with the Carbon Credit Trading System

Basis	Carbon Credit Trading System	Water Green Credit system/ Green credit Program
Act	A designated act like the Energy Conservation Act, 2022 [13]	No such act is presently there
Authorized agencies	Authorization to agencies for issuing carbon credit certificates [2, 13	The green credits would be granted by the ICFRE [20]. There is absence of any agency specifically designated for water credits
Methodology for calculation of credits	Methodology for the calculation of the value of each carbon credit certificate (CCCs), e.g., 1 Carbon credit certificate = 1 tonne of CO_2 equivalent reduction [2, 13]	The methodology for calculating green credits, which sets the benchmark for each activity/process, has been developed to guarantee the environmental effectiveness and transferability of green credits across various sectors but there is absence of such methodology specific to water green credits [20]. Since no value has yet been assigned for the value of 1 water credit certificate, it could take inspiration from the carbon credit certificate and assign the value of 1 water credit with 1000 cubic meters of water saved with specified water quality [1] for assessing the water quality
A ministry to oversee the mechanism	The Ministry of Power is designated for the Carbon Credit Trading Scheme [13]	The Ministry of Environment, Forest and Climate Change is designated to oversee the Green Credit Program [20]
A designated administrator	The BEE oversees the issuance of Carbon Credit Certificates [2, 13]	The ICFRE will serve as an administrator to the Green Credit Program, which will be responsible for the implementation, management, monitoring, and operation of the program [20]
Baseline Targets	Carbon emission intensity baseline targets for obligated entities (in terms of CO_2 equivalent per unit of output) to be set after considering factors such as available technology, and the cost of implementation for carbon reduction [2, 13]	The baseline targets for water credits are to be determined considering factors such as rainfall, groundwater availability, water quality, soil characteristics, current consumption levels, and the virtual or hidden water consumption in the production of goods, inculcating green, blue, and grey water consumption. Here, green water pertains to rainwater usage, blue water relates to surface water use, and grey water is the water utilized to dilute contaminants [7]

(continued)

Table 1 (continued)

Basis	Carbon Credit Trading System	Water Green Credit system/ Green credit Program
Sectors to be covered	The sectors covered include aluminium, chloralkali processes, fertilizers, cement, iron and steel, paper and pulp, petrochemicals, textiles, and petroleum refining sectors [13]	There is absence of a compliance mechanism that shrugs off the need for sector specification in the case of water green credits
Pollutants to be included	The pollutants initially covered include carbon dioxide and perfluorocarbons (PFCs) [13]	The Green Credit Program talks about water conservation as one of the objectives of the program on a broad spectrum, it doesn't specifically describe the pollutants that need to be covered for maintaining the water quality hitherto
Link to Nationally Determined Contributions	Setting up a short-term trajectory for GHG emissions reduction up to 2030 based on Nationally Determined Contribution commitments, i.e., currently to reduce carbon emissions by 45% by 2030, available technology, and expected abatement costs in different sectors [2, 13]	The Green Credit Program is an initiative under the LiFE Mission of the Government of India, which is linked to the Nationally Determined Contributions. So, Water Green Credit Program while not directly but indirectly linked to India's Nationally Determined Contributions through the broader scope of LiFE Mission [25]
Mechanism to issue credit certificates	The compliance mechanism to issue CCCs: entities are given obligatory emission intensity targets in CO_2 equivalent/ output for each specified compliance period, the entities that achieve beyond the targets are issued with the CCCs and those that don't achieve the targets are required to purchase and hand over the certificates for gap below the target. The entities with surplus certificates could trade them to other entities needing those certificates [2, 13]	The mechanism to issue Water credits: The entities could earn water credits just like carbon credits by voluntarily adopting water-saving measures. These credits can then be traded to those who need them for offsetting their excess water usage [7]

(continued)

Table 1 (continued)

Basis	Carbon Credit Trading System	Water Green Credit system/ Green credit Program
Mechanism for trading credit certificates	The mechanism for trading of CCCs: The certificates could be traded through power exchanges in India. Obligated entities need to register themselves on the national registry. Moreover, non-obligated entities could also trade the certificates. The scheme doesn't allow over-the-counter trading initially[13]	The Green credit registry and trading platform is developed by the ICFRE, with due consultation from experts, that would facilitate the registration and trading of green credits. The individuals and entities have to register themselves through the central government's website www.moefcc-group.in to obtain green credits. The green credit certificate would be granted after verification by the administrator [20]. The smaller companies could create tailored water-conservation projects and thus earn water credits that could be sold to bigger companies that need to offset their excess water requirements [1]
Voluntary mechanism	The voluntary mechanism for CCCs: The non-obligated entities can register in projects to reduce the GHGs to avail CCCs that could be traded in power exchanges. This will incentivize the non-obligated entities to reduce emissions. [2, 13]	The Green Credit Program, notified on 13th October 2023 by the MoEFCC, stated the deployment of a market-based mechanism to incentivize voluntary environmental actions across different sectors by stakeholders, including individuals, communities, the private sector, and companies. It focuses on two key areas, viz., water conservation and afforestation (Ministry of Environment, [20]
Authority for overseeing trading	A designated authority to oversee the voluntary trading mechanism [2]	No such authority has yet been assigned for the Green Credit Program
Digital Platform	Establishment of a supporting framework: The National Framework for Indian Carbon Market works through a national carbon credit electronic platform that assists entities to price their additional actions for GHG emissions reduction for decarbonization [2]	The Green Credit framework uses a user-friendly digital platform that will facilitate the registration of water conservation projects, their verification, and the grant of green credits [20]

(continued)

Table 1 (continued)

Basis	Carbon Credit Trading System	Water Green Credit system/ Green credit Program
Steering Committee	Constitution of a designated committee for overseeing the function of the Indian Carbon Market: The Central government of India constituted the National Steering Committee for the Indian Carbon Market for this purpose with member from different ministries and organizations under the chairmanship of secretory of Ministry of Power and co-chairmanship of secretory of Ministry of Environment, Forest, and Climate Change [2]	The green credit program will be supported by the inter-ministerial Steering Committee [20]
Functions of the Steering Committee	Specify the functions of the National Steering Committee for the Indian Carbon Market [2]	No such functions have yet been notified
Function of Administrator	Specify the functions of the administrator, i.e., the Bureau of Energy Efficiency [2]	No such detailed functions have yet been notified
Registry Operator	A designated registry operator for the Indian Carbon Market: The Grid controller of India operates the Indian Carbon Market [2]	No such operator has yet been notified
Function of Registry Operator	Specify the functions of the registry operator [2]	No such functions have yet been notified
Regulating body for the credit certificate market	A designated body for the regulation of trading activities in the Indian Carbon Market: The Central Electricity Regulatory Commission (CERC)	No such body has yet been notified
Functions of the regulatory body	BEE specifies the functions of the regulatory body for trading activities in the Indian Carbon Market [2]	No such functions have yet been notified
Eligibility criteria for approval of projects	BEE specifies the eligibility criteria for the approval of projects registered under the voluntary scheme and the phases involved in it [2]	A user-friendly digital platform will facilitate the registration of water conservation projects, their verification, and the grant of green credits [20]

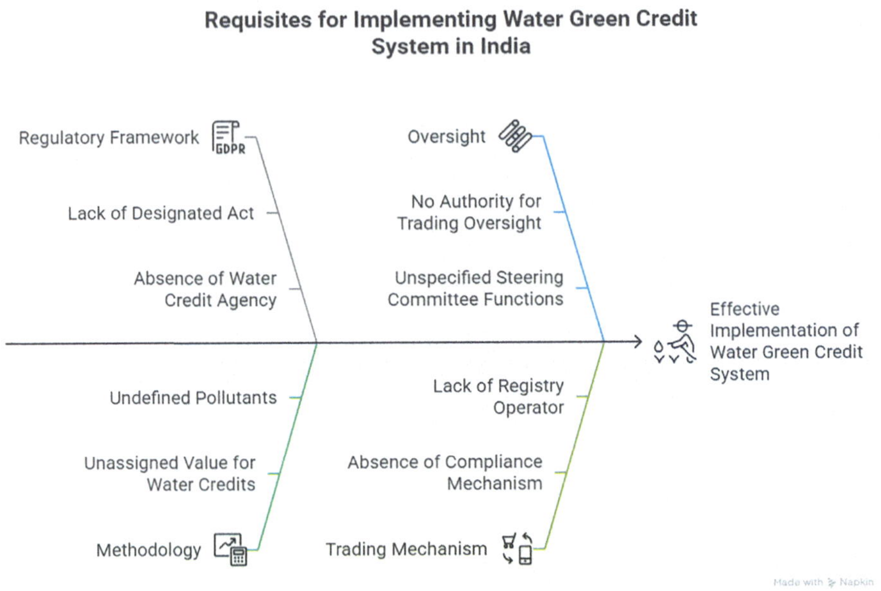

Fig. 3 Absence of requisites for implementing the Water Green Credit mechanism in India

4.3 The Implications of Implementation of Water Green Credits

Figure 4 briefly shows the prospective impacts of the Water Green Credit mechanism. Further, the following discussion highlights in detail the prospective implications of the Water Green Credit system in India:

1. It will incentivize farmers, cooperatives, and entities to adopt water-efficient practices [7].
2. It will promote water conservation practices in agriculture, like drip irrigation, precision technologies, and micro-irrigation. These techniques could reduce the irrigation cost by 45% and increase the yield by 114% in comparison to traditional methods [8].
3. It will provide financial benefits to farmers and support their livelihood by allowing them to trade the water green credits earned by practicing water-efficient practices [7].
4. It will secure water availability [7].
5. It will enhance crop yields in the agricultural sector [7].
6. Promoting water conservation will also boost rainwater harvesting, efficient water usage, and wastewater treatment [8].
7. Entities will be rewarded for their water-efficient practices through water credits [8].
8. Business entities could be held accountable for their water footprints [8].

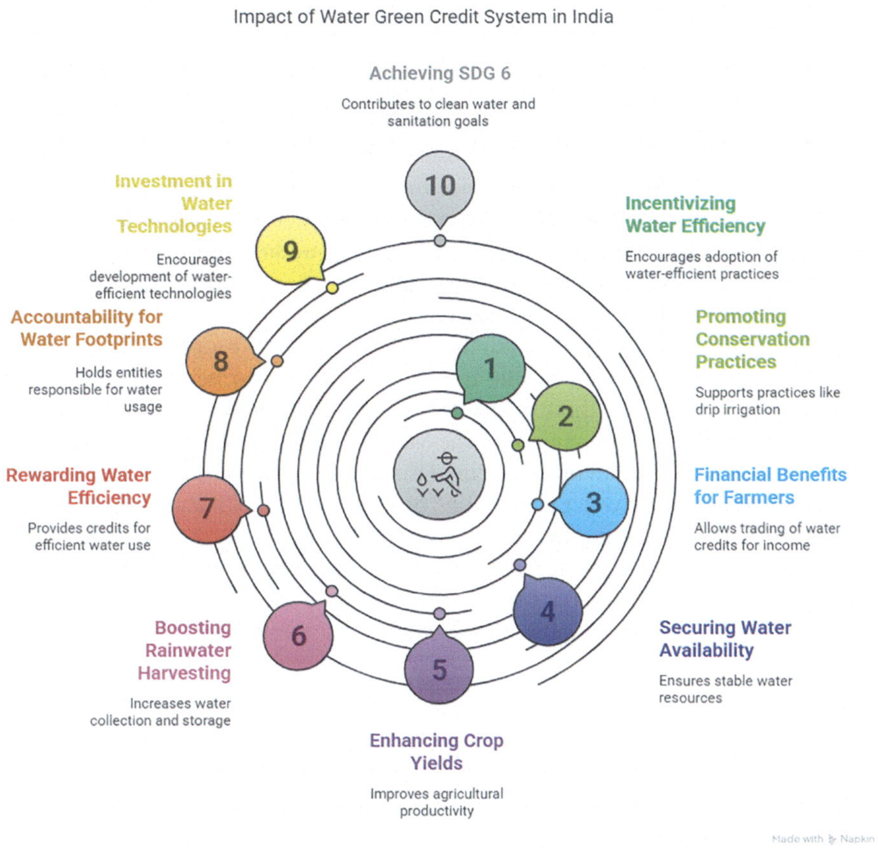

Fig. 4 Implications of the Water Green Credit mechanism

9. It would foster investment in water-efficient technologies and infrastructure [7].
10. It will assist in achieving SDG 6 for Clean water and sanitation [7].

5 Conclusion

This study highlighted the need, requisites, and implications of adopting the Water Green Credit mechanism in India by doing a review of related articles, reports, and grey literature. It is incumbent on India to adopt a revolutionary system for managing and ensuring the sustainability of water like Water Green Credit System due to its international commitments, rapid urbanization, water scarcity, depleting groundwater levels, increasing population, natural calamities like droughts, overuse in agricultural activities, and inefficient usage of water in industrial activities. Although the Ministry of Environment, Forest, and Climate Change has introduced the Green

Credit Program to focus on afforestation and water conservation under the Mission LiFE, there are still grounds to cover. A designated Act has to be assigned for the Water Green Credit system, a designated agency has to be authorized specifically for water credits, a methodology needs to be adopted to calculate the value of each water green credit, baseline targets for water green credit should be determined and informed to all stakeholders, a compliance mechanism should be framed for effective implementation, the pollutants that are considered for determining the water quality should be specified, authority to oversee the trading of water credits is to be sanctioned, a registry operator for registering water conservation projects should be sanctioned, a regulatory body for the regulation of the water credit market is to be authorized, the functions of the administrator are to be prescribed, and steering committee, registry operator, and the regulating body are to be clearly identified and specified. Working on these requisites aligned with the carbon market could lead to a successful implementation of the Water Green Credit system in India. Its implementation will result in incentivizing farmers and other entities to adopt water-efficient practices, secure water availability, promote rainwater harvesting and waste treatment activities, assist the farmers in earning extra income by trading the water credits earned, improve their crop yield, and will facilitate in holding the accountability of companies for their water footprints. While there are many benefits of implementing this system in India, many challenges are hindering its progress, as specified below:

1. It is intricate to demarcate the baselines for water credits due to the localized nature of water resources [7]. Unlike carbon credits, wherein the CO_2 is transferred to the atmosphere, in the water credit system, water is sourced from different areas sharing boundaries with each other [1].
2. There may arise differences in the accessibility of water credits between small and large entities [7].
3. There are different water rights and regulations across different states in India [7]. These states may have differences in the degree of water regulations that may lie anywhere between effective water regulation policies and ineffective water regulation policies [1].
4. There is a lack of a stringent verification process for ensuring bona-fide water conservation practices and thus the eligibility of attaining water credits [7].
5. The companies may suffer from the initial capital outlay for technology, labor, and other expenses for taking water efficiency measures [1].

The following recommendations are offered for the effective implementation of the Water Green Credit mechanism in India:

1. Factors such as rainfall, groundwater level, water quality, current water consumption levels, water scarcity level in the area, and virtual water consumption, including green water, blue water, and grey water consumption levels, should be considered for framing baseline targets for water credits. Moreover, the location and stage of use of virtual water should also be traced [7, 8].

2. The eligible water trading activities, measurement, and verification standards should be clearly defined, and a marketplace is to be created for the trading of water credits [7].
3. Equality in access to water credits between small and large entities should be promoted, precluding the wealthier entities from dominating the water credit market [7].
4. There is a need for uniform water rights and regulations across the different states in India for harmonized and effective implementation of the Water Credit system [7].
5. There should be active monitoring and evaluation schemes to track the reduction in water footprints and the quality of water for both the project implementer and credit provider [1].
6. Third-party audit and digital tracking systems should be used to verify the authenticity of the water conservation practices before granting water credits [1, 7].
7. There should be defined methodologies for valuing each water credit certificate [7].
8. The system should initially be placed in harmony with the highly water-regulated regions and poorly water-regulated regions, wherein the highly water-regulated regions that inculcate the big entities who are already complying with the regulations and want to achieve their climate goals as part of those regulations and to fulfill their corporate social responsibility can purchase water credits from the smaller entities in the poorly water-regulated regions thus incentivizing these regions to adopt water conservation measures [1].
9. The project implementers should create standard approved methodologies for a certain project, inculcating project design, activities, certification for the reduction of water footprints, and the water quality output [1].
10. There should be regulations on the greenwashing accusation of the entities claiming to follow the water conservation measures, but not actually doing so [1].
11. The water credit provider should be designated to act as a liaison between the big entities that want to purchase the water credit and the smaller entities that create the projects to earn water credits from the providers, which are sold to the big entities through the provider to offset their water consumption [1].
12. Government bodies, agricultural associations, private organizations, and water management agencies should share expertise with each other and work cohesively to align the water credit system with the local conditions of an area, and driving sustainability in water management [8].
13. There is a need to consider the industrial, agricultural, and ecological preservation perspective simultaneously in order to derive a significant impact of this system [8].
14. It is essential for the government to provide assistance to farmers, especially small farmers with smaller landholdings, by educating them about water conservation activities, optimal irrigation techniques based on the soil type. Furthermore, to implement the mechanism successfully, capacity-building programs

could be undertaken, and farmers should be provided with financial support for implementing water conservation activities [8].

15. Water conservation in the system could also be promoted through activities like rainwater harvesting for non-potable uses, and grey water recycling for toilet flushing which is currently in practice in the DLF Mall of India, Delhi [8].
16. Smart technologies like leak detection technologies, smart sensors in taps, smart meters, and acoustic sensors could be used to conserve water.
17. Water-efficient appliances in homes and industries could further reduce water consumption [8].
18. There should be fungibility across different sectors to ensure flexibility and equivalence in water conservation activities [8].
19. The government should utilize digital platforms for registration, issuance, and monitoring of water credits to foster transparency and efficiency [8].
20. The government should leverage the expertise, funding, and advocacy of international organizations and NGOs to intensify the water credit system [8].

6 Implications

6.1 Policy Implications

This research will direct the policy makers to foster the implementation of Water Green Credits in India by guiding them in filling the gaps in the mechanism for the necessary bodies to be constituted, the functions that need to be assigned to these bodies, and the standards to be set for the effective execution of it. The effective implementation of water green credits would ultimately assist in the achievement of SDG 6 related to clean water and sanitation, and SDG 15 for Life on Land that emphasizes the sustainable use of ecosystems.

6.2 Industrial Implications

The industry would benefit by getting aware of the basic tenets of the system that will assist the companies to be prepared for the changes that need to be incorporated for economical use of water which will be well within the cap under the mechanism, and take leverage of the water allowances by selling the extra allowances to those who need them.

6.3 Societal Implications

The society would become better aware of the system and could take the opportunity to execute water conservation projects, earn credits through them, and sell them to the companies that need to buy allowances for excess water usage.

6.4 Research Implications

Future research could keep a track of the progress of the system in India, compare the updates with other countries that have already adopted the system, search for the challenges in the implementation of the system, and the effectiveness of the system in terms of water saved, and efficient usage of water.

References

1. Al-Yacoub LK (2024, January 9) Water credits (II): a project-based approach. Smart Water Magazine. https://smartwatermagazine.com/blogs/laith-k-al-yacoub/water-credits-ii-a-project-based-approach
2. Bureau of Energy Efficiency (2021) Carbon market | Bureau of Energy Efficiency, Government of India, Ministry of Power. Beeindia.gov.in. https://beeindia.gov.in/en/programmes/carbon-market
3. Bureau of Reclamation (2020, November 4) Water facts—worldwide water supply. Bureau of Reclamation. https://www.usbr.gov/mp/arwec/water-facts-ww-water-sup.html
4. Chandra A (2024) The water-carbon nexus: is it worthwhile to generate carbon credits based on agricultural water management? Environ Res Lett 19(9):091004. https://iopscience.iop.org/article/10.1088/1748-9326/ad6339
5. DGB Group (2024, June 17) Revolutionising water use: proposal for water credits in India. Green.Earth; DGB Group. https://www.green.earth/news/revolutionising-water-use-proposal-for-water-credits-in-india
6. Droogers P, Immerzeel WW, Terink W, Hunink JE, van Lynden GW (2012) Water allocation in 2050. Green Growth and Water Allocation, 101
7. George A (2024, June 23) Water credits for improving water use efficiency in the agricultural sector. BusinessLine. https://www.thehindubusinessline.com/economy/agri-business/water-credits-for-improving-water-use-efficiency-in-the-agricultural-sector/article68320315.ece
8. George A (2025, March 25) How water green credits can further conservation. Hindustan Times. https://www.hindustantimes.com/opinion/how-water-green-credits-can-further-conservation-101742910939168.html
9. Hirani P, Dimble V (2019, October 4) Water pollution is killing millions of Indians. Here's how technology and reliable data can change that. World Economic Forum. https://www.weforum.org/stories/2019/10/water-pollution-in-india-data-tech-solution/
10. Hunink JE, Droogers P, Kauffman S, Mwaniki BM, Bouma J (2012) Quantitative simulation tools to analyze up- and downstream interactions of soil and water conservation measures: supporting policy making in the Green Water Credits program of Kenya. J Environ Manage 111:187–194. https://doi.org/10.1016/j.jenvman.2012.07.022
11. Hunink JE, Immerzeel WW, Droogers P, Kauffman S (2010) Green water credits target areas for the Upper Tana catchment, Kenya: biophysical assessment using SWAT. ISRIC World Soil

Information Wageningen. https://files.isric.org/public/documents/isric_report_2010_04_addendum.pdf
12. India Meteorological Department (2020) Hazard Atlas of India. Imdpune.gov.in. https://imdpune.gov.in/hazardatlas/droughtnew.html
13. International Carbon Action Partnership (2024, September 2) India adopts regulations for planned compliance carbon market. Icapcarbonaction.com. https://icapcarbonaction.com/en/news/india-adopts-regulations-planned-compliance-carbon-market
14. Kauffman S, Droogers P, Hunink J, Mwaniki B, Muchena F, Gicheru P, Bindraban P, Onduru D, Cleveringa R, Bouma J (2014) Green Water Credits—exploring its potential to enhance ecosystem services by reducing soil erosion in the Upper Tana basin, Kenya. Int J Biodivers Sci, Ecosyst Serv & Manag 10(2):133–143. https://doi.org/10.1080/21513732.2014.890670
15. Kayatz B, Harris F, Hillier J, Adhya T, Dalin C, Nayak D, Green RF, Smith P, Dangour AD (2019) "More crop per drop": exploring India's cereal water use since 2005. Sci Total Environ 673:207–217. https://doi.org/10.1016/j.scitotenv.2019.03.304
16. Mehla MK, Kothari M, Singh PK, Bhakar SR, Yadav K (2023) Water footprint assessment and its importance in Indian context: a meta-review. Water Science & Technology: Water Supply 23(8):3113–3127. https://doi.org/10.2166/ws.2023.174
17. Meijerink GW, Muchena FN, Njue E, Dent D (2007) Political, institutional and financial framework for Green Water Credits in Kenya. ResearchGate; Unknown. https://www.researchgate.net/publication/40100090_Political_institutional_and_financial_framework_for_Green_Water_Credits_in_Kenya
18. Ministry of Culture, Government of India (2021) Azadi Ka Amrit Mahotsav. Azadi Ka Amrit Mahotsav, Ministry of Culture, Government of India. https://amritmahotsav.nic.in/lifestyle-for-environment-life.htm
19. Ministry of Environment, Forest and Climate Change (2022) Mission life—lifestyle for environment. https://static.pib.gov.in/WriteReadData/specificdocs/documents/2022/nov/doc2022119122601.pdf
20. Ministry of Environment, Forest and Climate Change (2023) Notification issued for Green Credit Program (GCP) and ecomark scheme under life initiative to promote sustainable lifestyle and environmental conservation. Pib.gov.in. https://pib.gov.in/PressReleaseIframePage.aspx?PRID=1967476
21. Ministry of Information and Broadcasting (2017) Mission life—lifestyle for environment. Pib.gov.in. https://pib.gov.in/PressNoteDetails.aspx?NoteId=151925&ModuleId=3&reg=3&lang=1
22. Ministry of Jal Shakti (2017) Synopsis of Water Data in India | National Water Mission, Ministry of Jal Shakti, Department of Water Resources, RD & GR, Government of India. Nwm.gov.in. https://nwm.gov.in/synopsis-water-data-india
23. Muchena FN, Onduru D, Kauffman JH (2011) Analysis of financial mechanisms for Green Water Credits in the Upper Tana, Kenya. Green Water Credits... researchgate; ISRIC world soil information Wageningen. https://www.researchgate.net/publication/236215701_Analysis_of_Financial_Mechanisms_for_Green_Water_Credits_in_the_Upper_Tana_Kenya_Green_Water_Credits_Report_17
24. National Water Mission (2025, April 15) National water mission, ministry of Jal Shakti, department of water resources, RD & GR, Government of India. https://nwm.gov.in/sites/default/files/1.%20National-water-mission-%20%20%20water-use-efficiency.pdf
25. Niti Aayog (2022) Life | lifestyle for environment. https://www.niti.gov.in/sites/default/files/2022-11/Mission_LiFE_Brochure.pdf
26. Rodrigues LN, Agbossou E, Weikard H-P, Jaspers F (2015) Green Water Credits: institutional framework and compensation measures for Piedade watershed. Anais do III inovagri international meeting—2015, 1803–1816. https://doi.org/10.12702/iii.inovagri.2015-a196
27. Shisanya, C.A., Akombo, R. A., Cush, N. L., & Obando, J. A. (2014). Green Water Credits for Sustainable Agriculture and Forestry in Arid and Semi-Arid Tropics of Kenya. J Agri-Food Appl Sci 2(4)

28. The Hindu. (2024, July 12). India's population to peak in early 2060s to 1.7 billion before declining: United Nations. *The Hindu.* https://www.thehindu.com/news/national/indias-population-to-peak-in-early-2060s-to-17-billion-before-declining-united-nations/article68395623.ece
29. The Times of India (2019, June 19) "India has just 2.4% of world's land but 18% of its population" | Nagpur news—Times of India. The Times of India. https://timesofindia.indiatimes.com/city/nagpur/india-has-just-2-4-of-worlds-land-but-18-of-its-population/articleshow/69848388.cms
30. United Nations (2024) Sustainable development report 2024. Sustainable development report. https://dashboards.sdgindex.org/rankings

Waste Management for Phasing Out Mercury Added Devices

Sunil Sharma

Abstract Mercury-based medical devices, such as thermometers and sphygmomanometers, pose significant health and environmental risks due to their toxic content. The Minamata Convention, ratified by over 140 nations, mandates the global phase-out of mercury-containing products by 2020. Minamata convention has pushed the mercury-added devices out of usage. However, improper disposal practices—such as auctions, municipal waste dumping, or uncontrolled incineration—persist, particularly in developing countries like India. This chapter addresses two critical challenges: (1) eliminating mercury-based devices from healthcare systems and (2) managing end-of-life or broken devices safely. While existing guidelines for mercury disposal exist, this study focuses on tailoring solutions specifically for sphygmomanometers, which remain prevalent in Indian healthcare despite known hazards. Mercury-added devices such as thermometers and sphygmomanometers pose a health and environmental hazard due to their content. Hospital staff and cleaners are the primary persons in contact with these mercury-added devices. They are directly or indirectly exposed to mercury contamination. Commonly, mercury is released via these agents into wastewater and solid waste at the end of their shelf lives. This research details the disposal plan for medical devices containing mercury specifically sphygmomanometers on the principles of circularity.

Keywords Mercury · Waste management · Sustainability · Sustainable healthcare practices · Minamata convention · Mercury disposal · Circularity

1 Introduction

Phasing out mercury-based devices presents a significant global challenge due to the toxic nature of mercury and its widespread use in everyday items. Mercury is a potent environmental pollutant that poses serious risks to human health, even

S. Sharma (✉)
Lovely Professional University, Phagwara, India
e-mail: 85sunilsharma@gmail.com

V. Kandpal and T. Choudhury (eds.), *Blue Revolution: Sustainable Water Use and Circular Economy Innovations*, Urban Sustainability,
https://doi.org/10.1007/978-981-95-1916-3_4

in small quantities [3]. Exposure can occur through inhalation, ingestion, or skin contact, leading to severe damage to vital organs such as the brain, kidneys, and lungs. Mercury exists in three forms—metallic, inorganic, and organic compounds. Metallic mercury, once commonly used in thermometers, fluorescent lamps, and electrical switches, is particularly hazardous due to its ability to vaporize at room temperature. Inorganic mercury, often found in natural forms like cinnabar and metacinnabar, is equally concerning due to its persistence in the environment.

Recognizing these dangers, the Minamata Convention on Mercury—a landmark global treaty—has called for the phase-out of mercury-containing medical devices such as thermometers and sphygmomanometers [8]. Over 90 countries, including India, have committed to this global initiative. However, India remains one of the major contributors to mercury emissions, which were projected to reach 540 tonnes by 2020 [4]. This calls for urgent and coordinated action to cut emissions by at least 50%, not only to protect public health but also to safeguard the environment from long-term contamination.

Figure 1 provides a visual overview of global mercury consumption from 1850 to 2010, showing how the use of mercury has evolved across various industries. From 1850 to the early 1900s, the primary driver of mercury use was large-scale silver and gold mining. By the mid-twentieth century, mercury consumption surged again, reaching a second peak around the 1960s–1970s. This increase was due to its widespread use in various sectors like Chlor-alkali production, Vinyl chloride monomer and other chemicals, artisanal and small-scale gold mining (ASGM), medical devices, batteries, dental applications, wiring/measuring devices, and even explosives and pesticides. During this period, mercury was everywhere—from industrial production to everyday consumer products. After the 1970s, a gradual decline in mercury use across most sectors was observed, largely due to growing awareness of mercury's health and environmental impacts, regulatory action, the introduction of safer alternatives, and the global movement culminating in agreements like the Minamata Convention. Despite the reduction, mercury use in ASGM and certain chemical processes continues to this day, highlighting the ongoing challenge of fully eliminating mercury from global supply chains.

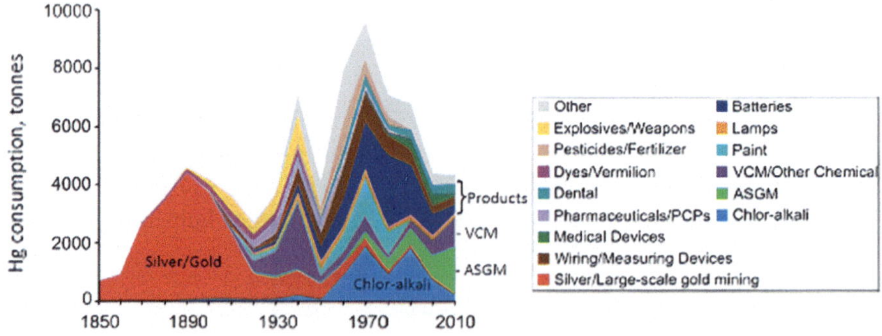

Fig. 1 Global Mercury consumption 1850–2010 [12]

2 Background

The Minamata Convention on Mercury is a global treaty adopted in 2013 to protect human health and the environment from the harmful effects of mercury. The Minamata Convention tackles mercury pollution by taking a lifecycle approach—addressing every stage from production and product design to trade, use, and disposal. It's not just about removing mercury from shelves; it's about transforming industrial practices, consumer behavior, and global trade for a safer, more sustainable future. The Minamata Convention on Mercury characterizes a mercury-added product as a "product or product component that contains mercury or a mercury compound that was deliberately incorporated" [7, 10].

The control measures under the Minamata convention take the life cycle approach for mercury to provide guidelines for supply, production and processing, emissions, storage, waste management, and managing contaminated sites. Globally many studies and reports have been published on the mercury supply chain [6], their usage in different industries, the contribution of mercury to ecological contamination [16, 18] and its reduction [25], regulations [21] methods for containing, recycling and reusing mercury [5] and immobilizing mercury in waste [15]. Many policies and guidelines have been set across nations for getting rid of the mercury problem. Developed countries are ahead of developing and underdeveloped countries in completely shunning and containing mercury usage. For example, Sweden has already decided that the use of mercury was phased out by 2000 [24]. In Australia, several states have arranged mercury-containing lamp drop-off points to collect household fluorescent lamp tubes for recycling. In India, estimated data is available for mercury and its usage in commercial products. Even, the number of thermometers and sphygmomanometers are approximated by the major suppliers [1]. Much research has also discussed the exposure and health threats to the ecosphere and human beings [3].

The implementation of the Minamata Convention's mercury phase-out faces significant challenges, particularly in balancing environmental goals with socioeconomic realities. In developing nations, the high cost of mercury-free alternatives, such as digital medical devices or cleaner industrial technologies, strains limited budgets, slowing adoption. Many regions lack infrastructure for safe mercury waste management, leading to improper disposal through incineration or landfill dumping, which perpetuates environmental contamination. Artisanal and small-scale gold mining (ASGM), responsible for nearly 40% of global mercury emissions, remains a critical obstacle due to its informal, unregulated nature and miners' reliance on mercury as a cheap, accessible method for gold extraction. Illegal mercury trade thrives in areas with weak governance, undermining global supply restrictions. Additionally, gaps in awareness and technical training among healthcare workers, waste handlers, and miners result in unsafe practices, such as inadequate spill responses or improper storage. Political and institutional inertia, compounded by competing priorities like economic growth or public health crises, often delay policy enforcement and funding. Technological gaps persist in sectors like chlor-alkali production, where mercury-free alternatives remain underdeveloped or costly. Finally, the transboundary nature of mercury pollution demands robust international cooperation, yet disparities in resources and commitment between high—and low-income nations hinder unified progress, leaving vulnerable populations—particularly in the Global South—exposed to mercury's toxic legacy. Addressing these challenges requires tailored financing, capacity-building, and inclusive policies that prioritize both planetary health and equitable development.

This study focuses on the resolution of two challenges, (1) eliminating mercury-based devices from healthcare systems and (2) managing end-of-life or broken devices safely. Despite growing global awareness about mercury's toxic effects, many healthcare facilities—especially in low—and middle-income countries—still rely on traditional mercury-based instruments such as thermometers and sphygmomanometers. These devices have long been favored for their accuracy, affordability, and familiarity among healthcare professionals. However, their continued use presents significant health and environmental risks, particularly in scenarios where breakage occurs and mercury is released into the environment.

The most common mercury-containing medical devices include mercury thermometers, which hold 0.5–1.5 g of liquid mercury for temperature measurement, and mercury sphygmomanometers, used for blood pressure monitoring, each containing 60–100 g of mercury in a calibrated glass column. Other devices include esophageal dilators and gastrointestinal tubes, which historically used mercury-weighted tips to aid insertion, and mercury-based dental amalgams for fillings, though these are increasingly replaced by composite materials. Additionally, mercury is found in thimerosal, a preservative used in some vaccines and antiseptics, though its medical use is now limited.

Distributed studies of mercury-containing devices and their disposal at healthcare facilities in India have been carried out by various researchers [11, 19, 20, 22]. Research also reports that many healthcare facilities in private and government hospitals are also mercury-free and phased out mercury long back [23]. Chalkidis et al. described the main methods employed for the recovery of mercury and the treatment of contaminated waste in industry with policies and challenges in a global context [5]. Despite guidelines and laws by the Indian government, mercury-containing thermometers/sphygmomanometers are still preferred by health professionals and there is an urgent need for source reduction, recycling, and waste minimization [11].

2.1 Why Sphygmomanometers Are a Priority for Disposal in Mercury Based Devices?

This work focuses on the disposal of mercury-added sphygmomanometers. Research suggests that staff and healthcare authorities are often unaware of the proper disposal of mercury-added devices, such as thermometers, when they are broken or condemned [9]. Mercury-added thermometers/sphygmomanometers are still preferred by health professionals [11]. Mercury usage in medical instruments is estimated to be 26 tons in India. The amount of Hg released through medical devices is approximately 8 tons. Detailed data regarding mercury consumption and its waste generation in society and industries already exist in developed countries [17]. However, there is a lack of research data and efforts for India. Concrete evidence can be provided by conducting research studies. More efforts and funding are needed in this regard, as India is the second largest emitter of Hg in the world [13].

2.2 Challenges with Mercury-Added Devices

Manufacturing healthcare tools with elemental mercury poses significant challenges. Regulators face challenges in monitoring the management of hazardous materials due to their large number and geographical distribution. Efforts to replace mercury in mercury-added devices in healthcare facilities face several challenges, including a lack of high-quality alternative products at affordable prices, a lack of standardization for alternative instruments, and a lack of affordability. Barriers to shifting from mercury to non-mercury goods include the standardization of alternative instruments and financial ramifications. The high expense of these items necessitates teaching healthcare professionals to utilize alternative alternatives. Healthcare professionals' attitudes towards transitioning from mercury to alternatives are also a concern. Most people consider mercury instruments to be the gold standard and resist the change to substitutes. The downstream problem is how to handle the garbage created by these instruments, as well as the product waste that becomes outdated with the advent of alternatives.

2.3 Cost Analysis for Replacing Sphygmomanometers from Hospitals

Sphygmomanometers contain the highest amount of Hg i.e. 60 gm.. This makes them a significant source of mercury in healthcare settings, especially when considering the sheer number in circulation. For example, with over 426,000 sphygmomanometers reported in public health facilities in India alone, the cumulative mercury load can exceed 25 metric tonnes. Refer to Table 1. In India, public health facilities include Sub-Centers (SCs), Primary Health Centers (PHCs), Community Health Centers (CHCs), and District/Sub-Divisional Hospitals (SDHs), with a sphygmomanometer (blood pressure monitor) being a common piece of equipment. This high volume underscores the potential environmental and health hazards if these devices are mishandled, broken, or improperly disposed of.

Mercury spills from damaged equipment not only pose immediate risks to healthcare workers and patients but can also lead to long-term contamination of hospital environments and nearby ecosystems. Therefore, accurately quantifying mercury

Table 1 Number of sphygmomanometers in health facilities in India

Type of public health facility	Sphygmomanometers	Numbers of HCFs in India	Total sphygmomanometers
Sub-centers	2	147,838	295,676
PHC	3	23,790	71,370
CHC	1	4761	4761
District and sub-divisional hospitals	40	1372	54,880
Total			426,687

content is essential for planning effective phase-out strategies, ensuring safe collection and storage, and supporting compliance with international commitments like the Minamata Convention. Even though phasing out of mercury-based sphygmomanometers has been a declared policy of the Government, there is no timeline-based roadmap for the same and the complete phase-out is still in process.

The large-scale presence of mercury-based sphygmomanometers in India presents not only an environmental challenge but also a potential opportunity for controlled mercury recovery. With each device estimated to contain between 80–100 g of mercury, and assuming a 50% recovery efficiency, approximately 50 g of mercury can be extracted per unit. Refer to Table 2. Given a total of 426,687 sphygmomanometers in use across public health facilities, the cumulative mercury content is substantial. Even with just 30% of these devices being accessed for recovery, over 6,400 kg of mercury could potentially be reclaimed. At the current estimated market rate of Rs. 4,000 per kilogram, this translates into a value of approximately Rs. 25.6 million. This highlights the importance of developing structured recovery and recycling mechanisms, not only to reduce mercury pollution but also to harness the economic value of end-of-life devices.

The transition from mercury-based to digital sphygmomanometers in India's public health facilities involves a significant upfront investment but is crucial for aligning with global environmental and health standards. To replace all 426,687 existing mercury sphygmomanometers, the total estimated cost at a unit price of INR 2,000 would be around INR 85.34 crore. In comparison, the cost of continuing with mercury-based devices, priced at INR 800 per unit, stands at approximately INR 34.13 crore. This results in a cost differential of INR 51.2 crore for

Table 2 Resale value of mercury sphygmomanometers

Amount of mercury in sphygmomanometer (g/unit):	Assuming 50% recovery@100 gm, mercury extracted (gm)	Total Hg sphygmomanometers (India)	At 30% access, the total amount of Hg recovered	Sale @ Rs 4000/kg
80–100	50	426,687	6400	25,601,220

Table 3 Cost of replacing all mercury sphygmomanometers with digital sphygmomanometers

Total digital sphygmomanometers required	426,687
Total cost for digital sphygmomanometers (unit cost @2000 INR)	85.34 crore
Total Cost for mercury added sphygmomanometers (unit cost @ 800 INR)	34.135 crore
Cost differential of shifting	51.2 crore

the shift. Refer to Table 3. While the initial financial outlay may appear substantial, the long-term benefits—in terms of improved safety, reduced mercury pollution, compliance with the Minamata Convention, and avoidance of costly environmental clean-ups—far outweigh the investment. Moreover, digital devices reduce the risk of breakage, require less maintenance, and support better record-keeping, making them a sustainable and health-conscious alternative for the future of India's healthcare system.

3 Mercury Added Device Collection, Disposal, and Reuse Plan

The model illustrated in Fig. 2 presents a structured flow for the phase-out and safe management of mercury-added medical devices, not limited to sphygmomanometers, in healthcare facilities. It begins with hospitals, where mercury-containing devices become either broken or redundant due to obsolescence or replacement by digital alternatives. These devices are then segregated into broken mercury devices, which are sent to mercury recovery units, and redundant but intact devices, which can be routed to medical device manufacturers for possible reuse, refurbishment, or safe dismantling. Additionally, mercury waste—residual elemental mercury or contaminated materials—is directed toward appropriate waste management channels. This includes Hazardous Waste Treatment, Storage and Disposal Facilities (TSDFs) for secure and environmentally compliant disposal, and Common Biomedical Waste Treatment Facilities (CBWTFs) for handling mercury waste per biomedical waste regulations. Finally, recovered mercury that meets regulatory standards can be sold to buyers and traders licensed to handle hazardous materials. This systematic approach ensures that mercury is removed from healthcare environments safely while promoting recovery, reuse where possible, and secure disposal, in line with the goals of the Minamata Convention on Mercury.

Austria, South Korea, and Germany implement either Extended Producer Responsibility (EPR) or deposit-refund policies to manage mercury waste. Under the EPR system, producers bear the cost of waste collection, packaging, transport, and recycling. In contrast, the deposit-refund scheme requires consumers to pay a refundable deposit (e.g., €1 per unit in Austria) when purchasing, which is returned upon returning the used product. Switzerland has established Producer Responsibility Organizations such as the Swiss Lighting Recycling Foundation (SLRS) and the

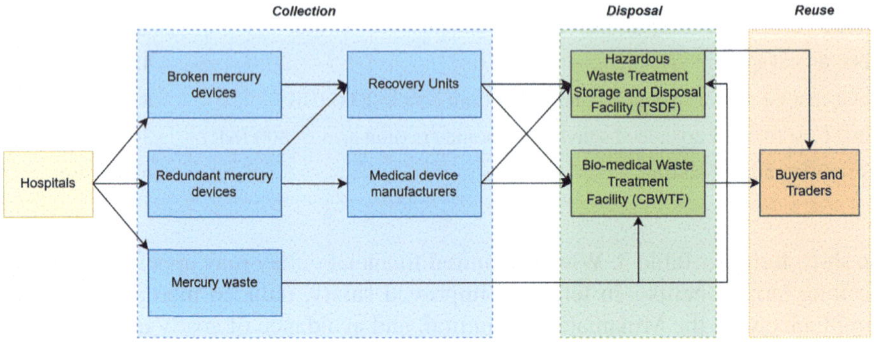

Fig. 2 Mercury collection, disposal and reusage plan

Swiss Foundation for Waste Management (SENS) to coordinate with lamp manufacturers, importers, and retailers to collect waste mercury lamps through stores and designated collection points [14].

We explain the role of each entity in the disposal and reuse supply chain as follows.

Hospitals: In this disposal plan, hospital staff collects the mercury-containing devices and categorizes them into three cartons. First carton contains broken mercury devices which have some mercury content in plastic bottles. The second carton consists of obsolete and working mercury devices. The third carton consists of mercury waste in mercury disposal kits. The first and second cartons are rich and can yield quality mercury, they can be sent to authorized recovery units. We assume the recovery units are state-owned and can handle transportation charges. However, collection from the entire district or a limited number of hospitals can be more feasible. The hospitals can put their request for disposal via mobile apps. Collected data from a region can help to chalk out a better plan. Therefore, it is imperative to take various stakeholders in confidence and inform them of the plan of action for collection via mobile app.

Alternatively, obsolete devices can be sent to medical device manufacturers when the delivery of medical supplies comes. A reverse logistics model can be used here. Medical device manufacturers and recovery units can recover the mercury.

Recovery Units: Recovery units recover the mercury and store it by distillation or roasting methods and reuse recovered mercury in the manufacturing of mercury. This step can be monetized where the recovery units can be given remuneration for facility usage. The waste, contaminated, and non-extractable mercury devices are further sent to Waste treatment and bio-medical treatment facilities for final disposal. Recovery units and medical device manufacturers can further send the recovered mercury directly to the buyers and sellers. They can also direct it to treatment facilities for further refining of mercury.

Treatment disposal facilities: The treatment facilities and disposal facilities can refine the mercury accordingly and sell or auction it to sellers and traders. Treatment Storage and Disposal Facility TSDF and Common Bio-medical Waste Treatment and Disposal Facility CBWTF are already functional in many states of India. Disposal option for disposal of mercury-bearing waste containing residual mercury and not economically viable for recovery, such waste needs to be disposed of through a nearby authorized Common Hazardous Waste Treatment, Storage and Disposal Facility (TSDF). The storage of mercury-bearing waste within the health care facility for a longer period should be avoided, and it should be sent to the authorized common bio-medical waste treatment facility (CBWTF). CBWTF should have a separate storage provision for the storage of mercury-bearing waste.

Buyers, sellers, and traders: At TSDFs and CBWTFs, buyers, sellers, and traders play pivotal roles in the responsible management of mercury, a hazardous substance. Sellers, often including healthcare facilities, laboratories, and industries, are responsible for handling mercury-containing waste in a safe and documented manner. Buyers, such as recycling firms or authorized mercury recovery units, procure mercury waste for safe extraction, reuse, or environmentally sound disposal. Traders act as intermediaries, facilitating the movement of mercury waste from generators to authorized treatment or disposal entities, ensuring compliance with environmental regulations. These stakeholders must adhere to strict guidelines for collection, packaging, transport, and documentation to prevent leakage, contamination, or exposure, thereby supporting safe and sustainable mercury waste management.

3.1 Cleaning up and Spill Disposal in the Mercury-Added Devices

The cleanup and disposal of these spills present complex challenges, requiring meticulous handling to avoid contamination of air, surfaces, and water sources. Moreover, improper or delayed cleanup can result in prolonged exposure, increasing the risk of neurological and developmental health impacts. Given mercury's persistence in the environment and its ability to bioaccumulate, effective spill management of mercury-containing devices remains a critical concern in ensuring public and environmental safety. When medical instruments containing mercury, such as thermometers and sphygmomanometers, become outdated, they become part of the solid waste stream and should be treated with caution.

- To improve mercury waste management in the healthcare sector, the government must implement the recommended measures.
- To manage mercury waste in the healthcare sector, CPCB guidelines should be followed.
- Manufacturers should provide a take-back system for discarded devices, and common biomedical waste treatment facilities should be used for incineration rather than individual hospitals.
- To prevent further exposure, only trained personnel should use mercury spill management kits.
- Inventorize and quantify the cumulative weight of medical devices and mercury.

3.2 Strategies for a Mercury-Free Indian Healthcare Sector

Developing and implementing strategies for a mercury-free Indian healthcare sector is crucial due to the significant health and environmental risks associated with mercury use. Mercury is still present in various medical instruments such as thermometers and sphygmomanometers, posing potential hazards to healthcare workers, patients, and waste handlers, especially in cases of breakage or improper disposal. In India, where the healthcare infrastructure is vast and diverse, the risk of mercury exposure is amplified by inconsistent handling practices and limited awareness. Furthermore, mercury released into the environment from healthcare facilities can contaminate air, soil, and water, ultimately entering the food chain. Transitioning to mercury-free alternatives not only aligns with global efforts under the Minamata Convention but also promotes safer healthcare practices, reduces long-term liabilities, and supports sustainable public health outcomes across the country. We suggest the following remedies for the same.

1. **Prevention of dumping of Mercury devices (Thermometers & BP monitors)**
   A strict check must be placed on the disposal and resale of mercury-based devices. Many outdated devices are discarded unsafely or enter second-hand markets, increasing environmental and health risks.
2. **Standardization of non-Mercury healthcare instruments**
   Guidelines and technical standards should be established to ensure all healthcare providers adopt reliable and accurate non-mercury alternatives, enabling consistency and trust in such instruments.
3. **Stop production of Mercury-based products & encourage domestic production of Mercury-free alternatives**
   Banning domestic manufacturing of mercury devices and promoting indigenous production of digital and aneroid alternatives will not only reduce dependency on imports but also support the local economy.
4. **Regulate export-import to stop India from becoming a dumping ground**
   Import regulations must be tightened to prevent India from becoming a hotspot for international dumping of mercury-containing healthcare waste.
5. **Dental amalgam to be phased down**
   As dental amalgam contains around 50% mercury, efforts must be made to gradually reduce its use by promoting safer alternatives such as composite resins, especially in public healthcare systems.
6. **Mandates for switching over to alternatives**
   The government should issue formal directives requiring public and private hospitals to replace mercury instruments with digital or other mercury-free technologies within a specified timeframe [2].

3.3 *Roadblocks to Mercury-Free Health Center*

Despite widespread knowledge of mercury's adverse effects, converting healthcare systems to be totally mercury-free remains a substantial issue, particularly in resource-limited areas. While the intention to replace mercury-containing medical devices is obvious, practical obstacles frequently impede progress. Many healthcare institutions continue to suffer owing to a scarcity of inexpensive alternatives, a lack of standard procedures, and inadequate training for healthcare workers. Many experts continue to use mercury-based devices out of habit or familiarity, despite the fact that safer alternatives exist. Furthermore, there is a general lack of understanding on how to securely remove or dispose of damaged mercury devices, which heightens the hazards. These difficulties, taken together, create a real-world gap between policy and practice for establishing mercury-free health facilities.

4 Conclusion

As required by the Minamata Convention, the phase-out of medical devices that contain mercury, especially sphygmomanometers, is an essential step in protecting environmental integrity and human health. This study emphasizes how urgently India, a significant mercury emitter in the world, must address the continued usage and inappropriate disposal of medical equipment containing mercury. Because sphygmomanometers are so common in Indian healthcare institutions and can contain up to 100 g of mercury each unit, they are a priority for phase-out. Cumulative mercury reserves surpass 25 metric tons, with over 426,000 devices in public health facilities alone. If improperly handled, these mercury reserves might pose serious environmental risks. Long-term advantages of switching to digital alternatives include lower contamination concerns and conformity with international standards, notwithstanding the initial expense (INR 85.34 crore for nationwide replacement). The potential production of 6,400 kg of mercury (worth INR 25.6 million) through cost recovery through structured mercury reclamation demonstrates the economic feasibility of circular disposal models.

Segregation, reverse logistics, and cooperation between hospitals, recovery units, and treatment institutions are the main focuses of the suggested disposal system. Mercury may be safely removed, reused, or disposed of in accordance with the principles of the circular economy by incorporating Extended Producer Responsibility (EPR) and utilizing already-existing infrastructure like as TSDFs and CBWTFs. Important tactics include phasing down dental amalgam, standardizing substitutes, controlling commerce, and outlawing the manufacture of mercury devices.

Phasing out mercury-based devices from healthcare settings is not merely a matter of switching to digital alternatives—it also requires overcoming barriers such as cost, availability of high-quality substitutes, and resistance to change due to a lack of training or awareness. Furthermore, many healthcare institutions lack established protocols for the safe collection, storage, and disposal of broken or obsolete mercury devices. Improper handling of such waste can lead to contamination of hospital premises and surrounding areas, posing long-term risks to both medical staff and patients.

To ensure a successful transition, healthcare systems must adopt a holistic approach that includes policy implementation, procurement of mercury-free alternatives, staff training, and awareness campaigns. At the same time, the development of secure mercury waste management systems—aligned with the Minamata Convention guidelines—is essential to prevent secondary contamination. By focusing on both the elimination of mercury use and the responsible handling of legacy devices, this study contributes to safer, more sustainable healthcare practices aligned with global environmental health goals.

Public health officials and solid waste management companies should be uniform in their approach to interdicting any mercury that otherwise would be added to the environment [9]. Awareness programs for the general public and education about the potential perils should be imparted. Proper clean-up of spills and compliance with hospital mercury management policy should be stressed. There is an urgent need for accurate information about mercury and its contaminant exposures in various parts of the ecosphere to make informed choices and decisions. Further, there is an urgent need to translate the policy guidelines and measures at the local level across the country. Replacements of mercury-containing thermometers and sphygmomanometers are available and are accurate and practical. They should be considered while phasing out mercury added devices.

References

1. Agarwal R, Wankhede K (2002) Mercury in India: usage and releases. https://toxicslink.org/wp-content/uploads/2023/08/2008_Mercury%20In%20India-1-33-2.pdf
2. Asayama K, Ohkubo T, Hoshide S, Kario K, Ohya Y, Rakugi H, Umemura S (2016) From mercury sphygmomanometer to electric device on blood pressure measurement: correspondence of Minamata convention on mercury. Hypertens Res 39(4):179–182. https://doi.org/10.1038/hr.2015.158
3. Budnik LT, Casteleyn L (2019) Mercury pollution in modern times and its socio-medical consequences. Sci Total Environ 654:720–734. https://doi.org/10.1016/j.scitotenv.2018.10.408
4. Burger Chakraborty L, Qureshi A, Vadenbo C, Hellweg S (2013) Anthropogenic mercury flows in india and impacts of emission controls. Environ Sci & Technol 130726132711009. https://doi.org/10.1021/es401006k
5. Chalkidis A, Jampaiah D, Aryana A, Wood CD, Hartley PG, Sabri YM, Bhargava SK (2020) Mercury-bearing wastes: sources, policies and treatment technologies for mercury recovery and safe disposal. J Environ Manage 270:110945. https://doi.org/10.1016/j.jenvman.2020.110945
6. Chen B, Wang XB, Li YL, Yang Q, Li JS (2019) Energy-induced mercury emissions in global supply chain networks: structural characteristics and policy implications. Sci Total Environ 670:87–97. https://doi.org/10.1016/j.scitotenv.2019.03.215
7. Coulter MA (2016) Minamata convention on mercury. Int Leg Mater 55(3):582–616. https://doi.org/10.5305/intelegamate.55.3.0582
8. Datta P, Mohi G, Chander J (2018) Biomedical waste management in India: critical appraisal. Journal of Laboratory Physicians 10(01):006–014. https://doi.org/10.4103/JLP.JLP_89_17

9. DiCarlo M, Ruck B, Marcus S (2002) How should a fever mercury thermometer be disposed of? a survey of those likely to be asked. Pediatrics 109(5):e71–e71. https://doi.org/10.1542/peds.109.5.e71
10. Evers DC, Keane SE, Basu N, Buck D (2016) Evaluating the effectiveness of the Minamata convention on mercury: principles and recommendations for next steps. Sci Total Environ 569–570:888–903. https://doi.org/10.1016/j.scitotenv.2016.05.001
11. Halder N, Peshin SS, Pandey RM, Gupta YK (2015) Awareness assessment of harmful effects of mercury in a health care set-up in India: a survey-based study. Toxicol Ind Health 31(12):1144–1151. https://doi.org/10.1177/0748233713488237
12. Horowitz HM, Jacob DJ, Amos HM, Streets DG, Sunderland EM (2014) Historical mercury releases from commercial products: global environmental implications. Environ Sci Technol 48(17):10242–10250. https://doi.org/10.1021/es501337j
13. Jetashree, Wicke B, Liang S (2024) Parsing India's mercury crisis: a socioeconomic take reveals the defining role of domestic consumerism. J Clean Prod 463, 142742. https://doi.org/10.1016/j.jclepro.2024.142742
14. Lee C-H, Popuri SR, Peng Y-H, Fang S-S, Lin K-L, Fan K-S, Chang T-C (2015) Overview on industrial recycling technologies and management strategies of end-of-life fluorescent lamps in Taiwan and other developed countries. J Mater Cycles Waste Manage 17(2):312–323. https://doi.org/10.1007/s10163-014-0253-y
15. Makarova A, Fedoseev A, Yakubovich L (2020) Research on green technologies for immobilizing mercury in waste to minimize chemical footprint. Pure Appl Chem 92(4):557–565. https://doi.org/10.1515/pac-2019-0813
16. Mukherjee A.B., Bhattacharya P, Sarkar A, Zevenhoven R (2009) Mercury emissions from industrial sources in India and its effects in the environment. In Mason R & Pirrone N (eds) Mercury fate and transport in the global atmosphere, Springer US, pp 81–112. https://doi.org/10.1007/978-0-387-93958-2_4
17. Mukherjee AB, Zevenhoven R, Brodersen J, Hylander LD, Bhattacharya P (2004) Mercury in waste in the European Union: sources, disposal methods and risks. Resour Conserv Recycl 42(2):155–182. https://doi.org/10.1016/j.resconrec.2004.02.009
18. Pacyna EG, Pacyna JM, Sundseth K, Munthe J, Kindbom K, Wilson S, Steenhuisen F, Maxson P (2010) Global emission of mercury to the atmosphere from anthropogenic sources in 2005 and projections to 2020. Atmos Environ 44(20):2487–2499. https://doi.org/10.1016/j.atmosenv.2009.06.009
19. Peshin SS, Halder N, Jathikarta C, Gupta YK (2015) Use of mercury-based medical equipments and mercury content in effluents of tertiary care hospitals in India. Environ Monit Assess 187(3):145. https://doi.org/10.1007/s10661-015-4311-2
20. Ramesh K, Ramesh M, Krishnan R (2019) Management and disposal of mercury and amalgam in the dental clinics of South India: a cross-sectional study. Journal of Pharmacy And Bioallied Sciences 11(6):151. https://doi.org/10.4103/JPBS.JPBS_280_18
21. Sharma BM, Bharat GK, Šebková K, Scheringer M (2019) Implementation of the Minamata convention to manage mercury pollution in India: challenges and opportunities. Environ Sci Eur 31(1):96. https://doi.org/10.1186/s12302-019-0280-3
22. Singh RD, Jurel SK, Tripathi S, Agrawal KK, Kumari R (2014) Mercury and other biomedical waste management practices among dental practitioners in India. Biomed Res Int 2014:1–6. https://doi.org/10.1155/2014/272750
23. Singh R, Rustagi N (2010) Mercury and health care. Indian J Occup Environ Med 14(2):45. https://doi.org/10.4103/0019-5278.72240
24. von Rein K, Hylander LD (2000) Experiences from phasing out the use of mercury in Sweden. Reg Environ Change 1(3–4):126–134. https://doi.org/10.1007/s101130000016
25. Wang L, Hou D, Cao Y, Ok YS, Tack FMG, Rinklebe J, O'Connor D (2020) Remediation of mercury contaminated soil, water, and air: a review of emerging materials and innovative technologies. Environ Int 134:105281. https://doi.org/10.1016/j.envint.2019.105281

The Evolving Oasis: Modernizing Irrigation in Rajasthan's Arid Regions

Anshuman Jaswal and Debjani Mukherjee

Abstract Rajasthan's irrigation system has changed dramatically since 2000, through infrastructure developments and technological interventions. Expansion of irrigation to drier areas has been primarily via modernization of canals and pumps, especially through the Narmada and Indira Gandhi canals, and expansion of agricultural production in previously dry areas. The expansion and use of more tube wells has increased capacity, but also raised questions about the sustainability of groundwater sources. Also important are traditional practices such as johads (percolation ponds/check dams) and the chauka system, where remarkable expansion of capturing groundwater flush has restored groundwater levels and adaptability. There are positive initiatives from government projects—Pradhan Mantri Krishi Sinchayee Yojana (PMKSY) and Rajasthan Agricultural Competitiveness Project (RACP) to encourage agricultural practices for better use of water and through improvements in micro-irrigation. Reduction of groundwater supplies, climate change, and farmer adaptability remain key unresolved challenges. This chapter will assess these changes in irrigation since 2000, highlighting changes due to infrastructure development and technological intervention, and how these changes have impacted agricultural sustainability and water resource management. This chapter will answer the following research question: "Since 2000, how have developments in infrastructure and technological interventions transformed irrigation practices in Rajasthan, and how have they impacted agricultural sustainability and water resource management states?" In this chapter, we will bring together empirical data, analysis, and case studies to help provide a rich and comprehensive perspective on the multiple ways that infrastructure developments and technological interventions have drastically transformed irrigation practices in Rajasthan. This chapter will provide useful insights for formulating sustainable water resource management strategies in arid and semi-arid

A. Jaswal (✉)
SVKM's NMIMS Deemed-to-Be University, Indore, India
e-mail: anshuman.jaswal@nmims.edu

D. Mukherjee
School of Media and Marketing, Symbiosis University of Applied Sciences, Indore, India
e-mail: debjani.mukherjee@suas.ac.in

V. Kandpal and T. Choudhury (eds.), *Blue Revolution: Sustainable Water Use and Circular Economy Innovations*, Urban Sustainability,
https://doi.org/10.1007/978-981-95-1916-3_5

regions of the world, contributing to the water security and agricultural sustainability debate.

Keywords Irrigation · Groundwater · Micro-irrigation · Water management · Water harvesting · Sustainability · Resilience · Climate change · Rajasthan

1 Introduction: Transforming Irrigation in Rajasthan

1.1 Background: Rajasthan's Agro-Climatic Conditions and Dependence on Irrigation

Rajasthan, the largest state of India in terms of area, shows a lot of variability in its water resources, 70% of which is classified as dry or semi-dry. The region receives limited and irregular rainfall, averaging 200–550 mm annually, due to which agriculture is basically dependent on the irrigation infrastructure. Historically, agricultural communities in Rajasthan used a network of dams and wells for water management [1]. Effective water harvesting systems have developed indigenously—johad (seepage ponds), nadis (village water bodies), and khadin (uncultivated water systems)—which allow water production based on environmental factors. However, contemporary population pressure, agricultural density, and water resource changes have compromised the effectiveness of these traditional methods, which have necessitated the implementation of modern irrigation techniques and management strategies to ensure agricultural stability in this water-stressed area.

Since 2000, the State has witnessed a dramatic transformation in water management, driven by large-scale infrastructure projects, scientific advancements, and systemic reforms [2]. Table 1 shows the major milestones in the State's irrigation transformation. This chapter examines how these changes have reshaped irrigation systems, their socio-economic impacts, and the major challenges of stability and equality.

1.2 The Water-Stress Challenge in Arid/Semi-Arid Regions

Rajasthan stands as an important landmark in the global water crisis in dryland areas, where rising demand, environmental stress, and institutional limitations threaten long-term stability. In Fig. 1, data from the Central Ground Water Board (CGWB) shows that almost 82% of the administrative blocks in the state fall in the overexploited category [3], as illustrated in Fig. 1. This classification reflects a massive shortage of water as a result of irregular drainage systems for long periods of time [4]. At the same time, some manifestations of water resource change—in particular,

Table 1 Key milestones in Rajasthan's irrigation transformation (2000–2023)

Year	Milestone	Impact
2006	Pradhan Mantri Krishi Sinchayee Yojana, PMKSY (Per Drop More Crop), launched	Micro-irrigation adoption surges
2008	Narmada Canal Phase I completed	Brings water to 2.5 lakh hectares in western Rajasthan
2016	Mukhyamantri Jal Swavlamban Abhiyan begins	21,000+ water structures built
2018	State Remote Sensing Centre, Rajasthan, GIS mapping	Improves water resource planning
2020	KUSUM Scheme solar pumps installed	Reduces grid dependency in 50,000 farms
2023	ERCP (Eastern Rajasthan Canal) approved	Will benefit 13 drought-prone districts

the increasing frequency and intensity of some drought events—have further weakened the reliability of existing water resources, which has increased shortcomings in water-dependent economic sectors and environmental systems.

Agriculture has become the main consumer of water, which accounts for nearly 90% of the total consumption in the state. However, most of this waste is marked by inefficiency, contributing to huge losses through external irrigation systems, traffic, and transport inefficiency [5]. Without the efforts to improve the efficiency of the earth's resources, the earth's resources are shrinking, the ever-increasing environmental protection practices and the ever-increasing pressure on agricultural practices are posing several threats to the economic stability of our communities and our overall ecological integrity.

In response to these challenges, the state has launched a multi-pronged strategy focused on modernizing basic infrastructure, adopting micro-irrigation technologies, and institutionalizing sustainable water management [6]. This integrated approach not only attempts to stabilise the state's water economy but also provides effective management of water-stressed areas, which are undergoing similar transitions toward resilience and sustainability [7].

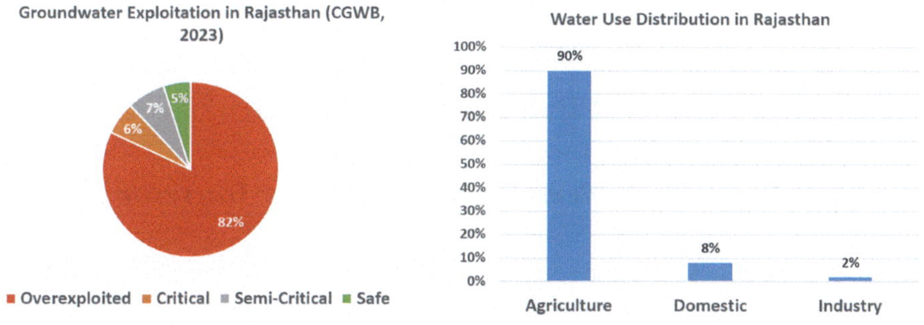

Fig. 1 Groundwater exploitation and water use distribution in Rajasthan

In response to the challenges related to water science, the kingdom has created a vast scientific framework that includes modernization of the basic structure, dissemination of shallow irrigation technology, and an improved water management system [6]. The objective of this integrated water management system is not only to stabilize the rural water science economy but also to generate sustainable development and stability in the water-stressed areas undergoing socio-economic changes in the direction of better sustainability and stability [7].

1.3 Research Question

This chapter addresses the central question: "How have infrastructural developments and technological interventions since 2000 transformed irrigation practices in Rajasthan, and what have been their impacts on agricultural sustainability and water resource management?".

1.4 Key Objectives

1. Analyze the role of major irrigation projects (e.g., Indira Gandhi Canal, Narmada Canal) in expanding water access.
2. Evaluate the adoption and impact of water-saving technologies (drip irrigation, remote sensing).
3. Assess socio-economic outcomes (crop yields, farmer incomes) and environmental trade-offs (groundwater depletion).
4. Identify policy lessons for sustainable irrigation in drylands globally.

Irrigation transformation in Rajasthan is a vital form of the global water-food-climate dynamic. Its solutions—and resolving tensions—provide a means to balance agricultural development with environmental limitations. By investigating this matter, we obtain information that can be applied from drought-stricken areas of India to dry areas of Asia, Africa, and the Middle East.

2 Historical Context of Irrigation in Rajasthan

2.1 Early Irrigation Systems: Traditional Water Harvesting

Rajasthan, India's most arid and semi-arid state, has historically been grappling with the problem of water supply. In response to its harsh water resource conditions and limited rainfall, communities across the region have developed a comprehensive series of traditional water management systems that reflect indigenous knowledge of hydro-engineering and sustainable resource management. Much before the advent of modern irrigation and basic facilities, these delicate systems created the foundations of water conservation and agricultural cultivation.

Noteworthy among these traditional systems was the Johad, a small-scale earthen dam which was constructed strategically to stop and maintain rainwater while also providing a facility for the protection of internal underground water reservoirs [8]. Khadins, is another innovative technique in which principles of rain fed farming are applied by stopping the flow of short term floods within agricultural areas, thereby making infiltration of crops slowly during the winter season and supporting seasonal crop farming. There were also baoris and beris, which were direct wells and small underground wells, which fulfilled the dual purpose of making water available for both drinking uses and irrigation. Similarly, rivers or village ponds played an important role in collecting water for domestic consumption, providing water to animals, and for overall irrigation purposes.

2.2 *Medieval and Colonial Period: Canal Systems Emerge*

During the medieval period, royal rulers constructed canals and ponds on a small scale [9]. However, in the nineteenth century, significant irrigation expansion took place under British colonial administration. Ganges Canal (later renamed as Rajasthan Canal) was conceived to divert water from the Himalayas, however, major construction started only after independence. The British also introduced revenue-driven irrigation policies, prioritizing cash crops over traditional subsistence farming. This led to a new phase of conventional irrigation farming, which would lead to a change in the agrarian practices. Despite these developments, a large part of Rajasthan remained dependent on rain and water.

2.3 *Post-Independence (1947–2000): Large-Scale Irrigation Projects*

In the post-independent period, there was a major shift in the irrigation policy of Rajasthan in which the reliance of traditional water systems was dissociated and a system of dams and canals was developed with the aim of increasing agricultural production and regional development. The Indira Gandhi Canal, which was formerly called Rajasthan Canal, was one of the most significant developments of the era and was only established in 1958. The Indira Gandhi Canal transferred water from the Sutlej and Beas Rivers in to portions of the Thar desert that are otherwise arid, signifying a huge advance in water supply system infrastructure. It was an achievement that transformed large areas of drought-prone land into agriculture land, especially in western Rajasthan.

The State was also involved in the Chambal Valley Project which was a joint collaboration with Madhya Pradesh, which involved building three multipurpose dams known as Gandhi, Rana Pratap and Jawahar Sagar. All of these dams were instrumental in the expansion of irrigation coverage in the south-eastern state and not only added to the agricultural productivity but also gave way to additional water and electricity production and flood control.

2.4 Challenges in the Late Twentieth Century

By the 1990s, the irrigation system of the state was under increasing stress and this was accompanied by a confluence of environmental, structural and socio-economic issues. First, groundwater had become the first source of irrigation in several regions; groundwater resources were being exploited to such an extent, serious shortages were registered in districts such as Jaipur, Jodhpur and Bikaner. For example, groundwater levels started to fluctuate seriously and rapidly, compromising not only the economy but possibly drinking water security.

At this time, canal-based irrigation facilities, especially in the command areas of the Indira Gandhi Canal, began demonstrating systematic inefficiencies. Poorly designed or inadequately constructed water drainage systems resulted in widespread waterlogging and soil erosion, which reduced agricultural production and left some parts of the land unsuitable for sustainable agriculture [10]. What further complicated these issues was the persistent concerns about equality. Access to river water remained unevenly distributed, and small and marginal farmers were often deprived of irrigation benefits due to environmental factors or institutional limitations, which further strengthened the existing rural inequalities.

These interlinked issues forced the government to redefine the irrigation policy of the State in the post-2000 era. The new approach focused on stability, technological modernization, and inclusive governance. The efforts encouraged water use efficiency through smart irrigation, improving canals, and enhancing participatory water management systems, to enable more ecologically viable and even-handed irrigation infrastructure in the state [11].

3 Infrastructure-Led Improvements Since 2000

The twenty-first century witnessed significant changes in the irrigation strategy of Rajasthan with a focus on modernization of infrastructure, expansion of canal networks, and integration of water conservation measures. Government initiatives, technological progress, and policy reforms played a vital role in improving water availability and agricultural productivity.

3.1 Major Irrigation Projects: Extension and Modernization of Canal Systems

Since 2000, the State Government has placed strategic emphasis on expanding and modernizing the canal system on a large scale to improve water supply efficiency and reduce transportation problems in its arid and semi-arid areas. Some of the State's major initiatives include the continued development of the Indira Gandhi Canal Project (IGNP), especially its second phase, which expanded the canal network in the western districts of Bikaner, Jaisalmer, and Barmer [12]. During this phase, an important technological innovation was implemented—concrete canal lining, which was a creative initiative that reduced seepage losses to a great extent and improved water transmission efficiency. Besides, the use of drip and sprinkler systems within the IGNP command area has further contributed to reducing water wastage while making agricultural systems convenient.

Another transformative initiative is the Narmada Canal project, implemented as part of the larger Sadar Sarovar project. The canal has made it possible to transfer water from the Narmada River to long-standing water-scarce districts like Jalore, Badaun, and Jodhpur. With irrigation coverage of over 2.5 lakh hectares, the project has played a vital role in converting dry lands into productive agricultural areas. Significantly, the Narmada Canal system has also been integrated into other irrigation schemes to promote water use efficiency. Currently, the state is undertaking the Eastern Rajasthan Canal Project (ERCP), which is under construction, to link the Chambal, Parbati and Kalisindh river systems and provided reliable water supply to 13 districts in Rajasthan. When completed, the ERCP is expected to irrigate approximately 560,000 hectares of agricultural land that will simultaneously increase agricultural productivity, as well as improve regional water security. It is an example of Rajasthan's commitment through infrastructure modernization and technology adoption for long term planning of water resources.

3.2 Minor Irrigation Schemes: Check Dams, Anicuts, and Water Harvesting Structures

In addition to investments in large-scale irrigation infrastructure, Rajasthan has consciously focused on decentralized water conservation methods to improve water availability at the local level and promote community use and management of resources. The main component of this has been the construction of check dams and anicuts at scale in the state. Small hydraulic structures like the Mahatma Gandhi National Rural Employment Guarantee Act (MGNREGA) and associated state schemes have made massive impacts to recharge groundwater and facilitate localized irrigation in water-scarce areas.

Collective efforts have been made through watershed development initiatives, specifically under the Integrated Watershed Management Programme (IWMP), which has focused on improving soil quality while restoring groundwater levels in rain-fed and ecologically sensitive areas. These programs have not only improved the water-holding capacity of agricultural landscapes but have also strengthened the resilience of rural communities to climate variability.

3.3 Public Infrastructure Investment Trends

In the last decade, Rajasthan has witnessed a significant increase in financial commitment towards the basic infrastructure of irrigation, which reflects the strategic priority of the state for water resource management. The budgetary allocation for irrigation development has also witnessed a substantial increase, from around ₹778 crores in the financial year 2010–11 to ₹9000 crores in 2023–24. This notable growth in financial investment highlights the government's importance of irrigation infrastructure as an important determinant of economic stability and rural economic well-being, especially in the context of the State's mainly dry and semi-dry climate.

In addition to domestic budgetary provisions, the state has also taken advantage of public-private partnerships and international development assistance to promote water infrastructure and services. In particular, projects supported by multidisciplinary agencies, particularly the Rajasthan Water Sector Livelihood Improvement Project (RWSLIP) supported by the Japan International Cooperation Agency (JICA), have played a vital role in expanding access to reliable irrigation to rural communities, thereby improving agricultural productivity and socio-economic development. These financial and institutional developments have been improved by progressive policy reforms. The Rajasthan State Water Policy, which was initially introduced in 2010 and later revised in 2019, is a step towards a significant overhaul of rural irrigation systems. The policy promotes the combined utilization of surface and groundwater resources. It accepts the need for appropriate and integrated water management strategies in a relevant manner.

3.4 Case Study: A Region Transformed by Canal Expansion—The Thar Desert (Bikaner-Jaisalmer Belt)

Before the beginning of the revolution, the Bikaner-Jaisalmer belt within the Thar desert was characterized by dry, marginal agricultural areas. Agricultural practices in this region were highly dependent on irregular and inadequate rainfall, due to which there were significant shortages in agricultural produce. The expansion of the Indira Gandhi Canal, one of the largest irrigation projects in India, was a transformational change that provided a reliable source of water to about 1.8 million hectares of land in the region. The introduction of canal-based irrigation facilitated a significant change in the crop pattern, making farming of high-consumption and valuable crops like wheat, sorghum, and cotton possible in a commercial manner. This infrastructural intervention led to remarkable growth in agricultural production, and in many areas, agriculture increased substantially, with farm incomes reportedly doubling in several areas. The shift from subsistence to market-oriented agriculture not only improved the economic condition but also contributed to broad rural development results, including an increase in food security and an increase in employment opportunities.

4 Technological Advancements in Irrigation

Rajasthan has taken up new irrigation technologies, since 2000, in order to combat water shortages, improve use efficiency, and increase agricultural productivity, because the problem of water shortages has affected the region for many years. The new technologies, from micro-irrigation systems to digital monitoring devices, have improved overall efficiency in water use, led to increased agricultural productivity, and provided a sustainable model for water conservation.

4.1 Micro-Irrigation Technologies: Drip and Sprinkler Systems

In an effort to optimise water use, particularly in its water-difficult areas, Rajasthan has actively promoted the adoption of drip irrigation techniques as a sustainable alternative to traditional irrigation systems. As a result, drip irrigation has gained significant momentum in delivering water directly to plant-root areas, thereby reducing losses caused by evaporation and surface runoff [13]. This method is especially effective in the cultivation of horticultural crops like citrus and pomegranate, as well as high-yield cash crops like cotton and guar. Drip irrigation is widely facilitated in adoption through targeted planning, especially through the provision of a subsidy under the Pradhan Mantri Krishi Sinchayee Yojana (PMKSY), which provides up to 50% financial assistance to small and marginal farmers [14].

In addition to this process, sprinkler irrigation has become a popular system particularly suited to crops such as wheat, sorghum and pulses. Sprinkler irrigation systems are just like natural rainfall; they provide uniform water distribution in agricultural lands, and they have particular relevance to the sandy soil profile of Western Rajasthan, which includes the districts of Jodhpur and Bikaner. In addition to water management, sprinkler irrigation systems are noted to be more energy efficient than traditional water irrigation processes, resulting in the reduction of groundwater depletion.

When the overall impact of these dry irrigation practices is examined, a cumulative effect can be observed. Research studies have measured about a 30 to 50% water saving in implementing dry irrigation methods, which resulted in an increase in yields of about 20–30%. Clearly, the impact of developing micro-irrigation methods will be an important aspect of increasing agricultural yields in ecologically sensitive areas while safeguarding important water resources [15].

4.2 Remote Sensing and GIS in Planning and Monitoring Irrigation

Increasingly advanced geospatial functionalities have resulted in changes to the irrigation management systems in Rajasthan that have increased operational efficiencies and environmental stability. A significant example of technology innovation in the state is the use of satellite-based canal monitoring in which Geographic Information System (GIS) mapping methodologies are utilized to monitor water flow variations within major irrigation schemes, with specific application to the Indira Gandhi Canal System. This technology enables detection of domestic leakages, excessive water flow, and loss of supply efficiencies, allowing for the better distribution of water to varying agricultural regions.

The water-use data collection of crop water requirements has also been enhanced by remote sensing technology, including the Normalized Difference Vegetation Index (NDVI). The use of these sensors enables in real time assessment of plant health, and in turn irrigated land, to optimize supply and curb wastage. In addition, fields of geospatial analysis have critical role in groundwater management, particularly in some contexts as mapping for areas of recharge potential. Hydrological modelling can help with site selection for construction of check dams, percolation tanks, and recharge structures.

A significant example of the evolution of these technologies is the application of data in the state through the State Remote Sensing Application Centre, Jodhpur (SRSAC). The SRSAC acts as significant institutional node encompassing both the historical and advanced technological applications of geospatial methodologies, providing real time support to both farmers and policymakers, and creating a means for data-informed decisions in irrigation and resource management.

4.3 Digital Soil Moisture Sensors and Mobile Apps for Farmers

In Rajasthan, innovative agricultural practices have improved accuracy and efficiency in the use of water in agriculture, particularly in areas of water scarcity. There has been a major innovation in agriculture in the area of Internet of Things (IoT) based soil-moisture sensors [16]. These technologies provide real-time information about soil moisture content and text message updates to farmers, which enables farmers to know what their irrigation needs are and allows farmers to make data-driven decisions at the farm. For example, in Rajasthan's Sikar and Nagaur districts. These technologies were used for grape and olive crops (which are water-intensive), and farmers using the sensor-based irrigation measures learned to stop over-irrigating, enabling them to save water and keep the plants healthy.

Mobile applications are becoming a complementary form of sensor-based technologies, such as showing irrigation recommendations based on each farmer. Examples include mKrishi, a service provided by Tata Consultancy Services (TCS), and Jal Sanchay, which is a service started by the Government of Rajasthan. These mobile applications are developed to provide advice for their users, including current weather forecasts, current water needs for the season, and a validated irrigation schedule through a smartphone screen. Mostly because of the smartphone/platform user interface, these tools are widely adopted by the leading-edge community of farmers.

Field-based reports suggest that when farmers use sensor-based irrigation technologies, in combination with mobile apps East Mi India Group Smart with agronomic advisory services, an estimated 20% of the water expenditures were reduced with no loss in yield. The confluence of new tools of digitized communication with agronomic application presents a significant opportunity for smart farming to contribute to water resource management in arid areas.

4.4 Impact on Water Use Efficiency and Crop Yields

The implementation of advanced irrigation technologies in the State has resulted in significant increase in water use efficiency, crop production and sustainability of groundwater resources. As shown in Table 2, water use efficiency under traditional irrigation methods was earlier limited to about 35–40%, mainly due to heavy losses caused by overflow and other hazards. The introduction of micro-irrigation systems along with drip and sprinkler technologies enhanced efficiency significantly, increasing it to a range of 75–85%. This change has not only optimised the use of water resources but has also played an important role in stabilising the groundwater levels.

The crop yields, which were earlier affected by the internal limitations of rain-fed agriculture, have also improved significantly. The technology interventions involving the soil moisture sensors, mobile-based advisory services, and micro-irrigation have facilitated a 1.5 to 2 times increase in yield across various crop categories. These

Table 2 Water use efficiency and crop yields before and after tech adoption

Metric	Before tech adoption	After tech adoption
Water Use Efficiency	35–40% (Flood Irrigation)	75–85% (Drip/Sprinkler)
Crop Yields	Low (Rain-fed)	1.5 × Increase
Groundwater Depletion Rate	1.5 m/year (1990s)	0.5 m/year (Post-2010)

benefits are especially significant in areas cultivating high-value horticultural and cash crops. The decline observed in the rate of groundwater depletion is also noteworthy. During the 1990s, as a result of unstable extraction systems, there was a depletion rate of about 1.5 m of demand per year in many districts. After 2010, after paying closer attention to sustainable irrigation systems, this loss has reduced to 0.5 m per year. This data underscores the significant effect of technology-driven water management strategies on Rajasthan's agrarian landscape [17].

4.5 Case Study: Adoption of Micro-Irrigation in Sikar

In the semi-arid districts of Rajasthan, the unviable cultivation of water-intensive crops like wheat and sorghum has contributed to a significant decline in the groundwater levels over time. In response to this growing crisis, both the state and central governments have implemented a series of targeted schemes intended to encourage viable food production techniques. These measures included adequate subsidies for adopting a drip irrigation system, especially for garden crops like kinnow and aloe vera, which are best suited to the conditions of the region. Complementary efforts by the Krishi Vigyan Kendras (KVKs) focused on capacity building by training farmers to improve their proficiency in the effective use of these irrigation technologies.

The results of these interventions have been significant. In areas using drip irrigation, there has been a 40% reduction in water use, while production has increased by 30% due to more accurate and timely distribution of nutrients. Equally significant is the economic consequences of this change: many farmers have shifted from farming low-yield cereals to farming high-yield horticultural produce, thereby improving their income. However, this change has not been without challenges. The high cost of drip irrigation systems remains a significant obstacle for small and marginal farmers, even if subsidies are available. In addition, maintenance issues such as clogging of drip lines keep hampering the final work, emphasizing the importance of technical support and infrastructural enhancements.

In the future, there will be a need for a multipronged strategy in water management of the state. To reduce the high cost of adoption of advanced irrigation innovations, the government must expand targeted subsidies for small and marginal farmers. Also, promoting technical literacy through rural training programmes is extremely important to enable farmers to effectively operate, maintain, and derive benefits from these systems. Moreover, integration of irrigation technologies with solar-powered pumps offers another way to long-term energy and water sustainability, especially in remote areas where access to electricity remains a challenge [18]. If these measures are implemented in coordination, it can be ensured that productivity improvements are spread across the poor areas of the state and distributed equally. Within the state, while implementing energy pumps and similar innovations, area-specific policies that assess the availability of water resources locally would be appropriate to optimise your resource use and prevent unexpected consequences [19].

5 Policy and Institutional Interventions in Rajasthan's Irrigation Sector

Since 2000, Rajasthan has implemented strategic policies and institutional reforms to enhance irrigation efficiency, promote water conservation, and support farmers. Government schemes, subsidies, and partnerships with NGOs have helped reshape the state's irrigation landscape.

5.1 Government Schemes for Irrigation Development

The State Government, in collaboration with central authorities and international development agencies, has launched a comprehensive set of irrigation-focused initiatives to enhance water use efficiency, expand irrigated agricultural land, and encourage ecologically sound agricultural techniques. The most important of these is the Pradhan Mantri Krishi Sinchayee Yojana (PMKSY), which has been launched under the guiding principle of "Per Drop More Crop". This flagship programme is designed to optimise water production by promoting the adoption of micro-irrigation techniques, including drip and sprinkler systems. In Rajasthan, this scheme has provided adoption of micro-irrigation facilities in more than 300,000 hectares of area between 2015 and 2023, supported by subsidies of up to percent for small and marginal farmers. The programme has given strategic priority to water-stressed districts like Jodhpur, Barmer, and Pali, where traditional irrigation methods are neither practical nor sustainable.

Complementing PMKSY is the Rajasthan Agricultural Competitiveness Project (RACP), an initiative supported by the World Bank with the aim of enhancing the economic viability of the country through integrated irrigation development. The project has supported the modernization of canal distribution systems, strengthened water management on farms, and encouraged the shifting of high-yield crops like horticultural crops and spices under efficient irrigation systems. Farmers in the project areas, including Kota and Bundi, have reported an increase in income of up to 20%. Apart from this, RACP has also promoted participatory irrigation governance through the organization of Water User Associations (WUAs), which has promoted community-led management of water resources.

At the ground level, the Mukhyamantri Jal Swavlamban Abhiyan (MJSA) has adopted a community leadership vision for water conservation. This initiative focuses on the construction of new check dams to restore traditional water conservation systems like johads and nadis, to enhance the local water resource supply. Between 2016 and 2020, more than 21,000 water conservation structures were constructed, which had a positive impact on groundwater availability in more than 12,000 villages.

5.2 Subsidies and Incentives for Micro-Irrigation

To accelerate the adoption of water-efficient technologies in water-stressed rural areas, the State Government has established a comprehensive support structure comprising financial incentives, technical assistance, and renewable energy integration. The main objective of this initiative is to create a tiered subsidy structure for various irrigation schemes. Small and marginal farmers are eligible for substantial subsidies, reaching up to 90% through joint state and central government funding, while other farmers receive support ranging from 50 to 60%. This differentiated model aims to ensure both equal access and scalable technology dissemination.

Apart from the direct financial subsidy, the state also provides interest-free loans for the purchase of irrigation equipment, thereby reducing financial constraints for the income-oriented farmers in its implementation. The Kisan Urja Suraksha evam Utthaan Mahabhiyan (KUSUM) scheme has further strengthened the stability initiative by providing a subsidy for solar-powered pumps. This not only reduces dependence on electricity for a connected grid but also links irrigation modernisation with broader energy transition objectives. The cumulative impact of these planned measures is considerable. Drip irrigation coverage in Rajasthan has increased tenfold, from about 50,000 hectares in 2005 to over 500,000 hectares by 2023. In crops with high water requirements like cotton and citrus, area-based studies have reported water saving of up to 40–60% without a concomitant reduction in yield levels. This significant change highlights the effectiveness of targeted incentives in catalysing behavioural changes and promoting sustainable use of resources in one of the most water-stressed states in India.

5.3 Role of NGOs and Farmer Producer Organizations (FPOs)

Non-governmental organizations (NGOs) and farmer producer organizations (FPOs) are playing a vital role in implementing sustainable irrigation and water conservation systems across the state. Their embeddedness within local communities and commitment to participatory development have enabled them to complement the state-led interventions, leading to the development of a more comprehensive, flexible, and decentralized model of water governance. One of the most prominent contributors is the Tarun Bharat Sangh (TBS), whose work in the Alwar district exemplifies the effective revival of indigenous water management systems [20]. By restoring more than a thousand johads, TBS has significantly enhanced groundwater restoration, revived local hydrological cycles, and strengthened the ecological validity of local water-engineering traditions. Similarly, the Watershed Organisation Trust (WOTR) has implemented water-efficient irrigation strategies in districts like Bhilwara and Rajsamand. This strategy emphasizes the integration of water-utilization efficiency with comprehensive water-availability adaptation requirements, and advances a model of irrigation that is both environmentally relevant and responsive.

In line with these efforts, Farmer Producer Organisations (FPOs) have played an important role in organising collective action among small farmers. By providing bulk purchase facilities for short-term irrigation equipment like drip and sprinkler systems, FPOs have helped reduce their own input costs, thereby improving access and affordability. In addition, FPOs also act as a platform for building technology capacity, providing training programs on accurate irrigation techniques, system maintenance, and efficient resource management. A special example of this is the Sujalam Sufalam FPO in Ajmer, where more than 500 farmers helped set up a community-managed drip irrigation network. The results of this initiative not only brought a reduction of 35% in water consumption but also increased the production of vegetable crops, showing dual benefits of environmental protection and increased agricultural production. Collectively, this initiative highlights the indispensable role of civic society and farmer-network institutions in transforming sustainable irrigation systems at local levels into real results.

5.4 *Case Study: PMKSY's Impact in Pali District*

The implementation of Pradhan Mantri Krishi Sinchayee Yojana (PMKSY) in Pali district provides one of the key studies which show how targeted irrigation strategies can effectively reduce water stress while simultaneously increasing agricultural production. The region has historically been facing the threat of depletion of water resources, particularly for cotton farming which has a high water consumption. Continuous extraction has led to a dangerous decline in groundwater tables, posing a major threat to both the practicality of crop production and the long-term economic stability. To address this important challenge, about 2500 farmers of the district were given a supplementary subsidy, which helped in adopting a specially designed drip irrigation system for cotton farming. These technological breakthroughs were accomplished through structured training programs designed to promote water-efficient agricultural practices and optimal use of irrigation technologies. The results of this integrated approach were significant. The participating farmers gave information about a reduction of 50% in the water consumption in each area, which was due to the precise delivery of moisture to the plant roots. Also, the average crop yield increased from 8 quintals per unit to 12 quintals per unit, which reflects the productive benefits obtained through better irrigation management.

6 Socio-Economic and Environmental Impacts of Irrigation Transformation in Rajasthan

Irrigation reforms in Rajasthan since 2000 have had profound socio-economic and environmental consequences, reshaped livelihoods, transformed the water system, and influenced gender roles in agriculture. This section analyses this dynamic impact.

6.1 Changes in Agricultural Productivity and Income Patterns

The changes in the irrigation environment of the state during the last two decades have brought about significant changes in the agricultural production and rural structures. At the root of these changes, there is a clear differentiation of high-yield, irrigated crops, including wheat, cotton, sour fruits, and various types of vegetables, from traditional, rain-fed subsistence crops like bajra and pulses. This change, facilitated by the expansion of canal-based irrigation and the widespread adoption of smart irrigation technologies, has significantly increased the crop yield. It has also enhanced economic benefits. In canal-command areas, especially under the Indira Gandhi Canal Project (IGNP), wheat yield has increased from about 2.5 tonnes to 4.5 tonnes per hectare, which highlights the productive benefits linked to reliable water access. Similarly, cotton farming under drip irrigation has shown improvement in yield by 20–30%, which shows the agricultural benefits of accurate and nutritious element application.

These improvements in agricultural performance have translated into tangible income growth. Many farmers in irrigated areas report a two to three times increase in income due to the shift to commercial crops and the ability to engage in multi-seasonal farming. Increased water availability has also strengthened market integration, which has helped farmers access Agricultural Produce Market Committees (APMCs) and, in some cases, international markets. For example, the expansion of irrigated guar cultivation in Barmer district has facilitated exports to the global pharmaceutical and oil exploration industries, highlighting the strategic potential of water-efficient agriculture.

Table 3 presents a comparative analysis of the differential impacts of irrigation reforms and technological interventions across two categories of farmers in Rajasthan: small and marginal farmers (owning less than 2 hectares) and large farmers (owning more than 4 hectares). The table examines four critical dimensions: access to water, adoption of technology, income change, and vulnerability.

Table 3 Impact on small vs. large farmers

Aspect	Small/Marginal Farmers (<2 ha)	Large Farmers (>4 ha)
Access to Water	Dependent on canal rotations & groundwater (often last priority)	Invest in private tube wells & drip systems
Adoption of Tech	Slow due to high upfront costs; rely on subsidies	Early adopters of sprinklers, sensors
Income Change	Moderate (10–20% rise) – Limited by landholding	Significant (50–100% + rise) – Economies of scale
Vulnerability	Higher risk during canal water shortages	More resilient due to diversified water sources

6.2 Effects on Groundwater Recharge and Depletion

The technological and structural interventions implemented in the state have yielded heterogeneous results in terms of groundwater stability. Positively, the developed water conservation initiatives, along with the construction of quick and comprehensive urban development programmes, have clearly improved groundwater status in specific areas. For example, in Alwar district, an increase of up to five metres in water level was recorded between 2000 and 2020, while a comparable positive trend was seen in Jhunjhunu. Besides, the establishment of canal-based irrigation systems in districts like Bikaner and Hanumangarh has reduced dependence on groundwater resources for economic purposes.

In contrast, significant challenges have arisen in other areas. In specific "dark areas" with special features of groundwater overexploitation, like Jaipur and Jodhpur, the unregulated proliferation and utilization of tube wells continue to drive unsustainable rates of groundwater extraction. Apart from this, there has been an increase in water supply and soil erosion as a result of inadequate water drainage infrastructure within the command areas of the Indira Gandhi Canal Project (IGNP), especially in the Sri Ganganagar district. The figures of the Central Groundwater Board show that in almost 70% of the administrative blocks of the State, the rate of groundwater extraction continues to be unsustainable, which highlights the necessity of more stringent regulatory structures and integrated water resource management strategies.

6.3 Gendered Impact and Labour Use Changes

Irrigation reforms in Rajasthan have brought about a range of complex challenges related to gender division of rural labour. In areas severely impacted by land degradation, women face additional burdens of water insecurity and often have to engage in arduous work that requires significant time and energy collecting water for household and agricultural use. This ongoing situation reflects the existing gendered disparities in rural water access and dimensions of labour.

Nevertheless, a range of targeted initiatives have begun to provide new pathways to women's empowerment in rural areas. NGO-led activities like the establishment of Self-Help Groups (SHGs) in Dausa district have been important in building women's technical skills particularly as they relate to the management and maintenance of micro-irrigation systems. These initiatives have fostered some incremental changes in women's roles that have allowed women increased involvement in better governance, and thus, improved decision making in rural areas. Similarly, increased involvement of women in higher-value horticultural pursuits including date farming in Jaisalmer is enlarging their contributions toward household income while challenging their economic autonomy.

The introduction and proliferation of drip and sprinkler irrigation systems have also altered rural labour configurations. By reducing the amount of manual irrigation labour required, these approaches have allowed women to repurpose their time toward higher-value pursuits and support activities. Improved reliability in irrigation has allowed for year-round farming in districts like Nagaur which has decreased the need for seasonal migration particularly for male workers, and thus ensured stronger domestic labour changes. Irrigation reforms have not solved the gender inequalities present in rural labour systems, but important changes have begun to promote a resurgence of rural labour. Changes outlined here promote greater opportunities to enhance women's economic participation and changes to inter-household labour within the context of an improving economy in the state.

7 Emerging Challenges in Rajasthan's Irrigation Sector

Despite significant progress, Rajasthan's irrigation system faces critical issues that jeopardize its long-term sustainability. From groundwater depletion to institutional inefficiencies, these issues require urgent policy attention.

7.1 Groundwater Over-Extraction and Unsustainable Use

Rajasthan is experiencing a severe groundwater crisis due to unsustainable extraction and usage, putting long-term agricultural production at risk. The Central Ground Water Board (CGWB) published a report in 2023 indicating that 82% of each administrative block in Rajasthan was either overexploited, or critical—which essentially means their resources are depleted. Jaipur, Jodhpur, and Nagaur districts have experienced declines in water levels from 0.5 to 1.5 m per year. The groundwater crisis can be analysed from several dimensions; there are structural factors and behavior factors in play. The number of unregulated tube wells in the state increased to over 1.5 million, allowing for a continuous pumping without further consideration, and the agricultural practices, of which are supported by large electricity subsides are incentivizing excessive groundwater extraction and pumping. Unfortunately, there are also agricultural practices that are environmentally unsustainable which intensify the situation—for example, environmentally unsuited rice and wheat-growing areas where the crops themselves use a large amount of water put additional pressure on water resources that are simultaneously being depleted.

The implications of the unsustainable practices are wide-ranging. First of all, the shallow wells are being drained requiring farmers to invest in deep bore wells—that typically are bored down to 150 to 300 m. The interplay of shallow wells, reduced depth of water and salinity are issues for farmers, as growing crops and farming has been made unsustainable and their can no longer replenish the nutrients in the soil [21].

A notable example is the Bikaner district, where groundwater levels fell by more than 10 m in the past twenty years creating a very bleak economic situation for the district. Salinity in soils is becoming an increasing problem, which one possible alternative could be to bring in and grow salt tolerant crop species [22].

7.2 *Climate Variability and Drought Resilience*

The water supply shifts are making Rajasthan's irrigation systems more vulnerable. Unpredictable rainfall patterns—in particular, the monsoon breaks and failures of 2018 and 2022—have created uncertainty for production, as well as lower resilience for agricultural processes. Similarly, the region has increased approximately 1.5 °C since 2000. This warming has increased the rate of evapotranspiration and contributes to the stress on limited water supplies [23].

These climatic trends have exposed structural weaknesses in the state's irrigation infrastructure. During the dry years, river networks—particularly the Indira Gandhi Canal—have recorded substantial reductions in water flow, with a decline of 30% in 2019. Such reductions have delayed water release schedules, which have disrupted sowing periods and have had a negative impact on crop production in many districts.

To bolster resilience, a broad and expansive strategy is necessary. Promoting climate-smart farming practices, such as integrating micro-irrigation systems with drought-tolerant seed varieties, provides a practical way to reduce the adverse effects of rain variability. Overall, expansion of water storage infrastructure, especially through initiatives like Eastern Rajasthan Canal Project (ERCP) has the potential to enhance inter-seasonal water security and stabilise irrigation supply in drought-hit districts (Fig. 2).

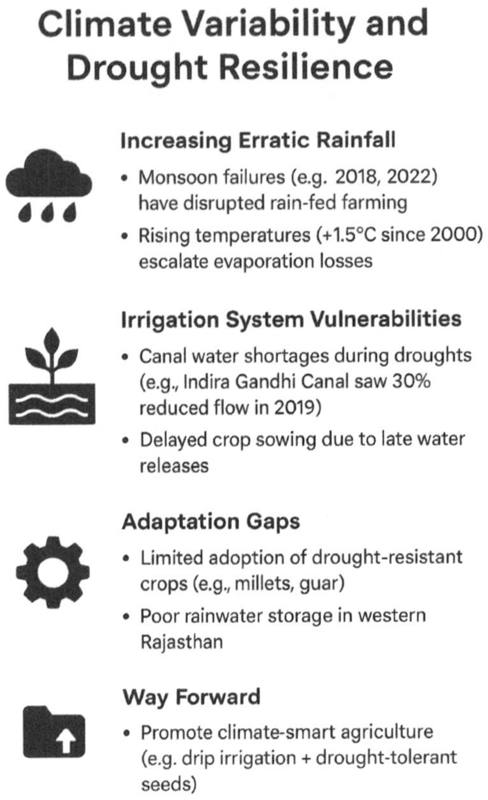

Fig. 2 Climate variability and drought resilience

7.3 Regional Disparities in Access to Irrigation Infrastructure

Rajasthan's irrigation landscape reveals marked regional disparities, as illustrated in Table 4 below.

Table 4 Access to irrigation infrastructure

Region	Irrigation Coverage	Key Issues
North-West (Thar Desert)	Low (20–30%)	Dependent on IG Canal; tail-end farmers face shortages
South-East (Kota, Bundi)	High (60–70%)	Better canal & groundwater access
Central (Ajmer, Jaipur)	Moderate (40–50%)	Groundwater depletion limits expansion

There are significant regional disparities in irrigation coverage across Rajasthan. The north-western Thar desert area, covering districts like Barmer and Jaisalmer, shows lower irrigation coverage rates (20–30%), depending mainly on the Indira Gandhi Canal. Farmers located at the end of these canal systems often have to deal with a shortage of water due to delays in supply and infrastructural management challenges. In contrast, south-eastern regions, including Kota and Bundi, benefit from high irrigation coverage (60–70%) due to a more reliable canal network and comparatively better water supply. Medium regions, including districts like Ajmer and Jaipur, show medium irrigation coverage (40–50%); however, fast-paced water supply presents a significant obstacle to future expansion.

These differences in irrigation access are related to significant differences in income. Farmers in high coverage areas (like Bundi) earn 2–3 times more than their counterparts in arid districts like Barmer. Unreliable irrigation and its resulting low crop yields have aggravated the rural crisis in western Rajasthan, leading to the migration of youth from areas like Jaisalmer and Jodhpur to urban centres in search of alternative livelihood. It has become critical for agricultural equity across diverse agro-ecological zones of the State to address through strategically targeted investment into irrigation infrastructure development and local water conservation initiatives.

7.4 Institutional Bottlenecks and Maintenance Issues

Rajasthan may have invested a lot of money into irrigation infrastructure, but ongoing institutional and operational challenges are contributing to inefficiencies in the use of water resources and limiting equitable access to farmers in Rajasthan.

- Weak Water Governance: systematic maladministration of the canal system creates inefficiencies with stolen water, and over- and under-availability of water for the last-mile user.
- Neglect of Infrastructure: Long-term disinvestment and neglect of infrastructure have caused much physical degradation. For example, segments of the Indira Gandhi Canal have apparently lost 30% of their capacity due to sediment. Furthermore, routine maintenance is generally delayed due to budget. When infrastructure is out of order, it limits the functionality and reliability of the system.
- Limited farmer involvement: Water User Associations (WUAs)- although interested in institutionalizing participatory irrigation management, have been operating locally with limited autonomy and legitimacy. In addition, considerable numbers of farmers do not know about, or do not appreciate the need for, efficient irrigation or water-saving technologies, which limits product adoption and, ultimately, impact at scale.

7.5 Case Study: Sri Ganganagar

The district provides a striking example of the challenges, where widespread waterlogging has occurred due to institutional inefficiency and poor water drainage infrastructure. Overall, more than 10,000 hectares of land have become saline, resulting in declining crop yields and compromising the livelihood of farmers. To address these inter-linked challenges, a dual strategy is required: first, physical rehabilitation of old irrigation infrastructure; and second, strong institutional reforms to enhance governance, improve inter-departmental coordination and enhance meaningful farmer participation in irrigation planning and management processes. This promotes long-term sustainability and parity of the irrigation systems of the State.

7.6 Policy Recommendations

To ensure long-term sustainability of irrigation in Rajasthan, a set of integrated policy interventions is required [24]. It is necessary to have strict groundwater regulation in areas with overexploited or "dark" areas. This includes mandatory installation of groundwater metering system and rational pricing of electricity to discourage indiscriminate water extraction. Also, in seriously affected blocks, a phased withdrawal from the cultivation of water-intensive crops is necessary to prevent further aquifer depletion and to encourage more viable cropping patterns [25, 26].

Climate adaptation measures should be promoted significantly to overcome the increasing risks arising from seasonal changes and increasing temperatures. The main objectives include expansion of crop insurance schemes to protect farmers from drought-related losses, increasing investment in rainwater harvesting infrastructure like check dams and watershed development structures to facilitate land reclamation and increasing overall water availability.

8 Rajasthan's Irrigation Sector: Important Lessons and a Future Roadmap

Rajasthan's two-decade irrigation transformation offers critical insights for sustainable water management in arid regions. This section synthesizes key lessons and outlines a roadmap for the future, with relevance for other water-scarce regions globally.

8.1 *Integrated Water Resource Management (IWRM) in Arid Zones*

Rajasthan's experience underlines the vital importance of Integrated Water Resource Management (IWRM), especially in arid and semi-arid areas, which are facing severe hydrological stress. A holistic approach has emerged as a key determinant of success, through schemes like the Mukhyamantri Jal Swavlamban Abhiyan, which demonstrated the supportive benefits of integrating canal irrigation, groundwater recharging strategies, and comprehensive watershed development within integrated structures. Importantly, the active participation of local communities has enhanced the long-term stability of water management efforts. In the future, Rajasthan should promote the IWRM model in all the districts, integrate sustainable water management, rainwater storage techniques, and sustainable land use systems on a wide scale. Adopting a "ridge-to-valley" perspective for watershed planning, which prioritizes upstream water recharging and downstream water conservation efforts, could improve water storage capabilities, improve land productivity, and help deal with growing challenges of climate variability. A balanced and comprehensive water strategy has become the key to ensuring long-term water security.

8.2 *Recommendations for Scaling Strategies*

Table 5 outlines scaling strategies for successful water management models in Rajasthan, highlighting localized innovations to enable wider policy replication across the state.

- **Drip Irrigation in Sikar:** The real success of this model was to employ Farmer Producer Organizations (FPO) to drive down initial costs and capably maintain maintenance in cooperation; expanding will simply require extending the reach of FPOs so that small and marginal farmers can access drip irrigation technologies in a more affordable and sustainable way.
- **Johad Revival in Alwar:** This model is an agriculture water recharge success story in the literature that clearly demonstrates the potential of using an ancient

Table 5 Strategies for scaling initiatives

Successful model	Scaling strategy
Drip Irrigation in Sikar	Expand FPO-led collective purchasing for small farmers
Alwar's Johad Revival	Replicate NGO-government partnerships in Jodhpur, Barmer
IGNP Canal Modernization	Apply lessons to Narmada Canal zones (e.g., sensor-based water release)
Solar Pumps (KUSUM Scheme)	Prioritize dark zone districts with solar-drip integration

water harvesting structure, a Johad, as an effective way to restore hydrological and ecological balance. The Johad experience in Alwar draws on partnership work with community organizing through NGOs, and could serve as a replicable example. Based on experience and success, the Johad model could be strategically adopted to sites in water-scarce districts such as Jodhpur and Barmer to promote sustainability.

- **Indira Gandhi Canal Project (IGNP) Improvement Model**: The extensive improvements to the IGNP, which included, among other technologies, sensor-based water release mechanisms, have vastly improved water delivery efficiency and minimized conveyance losses. The IGNP modernization model could be replicated and applied to existing canal systems, such as the Narmada Canal, for the purpose of improving equity in water distribution and performance of the overall irrigation system.
- **Deployment of Solar Pumps under the KUSUM Scheme**: The use of renewable energy in irrigation systems, such as solar pumps, promotes sustainable energy use in agriculture overall. Implementation in over-exploited "dark zone" districts allows for the option of coupling energy conservation with groundwater recharge. A more efficient targeted strategy would combine solar pumps with water used through micro-irrigation techniques.

8.3 Climate-Resilient Irrigation Systems

In order to meet the increasing challenges of water scarcity and climate uncertainty in arid and semi-arid areas, climate-resilient irrigation systems can be implemented through agronomic adaptation, niche creation of appropriate technological change, and infrastructure improvements. An adaptive measure to ameliorate the situation is crop diversification, where the focus moves away from traditional grains, such as wheat and rice, which are both high water-demanding crops, and instead embraces millets, pulses, and drought-resistant crops suited to the agro-climatic areas of Rajasthan.

Smart irrigation technologies are also enhancing water use efficiency. AI-based forecasting systems and apps (such as "Meghdoot" at India Meteorological Department (IMD)), identify regional and temporal levels of irrigation scheduling to provide agronomic farmers with real time local advisories, whilst soil moisture sensors and automated drip irrigation systems support the accuracy of water use efficiency and avoid over-irrigation at farm level.

8.4 Policy Implications for Other Indian States & Global Drylands

A number of localized initiatives, including building check dams and farm ponds, as well as watershed restoration initiatives, have brought life back to Rajasthan's traditional water management systems, creating a hybrid system that incorporates traditional water management with modern engineering [27]. The Government of Rajasthan has plans to address water security by 2030. The vision includes the objective of achieving 50% micro-irrigation uptake, ceasing groundwater over-extraction in regions in critical dark zones, and doubling the income of farmers through a transition to high-value, low-water using crops. Rajasthan's process of irrigation reform offers important policy and institutional insights not only for other Indian states but also for arid regions worldwide that wish to implement sustainable water management activities, given the climatic and hydrological stresses placed upon them and the associated climate adaptation challenges. The state's integrated and adaptive governance has resulted in a framework that can be replicated in transforming dryland agriculture.

States in India, such as Gujarat, can take advantage of the models for rainwater harvesting through wide-ranging networks of check dams and a community-centered groundwater governance framework in Rajasthan to support aquifer recharge and long-term water security. In Maharashtra, and especially in the drought-affected region of Marathwada, building on the potential for Water User Associations for irrigation, and scaling up of targeted subsidies for micro-irrigation technologies would provide a pathway forward. In Haryana and Punjab, where groundwater over-extraction is presenting an existential challenge, modernizing irrigation canal systems for agriculture, particularly enabling sensor-based water distribution, and real-time water management, provides a real opportunity to reduce unsustainable water use.

Beyond India, the adaptive practices in Rajasthan have significant relevance for dryland geographies across continents. Rajasthan could make a great contribution to international forums such as the UN Convention to Combat Desertification (UNCCD), where Rajasthan could showcase its institutional innovations in water management, participatory governance, and climate adaptability. In the Sahel region of Africa, johads (traditional water harvesting structures) might be a promising community-based solution to sequester water to address desertification. In the Middle East, where there is acute water scarcity, the model for traditional solar drip irrigation systems from Rajasthan presents opportunities to leverage renewable energy to achieve irrigation efficiency gains. Similarly, rainfed drylands in Latin America and the Andean highlands could also be nourished by an IWRM watershed-based blueprint that contemplated the Israeli revival of traditional hydrological systems in Alwar. Rajasthan's collaborative approach to capitalizing on expanding knowledge exchange would serve it well to share its know-how related to a IWRM and community-led water governance systems with other arid and semi-arid regions of the world, from the Sahel to the Middle East, and show that irrigation can be the springboard for advancing stronger climate resilience, social justice, and profound agricultural improvements.

References

1. Raju KV, Shah T (2000) Revitalisation of irrigation tanks in Rajasthan. Econ Polit Wkly 35(23):1930–1936. Retrieved from http://www.jstor.org/stable/4409359
2. Mahajan N (2020, March 30) Rajasthan is adopting a robust water management system for efficient irrigation. Egov. Retrieved from https://www.proquest.com/magazines/rajasthan-is-adopting-robust-water-management/docview/2384120749/se-2
3. CGWB (2024) Report on dynamic ground water resources of Rajasthan. Retrieved from https://www.cgwb.gov.in/cgwbpnm/public/uploads/documents/17084190408707263911file.pdf
4. Rosin R (1993) The tradition of groundwater irrigation in Northwestern India. Human Ecol 21:51–86. https://doi.org/10.1007/BF00890071
5. Dev SM (2016) Water management and resilience in agriculture. Econ Polit Wkly 51(8):21–24. Retrieved from http://www.jstor.org/stable/44004407
6. Deshpande V, Ahmad I, Singh C (2024) Sustainable groundwater management through micro irrigation: a critical review on challenges and solutions. J Landscape Ecol 17(1):16–34. https://doi.org/10.2478/jlecol-2024-0002
7. Birkenholtz T (2008) Irrigated landscapes, produced scarcity, and adaptive social institutions in Rajasthan, India. Ann Assoc Am Geogr 99(1):118–137. https://doi.org/10.1080/00045600802459093
8. Gupta S (2011, October) Demystifying 'tradition': the politics of rainwater harvesting in rural Rajasthan, India. Water Altern 4(3):347–364. Retrieved from https://www.water-alternatives.org/index.php/all-abs/146-a4-3-5/file
9. Chattopadhyaya BD (1973) Irrigation in early Medieval Rajasthan. J Econ Soc Hist Orient 16:298. Retrieved from https://www.proquest.com/scholarly-journals/irrigation-early-medieval-rajasthan/docview/1304344446/se-2
10. Pandey S, Mohapatra G, Arora R (2023) Groundwater quality, human health risks and major driving factors in arid and semi-arid regions of Rajasthan, India. J Cleaner Prod 427. https://doi.org/10.1016/j.jclepro.2023.139149
11. Rao PS (2005) Resource efficiency of ground water potential by tubewell irrigation in Rajsamand district of Rajasthan. Indian J Agric Econ 60(3). Retrieved from https://www.proquest.com/scholarly-journals/resource-efficiency-ground-water-potential/docview/201530380/se-2
12. Payal P, Poonam (2020) Role of irrigation in development: the Rajasthan experience. Int J Recent Technol Eng 8(6):2430–2435. https://doi.org/10.35940/ijrte.F7938.038620
13. Saini S, Roat M (2022) Irrigation pattern in Jaipur district of Rajasthan. Int J Innovations Res Anal 2(2):143–147. Retrieved from https://www.inspirajournals.com/uploads/Issues/544175492.pdf
14. WAPCOS (2018) State irrigation plan (SIP) for Rajasthan under PMKSY (2016–17 to 2022–23). Retrieved from https://pmksy.gov.in/mis/Uploads/2018/20180404041150904-1.pdf
15. Mohan G, Perarapu L, Chapagain S, Reddy AA, Melts I, Mishra R, et al (2024) Assessing determinants, challenges and perceptions to adopting water-saving technologies among agricultural households in semi-arid states of India. Curr Res Environ Sustain 7. https://doi.org/10.1016/j.crsust.2024.100255
16. Kaur A, Bhatt DP, Raja L (2024) Developing a hybrid irrigation system for smart agriculture using IoT sensors and machine learning in Sri Ganganagar, Rajasthan. J Sens. https://doi.org/10.1155/2024/6676907
17. Monir MM, Sarkar SC (2024) Analyzing post-2000 groundwater level and rainfall changes in Rajasthan, India, using well observations and GRACE data. Heliyon 10. https://doi.org/10.1016/j.heliyon.2024.e24481

18. Kishore A, Shah T, Tewari NP (2014) Solar irrigation pumps: farmers' experience and state policy in Rajasthan [India]. Econ Polit Wkly 49(10):55–62. Retrieved from http://www.jstor.org/stable/24479230
19. Jain P, Jain P, Jain P (2022) Implementation perspectives of solar energy irrigation policy in water deficient regions: a case of Rajasthan state in India. Local Econ 37(3):207–218. https://doi.org/10.1177/02690942221099400
20. Kashwan P (2006) Traditional water harvesting structure: community behind "community". Econ Polit Wkly 41(7):596–598. Retrieved from http://www.jstor.org/stable/4417829
21. Kumar M, Santra P, Panwar N, Moharana P, Pandey CB (2022) Whether canal command irrigation through pressurized irrigation system deteriorates soil properties in hot arid ecosystem of India? Geoderma Reg 28:e00459. https://doi.org/10.1016/j.geodrs.2021.e00459
22. Chaudhary V, Satheeshkumar S (2018) Assessment of groundwater quality for drinking and irrigation purposes in arid areas of Rajasthan, India. Appl Water Sci 8. https://doi.org/10.1007/s13201-018-0865-9
23. Verma AS (2025, March 6) Water crisis: decreasing levels in himachal dams affects irrigation supply to Rajasthan. Down to Earth. Retrieved from https://www.proquest.com/magazines/water-crisis-decreasing-levels-himachal-dams/docview/3174785046/se-2
24. Sangwan R (2016) From desert to oasis: regional water resource. Int J Eng Appl Sci 3(12). Retrieved from https://www.ijeas.org/download_data/IJEAS0312028.pdf
25. Bairwa A, Gupta A, Mishra V, Gupta K (2024) Physico-chemical and correlation study of ground drinking water of Tonk district of Rajasthan (India). Discov Water 4(23). https://doi.org/10.1007/s43832-024-00077-x
26. Kumar V (2015) Participatory irrigation management in Rajasthan—a case study. Indian J Agric Econo 70(3):384–385. Retrieved from https://www.proquest.com/scholarly-journals/participatory-irrigation-management-rajasthan/docview/1749188527/se-2
27. Narsimlu B, Prasad JV, Reddy AA, Chary G, Gopinath KA, Sridhar KB, Singh VK (2024) Catchment storage command relationship for sustainable rainfed agriculture in the semi-arid regions of Rajasthan, India. Sustain 16(10). https://doi.org/10.3390/su16103996

Water Privatization and Public Rights in the DRC: The Case of Community-Managed Water Systems (ASUREP) in Kinshasa's Peri-Urban Neighborhoods of Kisenso and Nsele

Hippolyte Ditona Tsumbu, Abhisekh Rodricks, and Pascal Sundi Mbambi

Abstract The Democratic Republic of Congo (DRC) is home to more than half of Africa's freshwater resources, yet it is a striking irony that millions of its people still struggle to find clean drinking water. This issue isn't due to a lack of water; rather, it is a result of ongoing governance problems, insufficient investment in infrastructure, rapid urban growth, and deep-rooted corruption. In Kinshasa and other semi-urban cities, the challenges are particularly severe, especially in the outskirts, where government services lack proper reach, as the current infrastructure struggles to keep up with the rapidly growing population. To address this problem, decentralized initiatives like Associations d'Usagers de l'Eau Potable (ASUREP) have stepped in to create community-driven mini water networks. This article takes a closer look at the ASUREP model in the slum areas of Kinshasa, highlighting its effectiveness in improving water access, social unity, and strengthening local governance. Although the DRC's legal and institutional frameworks acknowledge water as a basic right and consider integrated water resource management, there are still significant gaps in how these policies are put into action. The findings suggest that community-driven approaches like ASUREP, along with necessary regulatory changes and local empowerment, could pave the way for sustainable water management in the rapidly expanding urban areas of the DRC.

H. D. Tsumbu
School On Integrated Management of Tropical Forests and Lands (ERAIFT), University of Kinshasa, Kinshasa, Democratic Republic of the Congo

A. Rodricks (✉)
University of Antwerp, Antwerp, Belgium
e-mail: abhisekhrodricks@gmail.com

P. S. Mbambi
Faculty of Social, Political and Administrative Sciences, Université Pédagogique Nationale (UPN), Kinshasa, Democratic Republic of the Congo

V. Kandpal and T. Choudhury (eds.), *Blue Revolution: Sustainable Water Use and Circular Economy Innovations*, Urban Sustainability,
https://doi.org/10.1007/978-981-95-1916-3_6

1 Introduction

The Democratic Republic of Congo (DRC) is constantly faced with remarkable contradictions when it comes to water resources. Despite having one of the largest reserves of fresh water in the world, a large proportion of its population has limited access to clean and safe drinking water. With the mighty Congo River, as well as a network of abundant lakes and tributaries, the DRC holds more than half of Africa's water resources [1, 2]. However, only 50% of its citizens have reliable access to clean drinking water, revealing a systemic deficiency that affects millions of people [3, 4]. This failure is not the result of a physical shortage of water, but rather, of severe infrastructural limitations, governance challenges, and persistent corruption [5]. In both urban and rural areas, the lack of functional water supply infrastructure, combined with poor management practices, constitutes a major barrier to accessing safe water. This problem has been exacerbated in recent years by rapid urbanization, which is putting increased pressure on already overstretched public water services [3, 4, 6].

Beyond a situation marked by years of underinvestment, and an almost succession of wars between several states on its territory between 1997 and 2006, Trefon and De Putter argue that the difficulties of access to drinking water in the DRC also result from poor state management and chronic underinvestment [3, 4]. In addition, urbanization, driven in part by rapid population growth [5, 6], has not been accompanied by the systematic development of basic infrastructure [6]. The situation of access to drinking water has dramatically deteriorated during the first decade of the twenty-first century in the peri-urban areas of the country's large cities, particularly in the capital, Kinshasa.

In 2016, this megacity had 12 million inhabitants. It could be, in 2030, the largest city in Central Africa with 24 million inhabitants [5, 6]. The low investment by the central and provincial governments of the city of Kinshasa in creating water distribution infrastructure has not kept pace with the demographic explosion of the city's population. As in other large African cities, Kinshasa's peri-urban neighborhoods were formed without urban planning and without considering the topographical constraints that affect the development of the drinking water distribution network infrastructure [7]. The risk, as a recent World Bank study highlights, is that Kinshasa will become the largest slum in Central Africa if nothing is done to improve access to drinking water and sanitation in the peripheral areas [7].

To meet the considerable challenge of providing drinking water in these peri-urban neighborhoods, a professional community management framework emerged in 2005: the Drinking Water Network User Associations (Associations d'Usagers de l'Eau Potable, ASUREP), which are neighborhood associations that autonomously manage mini-water supply networks. Eijkelenburg and Vanderstichele [8] believe that this formula of community management brings encouraging results in most interventions in African cities. Likewise, Baron and Bonnassieux indicate that this type of community management model has been successful in Burkina Faso, as well as in many other West African countries. The authors suggest that this type of community management initiative be advocated and disseminated internationally over the coming decades, particularly in large urban areas in African countries [9].

The seriousness of the drinking water situation in major African cities has sparked a debate about the best approach to address these challenges. Essentially, the core of this argument revolves around the question of whether water should remain a public service or whether privatization can offer some improvements. Proponents of privatization believe that private sector involvement would inject a capital influx into the water infrastructure system, improve operational efficiencies, and eliminate the constant mismanagement of resources that characterized state-owned enterprises [10]. With the profit incentive, they believe the private industry will be incentivized to operate more efficiently, waste less water, and serve more customers where they currently are not. Proponents of privatization believe it could foster innovation and accountability in the provision of water services, relieving some of the burden on the government and its limited resources.

However, those who oppose privatization argue that it could further marginalize vulnerable communities, transforming water from a basic human right into a commodity available only to those who can afford it [11]. They argue that if private companies prioritize profit over public welfare, rates could rise, making water unaffordable for many people in a country with high levels of poverty. The experience of other African countries, such as Ghana and South Africa, has shown that privatization can lead to a negative reaction from the population, often due to increased prices and insufficient improvements [12]. Additionally, the ethical consideration of water as a fundamental right is supported by international (and even national in the case of the DRC) frameworks such as the United Nations Sustainable Development Goals, which advocate universal and equitable access to water and sanitation [13]. The critical trend (Water and sanitation—United Nations Sustainable Development Goals, no date) to privatization argues that in a country like the DRC, where poverty is widespread, water privatization could exacerbate inequalities, restrict access for the poorest, and potentially intensify social fractures [14].

This complex debate around water privatization and public rights raises critical questions for the future of the DRC. It is therefore important to find a solution that balances the need for investment with the principle of universal access to water. Given the considerable stakes, any initiative to reform water governance must prioritize efficiency and equity, while ensuring that the right to water is protected for all, regardless of income or geographic location [15].

2 Water as a Natural Resource in the Democratic Republic of Congo (DRC)

As discussed earlier, the Democratic Republic of the Congo (DRC), with its vast rivers, abundant rainfall, and large groundwater reserves, stands out as one of the country's most significant natural resources. The Democratic Republic of the Congo has roughly 62% of the Congo Basin, the second largest river basin in the world after the Amazon [16]. This river basin includes the Congo River (which is the second largest river in Africa and the deepest in the world), and not only is water abundant, but it also plays an important role in the ecological systems, economy, and banking of the country.

Surface water resources include more than 30 major rivers, lakes, and streams and provide a critical source of water for agriculture, domestic, and industrial purposes. The Congo River alone has enormous potential for hydroelectric energy production, especially through the Inga Dam project [17]. If fully developed, it could supply electricity not only to the entire DRC but also to neighbouring countries, transforming the region's energy landscape.

Rainfall is another important component of the DRC's water resources. The country receives an average of 1070 mm of rainfall annually, with higher amounts in the northeastern regions [18]. Precipitation makes a significant contribution to replenishing rivers, lakes, and underground aquifers, whose availability supports a rich biodiversity and enables farming in many parts of the country year-round [19]. Groundwater is also equally important, representing just under 47% of the country's internal renewable water resources. Groundwater supplies clean water to kin, especially in rural areas where seasonal surface water may or may not be contaminated. Water as a natural resource has relevance and significance in many areas of the Congolese economy [20]. In agriculture, for both crop irrigation and livestock. In fisheries, the Congo River and tributaries also provide a largely available protein source for the population, as well as income for all fishing communities [21]. Water is also important in transport, particularly in very rural areas where upland roads do not enhance mobility. Rivers constitute natural highways, linking towns, transporting goods, and facilitating trade to promote economic development and social justice.

Water in the Democratic Republic of Congo is not only an economic benefit but also has incredible ecological benefits. The Congo Basin harbours one of the two largest tropical rainforests, which acts as a huge carbon sink, influencing global climate as an ecosystem process. The wetlands and aquatic ecosystems in the Democratic Republic of Congo have experienced high biodiversity, but declining habitats have led to a decline in visual acuity, resulting in the report of endless varieties of plants and animals (which are, as genetically re-synthesized "species", often not acknowledged or referenced as endemic).

However, here is the painful truth: despite all this natural wealth, millions of people in the DRC lack safe water to drink or places to go to the bathroom. Around 50 million people live without access to clean drinking water, and up to 90% of the population doesn't have proper sanitation [22]. That is not just inconvenient, it is dangerous. Impoverished water and hygiene are among the leading causes of illness and death in the country.

3 Legal -Institutional Frameworks for Water Management in the DRC

The Democratic Republic of Congo (DRC) has a complex legal framework regarding the right to water, influenced by international agreements and regional legislation. As a signatory to the International Covenant on Economic, Social and Cultural Rights (ICESCR) treaty, the DRC is dedicated to ensuring that all citizens, particularly the marginalized communities, can access safe drinking water over time. Important treaties, such as the Convention on the Rights of the Child and General Comment 15 of the ICESCR, connect the right to water to essential human rights such as health and a decent standard of living. On the domestic front, the DRC's 2006 Constitution acknowledges the right to drinking water in Article 48. Nevertheless, challenges such as political instability, inadequate governance, and limited resources have impeded progress. In 2016, a significant water law was enacted, introducing reforms geared towards integrated water resources management (IWRM) and decentralized governance in line with constitutional directives (refer to RDC, loi no 15/026 du 31 décembre 2015 relative à l'eau). This law emphasizes the role of local user associations, cost-based tariffs, and management by provincial and territorial authorities.

3.1 Key Provisions of the Law Include

- Associations of water users: Article 32 allows local associations of users to manage water resources.
- Public Water Services (PWS): Articles 77–78 permit decentralized authority to establish PWS services, while allowing third parties to manage them.
- Quality standards: Article 80 requires compliance with standards of potability issued by public health authorities.
- Pricing: Articles 84–86 demonstrate that tariff prices are expressed transparently and equitably, in accordance with cost-recovery pricing rules for regulated fees.

This framework underscores water's status as a fundamental right while balancing affordability, sustainability, and quality standards.

4 The Experience of ASUREP in the Peri-Urban Neighborhoods of Kisenso and Nsele, in the City of Kinshasa

In the Democratic Republic of Congo (DRC), the drinking water and sanitation sector is marked by a fragmented institutional framework involving a multitude of stakeholders. The National Committee for Water, Hygiene, and Sanitation Action (*Comité national d'action de l'eau, de l'hygiène et de l'assainissement*, CNAEHA) is responsible for coordinating the actions of all involved parties. However, despite the establishment of CNAEHA, the institutional framework continues to suffer from a lack of harmonization in interventions and poor communication among stakeholders—factors that undermine both the effectiveness and efficiency of efforts and resource utilization.

Even with progress made in strengthening the legal and institutional frameworks, significant challenges remain in the provision of drinking water, especially in large cities like Kinshasa, where infrastructure development has failed to keep pace with rapid population growth, particularly in neighborhoods such as Kisenso and Nsele.

Before delving into the ASUREP model in Kinshasa, let us briefly revisit its origins. In response to the lack of infrastructure and investment in underprivileged areas of Kinshasa, a decentralized water supply initiative was launched in the mid-2000s. This innovative system involves small, autonomous neighborhood-based water supply networks designed to serve communities that are poorly or not at all covered by the national water utility, REGIDESO. The ASUREP initiative in Kinshasa was implemented in several phases and brought together various actors. It unfolded in three main stages: Kin-Est (2007–2010), the *Promotion de modalités Innovantes pour l'Accès à l'Eau Potable* (Promotion of Innovative Modalities for Access to Drinking Water) project—PILAEP I (2010–2014), and PILAEP II (2014–2020). The initiative was launched in 2006 by the former Belgian Technical Cooperation agency, in partnership with the *Action pour le développement communautaire en milieu rural* (ADIR), a Congolese NGO [23]. Site locations were carefully negotiated with REGIDESO to ensure that the decentralized systems would not conflict with the national utility. Founded in 1999, ADIR is a non-governmental organization focused on improving water access and sanitation infrastructure in both peri-urban and rural communities. It has played a key role in managing and implementing the Kinshasa water supply project [23]. The primary aim was to ensure that underprivileged neighborhoods on the outskirts, which often lack essential infrastructure, could finally gain access to clean water.

The initial phase was supported by DfID and the European Union over a four-year period. To date, 52 autonomous mini-networks have been established in Kinshasa's peripheral neighborhoods (Saint-Macary et al. 2023) as part of PILAEP I and PILAEP II.

Alongside the development of the ASUREPs, a federation was created—FEDASU (*Fédérations des ASUREPs*, Federation of ASUREPs)—which officially came into existence on August 14, 2012. FEDASU oversees ASUREPs operating both in

Kinshasa and in the province of Kongo Central. The federation is led by representatives from four ASUREPs in Kinshasa, who together form the General Assembly (with the president chosen from among them). Each representative serves as the president of their respective ASUREP. A rotation system among these four ASUREPs is intended to ensure the smooth governance of FEDASU, although the current president has remained vague on whether this rotation is being implemented in practice.

FEDASU's executive secretariat is managed by the local ADIR, which is responsible for centralizing financial records and balance sheets at its headquarters [23]. The federation's goals are to ensure the long-term sustainability of the service, provide management support, and mediate in cases of conflict, whether between ASUREP members or among different ASUREPs. It also offers legal assistance, monitors water quality across all ASUREPs in Kinshasa, and carries out a regulatory and support role for its members.

It is important to note that at this stage, all prices are deliberated with the mandatory support of FEDASU. No ASUREP can set a higher price without prior consultation. Thus, FEDASU has standardized the price of water for all ASUREPs, contrary to what was initially planned in terms of pricing based on the true cost of water and decision-making independence on the price of water for each ASUREP. This highlights that the decision to standardize the price of water can be imposed by FEDASU.

Furthermore, it should be noted that FEDASU receives 10% of the monthly revenue generated by each ASUREP, without passing on this charge to the price of the service. Finally, a solidarity fund, called *Caisse de solidarité* (CASOL), has been set up. This is funded by 3% of the revenue of the ASUREPs and is used to finance anti-erosion actions in neighborhoods exposed to risks.

Our investigation focused on 17 neighborhoods in the city of Kinshasa. The surveys were conducted in three separate municipalities: two neighborhoods in the municipality of Kisenso (Mbuku, Libération), eight neighborhoods in the municipality of Kimbanseke (Disasi, Bikuku, Elonga, Kika, Mangana, Ngandu, Ngamanzita, Way-Way), and seven neighborhoods in the municipality of N'Sele (Maba, Mata, Mikonga, Mimpa, Muke, Ngamaba, Ngina-Mpasa). These neighborhoods are located in municipalities among the least served by the national REGIDESO network. In general, these municipalities are neglected by state services, particularly with regard to the supply of drinking water. The shortcomings of the public sector are compensated by the involvement of the informal sector, in particular the ASUREP (*Associations d'Usagers de l'Eau Potable*), as well as other small private or religious operators that indicate the location of the ASUREPs which provide drinking water in the sites studied in the city of Kinshasa.

In Kinshasa, several neighborhoods that are located in the outer municipal areas face considerable obstacles regarding access to essential services like water and electricity [24]. Etienne (2003) characterizes these areas as peri-urban, disadvantaged, or slum neighborhoods, emphasizing the ongoing issues related to water supply, electrical connectivity, and sanitation facilities. These regions serve as a transitional zone between the rural and urban areas, highlighting the unique challenges that come with this position. Furthermore, Etienne (2003) notes that these peripheral communities reflect the broader development trends evident in many African cities, illustrating a common narrative of struggle and adaptation within the urban landscape. Typically situated on the outskirts of urban centers—hence the term peripheral—these neighborhoods are often marked by poverty and informality, frequently lacking legal land tenure [25, 24] (Bogaert and Halleux 2015; Mavakala 2019; Messina 2020; Olivier de Sardan 2012; Sambieni 2019).

Lelo and Tshimanga [24] explain that the issue of limited access to drinking water arises either from irregular distribution of the resource or from a lack of infrastructure within the national water supply network. In this context, women and children not only lose valuable time but also expend significant energy performing water-related tasks, such as collecting water from unprotected or poorly protected sources and transporting it over long distances in cans and buckets. This situation reflects a deep level of precariousness and poverty that affects every aspect of life for the people, including essentials like transportation, education, and healthcare [24].

It is important to note that these peri-urban neighborhoods have emerged because of the expansion of central municipalities and rural-to-urban migration. Many people aspire to settle in Kinshasa, where living conditions, economic opportunities, and security are perceived to be better than in the remote areas. These peripheral areas primarily accommodate migrants from the provinces, especially those coming from nearby regions such as Kwilu, Kwango, and Kongo Central.

5 Successes and Limitations of the ASUREP Project

5.1 Technical Implementation

From a technical standpoint, the implementation of the water networks proved to be generally satisfactory; they are functional and of good quality, although some aspects could still be optimized. The simplicity of the installations made it possible to mobilize residents, under direct management, for construction work. This approach encouraged community ownership of the network without compromising the quality of the infrastructure [26]. However, outsourcing certain tasks to local private companies led to delays in the execution of the work.

5.2 Improved Access to Water for the Population

The improvement in access to water for the populations of neighborhoods targeted by the Project is undeniable. The time spent on water collection has been significantly reduced, reaching an average of thirty minutes per day [27], while the daily per capita water allocation increased from four to ten liters [27]. These improvements benefited the 318,842 people who gained access to the thirteen mini-networks established before 2012 [27].

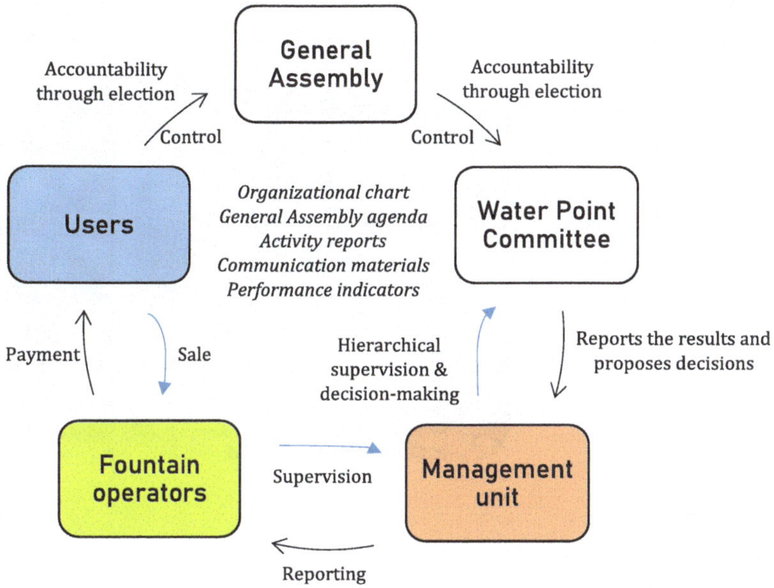

Fig. 1 Chart of the local governance framework [28]

5.3 Social Acceptance (Alternative Water Sources Abandoned)

ASUREPs introduced a new model of local governance in the peripheral neighborhoods of Kinshasa, contributing to constructive and non-confrontational vigilance, thus fostering moderate social contestability [28]. Indeed, the decentralized and participatory, or even community–based, management model they promote helps bring residents together around a shared vision of service provision [26]. ASUREPs thereby become local actors within their communities, contributing to the development of local skills through job creation and the assumption of responsibilities by their members. Moreover, due to their representative and democratic nature, including democratic oversight (Fig. 1 [26, 28]), they demonstrate that such a governance model is feasible in the slums of developing countries.

5.4 Economic Viability of ASUREPs

ASUREPs record an average annual turnover ranging between USD 70,000 and 120,000 [29]. These revenues, generated through water sales, not only enable the remuneration of staff and the maintenance of infrastructure but also allow for the funding of local public goods, such as the construction of public toilets, as is the case

Fig. 2 Simple technology adapted to local demand and payment capacity (Ditona 2020)

with the ASUREP in Mikonga, located in the N'sele municipality (Fig. 2) [28]. As a result, ASUREPs become financially autonomous, which allows them to ensure the long-term sustainability of water services [26]. Initially, their investments remain focused on the water sector. However, it is not inconceivable that, in the future, they may diversify their activities into other areas to improve living conditions in the neighborhoods, thus initiating a virtuous cycle of development [23].

Appropriate Technology in Light of Local Management Capacities and Effective Demand

The establishment of mini-networks managed by community-based associations proves particularly well-suited for the targeted neighborhoods. In a context marked by a weak state presence and an overburdened conventional water operator, this management model responds effectively to the local institutional situation, especially given the population's limited willingness to cooperate with state authorities [27].

Although service quality remains limited, residents still need to travel to access water—it appears to align with users' expectations. Users are willing to pay the requested price, suggesting that they perceive the service offer (fountain opening hours, community-based management, payment methods) as fair relative to their contribution [23].

The cost of the service is thus consistent with the effective demand—that is, the beneficiaries' capacity and willingness to pay [27]. Accordingly, the mini-network represents an appropriate technology that aligns with local management capabilities and the economic realities of the residents [25].

5.5 Water Quality: A Persistent Challenge

Water quality remains a critical issue for the provision of services. To ensure its safety, regular bacteriological analyses are required. However, the high cost of sending samples to specialized laboratories exceeds the financial capacity of the ASUREPs. To address this issue, the former *Coopération Technique Belge* (CTB), now ENABEL (the Belgian Development Agency), proposed a pragmatic solution: providing the management unit with bacteriological testing kits that are easy to use and suited to local capacities.

5.6 Risks of Conflicts of Interest, Political Co-Optation, and Takeover by "Local Leaders"

Although user associations generally function satisfactorily, vigilance is necessary regarding the risks of political co-optation and takeover of ASUREPs by certain local actors [25]. In some cases, statutes and internal regulations are either not respected or misinterpreted. Certain neighborhood leaders have attempted to take control of the ASUREPs for personal gain. More broadly, individual delegates often show a tendency to seek dominant positions within the organization. In other instances, attempts at ethnically motivated groupings have been observed within associations. These scenarios highlight genuine risks of conflicts of interest and power struggles over control of the structures.

5.7 Ongoing Need for Support—Limited Capacities

The project requires sustained support over several years, which is not always feasible for development agencies [25]. When construction work is delayed, the timeframe allocated for follow-up is correspondingly shortened. However, thorough follow-up is essential to enable ASUREPs to achieve full autonomy. Specific key competencies necessary for long-term sustainability, such as socio-economic analysis, as well as accounting, financial, and technical audits, require significant time to be fully developed.

6 Lessons on Strategy and Methodology for Universal Water Access Projects

This case study provides valuable insights into the strategies and methods employed in projects aimed at achieving universal access to water. It highlights both the achievements and the challenges faced by the ASUREP model, along with their underlying causes, offering insights into the factors that contribute to a project's success or, conversely, to the difficulties stemming from overlooked aspects.

Below, we present a table (Table 1) summarizing the strengths and weaknesses of the ASUREP model. For each component, we analyze the strengths, weaknesses, opportunities, and risks associated with ASUREPs.

Table 1 Summary table of a SWOT analysis of the ASUREP framework

ASUREP Component	Strength	Weakness	Opportunity	Threat
Governance is structured by the users	Creates a sense of belonging and responsibility	Requires a motivated management	Allows users to understand that good governance brings good services	Exploitation for political purposes
Professional management of operations	Service guarantee, Integrated proactive maintenance, and Network management by professionals	Can have a negative impact on rates; Limited influence over technical choices	Creation of sustainable jobs; Regular and structured collection of operational data	Conflicts between the management committee and the users' assembly Hijacking of the service to benefit financial performance at the expense of users
Adapted infrastructure	Professional design, Proper sizing, and Sustainability	Need for expropriation	Demonstrates the benefits of constructing sustainable structures	Utilization of infrastructure for purposes other than those intended by the ASUREP, potentially leading to premature deterioration

7 Impacts of This Type of Intervention

7.1 *Economic Effects*

Each ASUREP has enabled the creation of approximately 30 permanent jobs, the majority of which are held by women (fountain operators), providing them with stable and meaningful income.

The reduction in time spent collecting water relieves significant time capital, which households can dedicate to income-generating activities or improving their quality of life. The water services established are economically viable and autonomously managed. Experiences from the ASUREP program demonstrate that these associations generate substantial savings capacity, amounting to up to 30% of turnover during the initial years. These savings are used to renew infrastructure, enhance service reliability (through the purchase of pumps and backup generators), or support community initiatives (multipurpose halls, internet hubs, waste collection services).

The pricing for the service is designed to cover all costs completely. This pricing is recommended by the Board of Directors of each ASUREP and approved by the General Assembly. Typically, the tariff is set at 50 Congolese Francs for a 25-L container, equivalent to approximately €1.60 per cubic meter. While this might sound steep, it remains within reach for most households, all while ensuring a dependable and top-notch service.

7.2 *Environmental Effects*

ASUREPs play a significant role in protecting the environment, especially by enhancing hygiene around public fountains and homes, as well as by establishing more efficient sanitation systems. When water is distributed through public standpipes, it creates very little wastewater, which the soil typically absorbs naturally.

7.3 *Social Effects*

ASUREPs have a substantial social impact, particularly in terms of gender issues. By reducing the burden of water collection, they significantly lighten the workload of women and children, who are traditionally responsible for this task. This promotes their economic and social inclusion through the time that is freed up.

General health conditions also improve, with a noticeable decrease in waterborne diseases, epidemics, and unsanitary conditions.

7.4 Institutional Effects

In a context where public water institutions are weak, ASUREPs offer a model of local, autonomous, and participatory governance. The key stakeholders involved include:

- The ASUREPs themselves;
- The Federation of ASUREPs (FEDASU);
- The Provincial Water and Sanitation Action Committee (CPAEA);
- The relevant municipal authorities.

The ASUREP model promotes the emergence of innovative institutional solutions, combining community management with that of historical operators such as REGIDESO. It also encourages the establishment of regulatory mechanisms for pricing and water quality.

8 Sustainability of the ASUREP Concept

The sustainability of the ASUREP model is based on a participatory and autonomous service approach that directly empowers its beneficiaries. In the event of a malfunction, users turn to their community's management unit rather than waiting for intervention from the national operator. The technicians and managers, recruited from the neighborhood, are both accessible and accountable to the General Assembly.

The goal is for these REGIDESO service models not only to coexist but also to complement each other, thereby fostering a more integrated approach to water provision. As REGIDESO can offer both an affordable and reliable supply of water in sufficient quantities, ASUREPs may eventually transition to purchasing bulk water directly from REGIDESO. From a financial perspective, ASUREPs exhibit substantial resilience. Their pricing structures are designed to ensure that all operational costs are covered, and revenues are collected on a daily basis. This consistent cash flow secures their autonomy and underpins their long-term viability as providers of essential water services.

In summary, the ASUREP model has proven technically successful in establishing water distribution networks for users across nearly all Central African countries where it has been implemented, particularly in the districts of the peripheral municipalities of Kinshasa, in the DRC. Available data show that the approach is functional and delivers strong results [8]. Its simple technology enables the development of branched standpipe networks or straightforward systems involving drilling, pumping, and storage, combined with standpipes,spring catchments and gravity-fed supply systems at the municipal or local authority level have also been effectively implemented. These bodies have successfully taken charge of the networks without sacrificing water quality. The improvement in water access for residents in the project's focus areas is clear, as people are spending significantly less time fetching water. This has led to a boost in social cohesion and economic stability, fostering trust and peace within communities while also ensuring that the infrastructure is well-maintained.

9 Conclusion

The Democratic Republic of Congo (DRC) illustrates a profound dilemma; despite its abundant freshwater resources, millions of its citizens lack access to clean drinking water. This is not due to a lack of resources, but rather a result of deep-rooted governance issues, insufficient investment in infrastructure, and ongoing socio-political turmoil. The situation has been exacerbated by rapid urban growth, particularly in cities like Kinshasa, where the surge in population has outpaced the capacity of the water distribution systems. The DRC's legal framework, particularly the 2006 Constitution and the 2016 water law, lays down a solid groundwork for water governance and access. These laws recognize water as a fundamental human right and promote principles like integrated water resources management (IWRM) and decentralized governance. They support user associations, advocate for cost-based tariffs, and set standards for water quality.

However, while these frameworks are promising, their implementation is hindered by weak institutional capacity, a lack of funding, and the fragmented nature of the water management sector. Initiatives like ASUREP in Kinshasa showcase innovative solutions to tackle these issues. ASUREP's community-driven, decentralized approach demonstrates how self-sufficient networks can improve water access in underserved areas. By encouraging local ownership, reducing water collection times, and building sustainable infrastructure, ASUREP has made significant strides in peri-urban neighborhoods.

Its achievements highlight the significance of empowering communities and weaving localized solutions into broader national strategies. Still, challenges persist, including the need for standardized pricing and improved coordination among various stakeholders. The ongoing discussion about privatizing versus publicly managing water resources adds another layer of complexity. While privatization could bring in essential funding and improve operational efficiency, it also raises concerns about accessibility and equity.

References

1. Partow H (2011) Water issues in the democratic republic of the Congo (IRC Wash—Country Paper). London: IRC International Water and Sanitation Centre
2. Inogwabini BI (2020) The changing water cycle: freshwater in the Congo. Wiley Interdisc Rev Water 7(2):e1410. https://doi.org/10.1002/wat2.1410
3. Ditona Tsumbu H, Cammaerts D, Adant I, Nzuzi FL, Delière JF, Partow H (2021) Water issues in the Democratic Republic of the Congo: challenges and opportunities. Water true-cost to

manage social contestability? Users' perceptions about community-managed water standpipes networks in Kinshasa, Democratic Republic of Congo. Desalination Water Treat 260:321–332. https://doi.org/10.5004/dwt.2022.28619

4. Moe CL, Rheingans RD (2006) Global challenges in water, sanitation, and health. J Water Health 4(S1):41–57. https://doi.org/10.2166/wh.2006.019
5. Richards JA (1999) Remote sensing digital image analysis: an introduction. Springer-Verlag
6. United Nations Children's Fund (UNICEF) (2025) Global Annual Results Report 2024: Goal Area 4-Every child lives in a safe and clean environment. New York: United Nations Children's Fund.
7. Achore M, Bisung E, Kuusaana ED (2020) Coping with water insecurity at the household level: A synthesis of qualitative evidence. Int J Hyg Environ Health 230:113598. https://doi.org/10.1016/j.ijheh.2020.113598
8. Eijkelenburg A, Vanderstichele G (2009) Enjeux et perspectives de la gestion de l'eau potable en milieu rural: Expériences de la coopération belge dans le domaine de l'hydraulique rurale et périurbaine en Afrique. Enabel.
9. Baron C, Bonnassieux A (2013) Gouvernance hybride, participation et acces a l'eau potable: Le cas des associations d'usagers de l'eau (AUE) au Burkina Faso. Ann De Géogr 693(5):5–22. http://www.jstor.org/stable/24569678
10. Gashaw, B. (2023) Water privatization: A four nation case study [Report]. Stanford Digital Repository, Stanford University
11. Effah Ameyaw E, Chan APC (2013) Identifying risks of public-private partnerships (PPPs) in water supply project management in Ghana. J Facilities Manage 11(2):152–182. https://doi.org/10.1108/14725961311314651
12. Goldman M (2007) How the politics of 'Water for All!' became hegemonic: the power of the World Bank and its transnational policy networks. Geoforum 38(5):786–800. https://doi.org/10.1016/j.geoforum.2005.10.008
13. Vickers J, Yarrow G (1991) Economic perspectives on privatization. J Econ Perspec 5(2):111–132
14. Baer M, Gerlak AK (2015) Implementing the human right to water and sanitation: A study of global and local discourses. Third World Q 36(8):1527–1545. https://doi.org/10.1080/01436597.2015.1043993
15. Ioris AAR (2012) The persistent water problems of Lima, Peru: neoliberalism, institutional failures and social inequalities. Singapore J Tropical Geogr 33(3):335–350. https://doi.org/10.1111/j.1467-9493.2012.00454.x
16. Harrison IJ, Brummett R, Stiassny ML (2016) The Congo River Basin. In: The Wetland book, vol 1, pp 845–852. Springer. https://doi.org/10.1007/978-94-007-6173-5_92-2
17. Oyewo AS, Farfan J, Peltoniemi P, Breyer C (2018) Repercussion of large scale hydro dam deployment: the case of Congo Grand Inga hydro project. Energies 11(4):972. https://doi.org/10.3390/en11040972
18. Dyer EL, Jones DB, Nusbaumer J, Li H, Collins O, Vettoretti G, Noone D (2017) Congo Basin precipitation: assessing seasonality, regional interactions, and sources of moisture. J Geophys Res: Atmospheres 122(13):6882–6898. https://doi.org/10.1002/2016JD026282
19. Shang M, Xie J (2024) Agricultural sustainable development: soil, water resources, biodiversity, climate change, and technological innovation. Adv Resour Res 4(2):181–204
20. Nteranya JN (2024) Natural resources use in the Democratic Republic of Congo: State and opportunities for sustainable development. In The Palgrave encyclopedia of sustainable resources and ecosystem resilience, Springer, pp 1–2. https://doi.org/10.1007/978-3-030-67776-3_66-1
21. Béné C, Steel E, Luadia BK, Gordon A (2009) Fish as the 'bank in the water'–Evidence from chronic-poor communities in Congo. Food Policy 34(1):108–118. https://doi.org/10.1016/j.foodpol.2008.07.001
22. Zeufack A, Jha SK (2024, June 10) Democratic Republic of Congo: in the water sector, taking stock of challenges leads to new solutions. World Bank Blogs. https://blogs.worldbank.org/en/africacan/drc-water-sector-taking-stock-challenges-leads-to-new-solutions-afe-0624

23. Hydroconseil (2017) Évaluation finale: Projet PILAEP 1. Rapport final du Projet d'Amélioration de l'accès à l'eau potable pour les populations des quartiers périphériques de Kinshasa et de trois communes du Bas Congo
24. Tshimanga Mbuyi C, Nzuzi FL (2004) Pauvreté urbaine à Kinshasa. Kinshasa Gombe (République démocratique du Congo): Cordaid
25. Lemenager M (2013) Soutenir des projets AEP dans les quartiers périphériques d'un pays en crise: Gestion communautaire AEP à Kinshasa. AFD
26. Mouzon JL, Ebner V, Waut E (2010) Alimentation en eau potable et assainissement des quartiers périphériques urbains et des zones rurales. CTB
27. Denormandie C (2014) Eau potable et assainissement dans les villes du Sud: La difficile intégration des quartiers défavorisés aux politiques urbaines. Adapter les interventions dans les quartiers périphériques dans un contexte de crise: Exemple de Kinshasa. Collect Débats Et Controverses 8:62–68
28. Ditona Tsumbu H (2024) Contribution à l'analyse de la gouvernance décentralisée et participative de l'eau potable: la contestabilité des ASUREP en zone périurbaine de Kinshasa (RDC) [Doctoral dissertation, Université de Liège & ERAIFT]
29. Mott MacDonald (2016) PILAEP II, Diagnostic organisationnel et économique. AFD

Water Use in Palace Gardens

Hussein Mazyed and Gökçen Firdevs Yücel Caymaz

Abstract This study aims to examine the role and impact of water in architectural design. The research focuses on five palace examples and investigates the interrelationships between them and how their different functions are reflected in their architectural design. The integration of water into architecture is analyzed in terms of how it adds vitality to the structure and contributes positively to the surrounding environment. The study explores five palaces from different historical periods and religious contexts: **the Alhambra and Generalife, Chehel Sotoun, Versailles Palace, and Taj Lake Palace**, with attention given to the symbolism and significance of water within each. In some cases, water has been used to create spaces in harmony with nature. Additionally, the religious, cultural, aesthetic, and psychological meanings of water in Islam, Christianity, and Judaism are discussed. As a research method, these palaces were selected based on their historical significance and their embodiment of the architectural vision of their respective eras. Key design elements were identified and analyzed for their architectural relevance. The results were organized in a comparative table that examined the presence or absence of water elements according to several criteria: form, function, lighting, static and dynamic movement, color usage, the four classical elements of water, and the type of environment. The discussion further elaborates on the nature and symbolic role of water in architectural expression across these examples.

Keywords Water · Architecture · Alhambra Palace · Generalife · Chehel Sotoun · Palace of Versailles · Taj Lake Palace

H. Mazyed · G. F. Y. Caymaz (✉)
İstanbul Aydın University, Istanbul, Turkey
e-mail: gokcenfyucel@aydin.edu.tr

H. Mazyed
e-mail: habdulkadhimmazyed@stu.aydin.edu.tr

V. Kandpal and T. Choudhury (eds.), *Blue Revolution: Sustainable Water Use and Circular Economy Innovations*, Urban Sustainability,
https://doi.org/10.1007/978-981-95-1916-3_7

1 Introduction

The Alhambra Palace in southern Spain, the adjacent Generalife, Chehel Sotoun in Iran, the Palace of Versailles in France, and the Taj Lake Palace in India all exemplify the critical role of water in historical architectural and landscape design. These sites demonstrate how water was not merely a practical necessity but a central design element used to shape spatial aesthetics, symbolic meaning, and environmental interaction [1, 2].

Water is one of the most effective media through which the essence of nature can be conveyed in built environments. It fosters the creation of spaces that evoke tranquility, harmony, and sensory richness [3]. In addition to its visual and acoustic qualities, water has been shown to contribute significantly to psychological comfort by reducing stress, enhancing perception, and regulating microclimates [4].

Throughout history, water has played a foundational role in the development of civilizations. Societies that formed around water sources utilized these resources not only for agriculture, drinking, and industrial purposes, but also for architectural and urban symbolism [5]. The integration of fountains, pools, artificial canals, and other water features reflected not only technological advancement but also cultural values, aesthetic aspirations, and spiritual ideals.

Within the architectural composition, water frequently acts as a metaphor for purity, transformation, and divine presence. This symbolic association is deeply rooted in the fluid, reflective, and life-sustaining characteristics of water, which make it a timeless and universal design element [3, 6].

In many religious traditions, water is considered a symbol of purity and cleansing. In the three major religions, Judaism, Christianity, and Islam, water is associated both literally and symbolically with important events in human life. These include baptism (in Christianity), immersion in a pool of water before marriage (in Judaism), and washing before prayer (in Islam). Water also has special significance in Islam, where it serves an essential function in many religious rituals.

We created every living thing from water' (Al-Anbiya: 30).

"He sent down water (rain) from the sky to cleanse you" (Al-Anfal 8:11). It symbolizes personal cleansing, both physically and spiritually.

Seen the water that you drink ? Did you send it down from the clouds, or did We send it down?" (Vâkia) Surah: 68).

The biblical narratives (Kutsal Kitap)were concerned with "the first and most important act of creation: water".

Water holds profound cultural significance across numerous societies, playing a central role in rituals, ceremonies, and artistic expression. From ancient times to the present day, water has been symbolically linked to the fundamental stages of human life—birth, marriage, and death—representing themes of purification, transition, and renewal [5].

In Arabic poetry and classical literature, water is frequently used as a metaphor for beauty, love, and emotional depth, reflecting the intimate relationship between culture and natural elements. The mirror-like qualities of water often symbolize introspection, romanticism, and harmony with the divine [7].

Water also serves as a cultural agent in architectural design. Its integration into buildings and urban spaces is not merely aesthetic, but often represents shared community values, spiritual beliefs, and collective identity [6]. Architectural water features—such as fountains, pools, or flowing channels—can thus become symbolic markers of cultural memory and local traditions.

By incorporating water into architectural spaces, designers communicate the cultural identity of a place, reinforcing emotional ties between people, landscape, and tradition. In many cases, the presence of water in the built environment serves to honor ancestral heritage while fostering continuity between the past and present.

The integration of water in architectural design plays a vital role in shaping aesthetically engaging environments. Water elements such as fountains, reflective pools, streams, and even small lakes can transform the visual character of a space by introducing movement, sound, and light reflections, thereby creating atmospheres of tranquility and elegance [3].

Beyond visual enhancement, water serves as a powerful medium for emotional and symbolic expression within the built environment. It contributes to spatial harmony and can evoke feelings of calmness, intimacy, or grandeur depending on its scale and composition [5]. The sensory experiences associated with water—such as the sound of flowing water or the sight of rippling reflections—have been shown to reduce anxiety and promote emotional well-being [4].

From a theoretical standpoint, the inclusion of water in design aligns with the principles of biophilic design, which emphasize human affinity for natural elements. Water features help foster a deeper connection between users and nature, encouraging restorative experiences and enhancing the quality of interaction between people and their surroundings [8]. Consequently, the aesthetic and psychological dimensions of water in architecture extend far beyond ornamentation. They reflect a holistic approach to design that seeks to enrich the human experience through sensory, symbolic, and ecological engagement.

The presence of water is often perceived as a source of peace and relaxation, with a significant impact on an individual's psychological state. The auditory stimulation of flowing water or the visual presence of ponds can facilitate the relaxation of the nervous system and reduce stress [4]. Water is not only essential for physical life but also plays a crucial role in mental and spiritual well-being. A child's early dreams frequently involve organic themes, with water emerging as a prominent element symbolizing emotional depth and subconscious connections.

The empirical literature has found that "people have strong positive reactions to water in the ground in their dreams."

Furthermore, the incorporation of water features into architectural spaces has been shown to evoke feelings of calm and serenity, creating peaceful environments conducive to relaxation. The presence of water in design has been linked to stress relief and improved mental well-being, producing positive psychological effects [3]. Additionally, water engages multiple senses: visual, auditory, and tactile—thereby enhancing the overall spatial perception and enriching the user's experience.

2 Research Method

2.1 Research Areas

The properties of the water element in four different civilizations and three different religions are examined, with a particular focus on the use of water in architecture. Examples from Islam, Christianity, and Hinduism are presented for evaluation.

The Alhambra and Generalife in Spain, which reflect the historical importance of water use. The pools and aqueducts in the Alhambra and Generalife are emblematic of Islamic architectural design. A similar example is the Shil. It can also be seen in the Sotun Palace. The Versailles Palace in France is notable for its use of water. The large pools and use of water add to the grandeur of the palace. The Taj Lake Palace in India is located on water, and water is used as an important element in architectural design (Table 7.1).

Alhambra and Generalife

The Alhambra and the Generalife are historical palaces in Spain. One of the defining characteristics of these structures is that they are surrounded by water. Muhammad Ibn Yusuf al-Nusayri was the architect responsible for laying the foundations of the Alhambra. He also laid the foundations of the new castle and opened an irrigation channel to water the castle. After the death of Ibn al- Ahmar, his son Muhammad II (reigned 1273–1302) became the Nasrid. He brought a new cultural, administrative, and architectural vitality to his Sultanate. With the reign of Muhammad V (1354–1362), Nasrid architecture reached its highest level of development and wealth. The

Table 7.1 Water use in architectural structures

Sample Name	Country and City	Year	Deperiod	Purpose of Water Use	Coordinates
Alhambra	Granada, Spain	1238–1273	Nasrid Dynasty	Visuality, Worship	37.17610046533256, –3.5880924418856703
Generalife	Granada, Spain	1302–1309	Nasrid Dynasty	Visuality, swimming	37.178048169871666, –3.585487211951105
Chehel Sotoun	Isfahan, Iran	Sixteenth century	Safavid Dynasty	Worship, mystical meaning	32.65741309936842, 51.67169120955046
The Palace of Versailles	Versailles, France	1623–1715	Louis XIII	Visuality	48.805179513748264, 2.120161505624334
Taj Lake Palace	Udaipur, India	1746–1743	Royal Dynasty of Mewar	Visuality, mystical meaning	24.575512708563462, 73.67992646311397

pools and water channels inside the Alhambra reflect the architectural style of the Islamic period.

- The Court of Lions and its central fountain

Fountains were a common feature of Islamic gardens, both public and private [9]. The fountain consists of twelve lions arranged in pairs to form a circle around a basin. The lions' mouths act as fountains from which water is released. Although the lions have different designs, some researchers suggest that they can be grouped into four categories, each with different characteristics. This suggests that they were created by four different artists, each with their own unique patterns. All the tile elements are connected to the fountain. From the central fountain, four springs radiate out on the four sides of the palace, symbolizing the four rivers of paradise mentioned in the Quran [2]. Constructed of multicolored white marble, the original fountain creates an aesthetically pleasing ambiance and an auditory impression of birdsong, creating a sense of joy and vitality. In addition, the sound of flowing water can be heard in the surroundings, contributing to the sensory and psychological well-being of visitors (Bachelard 1983; Zube et al. 1974).

- Patio of the Arrayanes Alhambra

Myrtus plants (Myrtus communis L.) are placed on either side of the pool. Two large hedges of *Myrtus communis* form a distinctive landscape feature. The only plants in this garden are *Myrtus communis* [9]. The water for the pool is supplied by two marble basins located at each end of the garden. The pool and the sky are integrated into the garden, providing natural elements and reflecting the sky in the water.

The integration of water, vegetation, and open sky within the palace symbolically and physically represents the natural environment, thus contributing to the creation of a microclimate, humidity, ventilation, and aromatization [2]. Water acts as a magnificent mirror where everything becomes double-dimensional. The pool reflects the entire surrounding environment, creating the impression of an infinite and limitless space.

- Salad of the dead brisket Hermanas

The water channel originates from this area and then flows into another, more dominant space, the Lions' Courtyard. It provides the overall composition of the building, connects the different structures, and determines the forms of relations between them [9].

- The Partial Palace

The architectural feature in question is a large central pool surrounded by a five-arched portico. Built during the reign of Muhammad III (1302–1309), this structure is known as the Partial Palace within the Alhambra complex. It is believed to be the oldest surviving partial palace in the complex [9].

The Generalife is located on a high, autonomous hill at the northeastern end of the Alhambra, commanding a wide view of the complex. It also served as a place of rest and entertainment for the princes and princesses of the Banu al-Ahmar dynasty, providing a respite from the hustle and bustle of the palaces. The palace was owned by Sultan Muhammad III. The palace is rectangular in shape, with a rectangular pool of water in the middle and fountains on each side. It is surrounded by a collection of trees, shrubs, and rose bushes of various shapes and colors. The gardens are distinguished by the presence of waterways and waterfalls that set them apart from other similar structures [2, 9].

- EL Patio del Cipres de la Sultana

This patio is dominated by a pond, which occupies more than two-thirds of the available space. The pond is surrounded by 38 water fountains, two square plant beds, and a stone fountain above a small central basin. The pond is encircled by trees, which create aesthetically pleasing reflections in the water. The proximity of the pond to the palace is thought to have been designed to provide a source of water for the palace and to offer a place for relaxation during hot weather. Additionally, there is evidence in the basement that may indicate the location of the Generalife Palace's bathroom [2, 9].

- The Patio of the Acequia

The Patio de la Acequia, 48.70 m long and 12.80 m wide, is the most important component of the Generalife. The courtyard is divided in two by a channel that carries water from the Sakiyeh to the Al Hamra. This channel is surrounded by several small springs and ends in two stone pools. The area is furnished with various plants that reflect the prevailing aesthetic preferences of the period.

Chehel Sotoun

The function of this building in Isfahan, Iran, was to serve as a space for meetings and entertainment. In Persian, *Chehel Sotoun* means "Forty Columns." Construction began during the reign of Shah Abbas I and was completed in 1647 under Shah Abbas II. The palace is located in the center of the Chehel Pillar vineyard, which covers an area of approximately 67,000 m^2. Tall trees in its garden and a large reflective pool in front of the building are among its most notable features. The name "Palace of Forty Columns" refers to the visual effect created by the reflection of the twenty physical columns in the pool, giving the illusion of forty columns [10].

The Palace of Versailles

The Palace of Versailles is a renowned historical building located in France. It served as the principal residence of the French monarchs from the reign of Louis XIV to Louis XVI. The palace is especially noted for its innovative and symbolic use of water. The extensive pools, lakes, and canals within the palace grounds significantly contribute to its impressive visual and spatial composition [11].

The presence of water emphasizes the architectural features of the palace and reflects the grandeur of its Baroque design. Versailles is now a world-famous tourist destination, celebrated for its elaborate integration of water features into both the palace and its gardens. The gardens, designed by André Le Nôtre, include a wide range of water elements such as ponds, streams, fountains, and large lakes, and represent the peak of formal French garden design [12].

Historically, the Gardens of Versailles are significant because they demonstrate the use of landscape architecture as a political tool. The formality and grandeur of the gardens symbolized King Louis XIV's absolute control over nature and power over his domain, blending art, architecture, and political ideology into one unified landscape.

writes Tony Spawforth, a professor at Newcastle University, in his book "Versailles: A Biography of a Palace" (St. Martin's (The Press, 2008). effects Were the work of engineers whose machines made Versailles a hydraulic as much as an artistic "wonder".

- Fountains of the fight of the animals

Cornelius van Designed by Cleef and depicting a lion overthrowing a wolf, the fountain's water flows into the upper basin in each section, while the water of the defeated animal flows into the lower basin.

- Latona's fountain

Latona's Fountain, located in the Gardens of Versailles, is inspired by the mythological narrative found in *Metamorphoses* by the Roman poet Ovid. The story tells of Latona, the mother of Apollo and Diana, who pleaded with Jupiter to protect her children from the cruelty of the Lycian villagers. In response, Jupiter punished the villagers by turning them into frogs and lizards. The fountain captures this mythological moment through sculptural representations and is a symbolic reflection of divine justice and transformation, integrating myth and water in a theatrical landscape (Thacker 1979).

- Le bassin d'Apollon

The Apollo Fountain (*Le Bassin d'Apollo*) is one of the 55 water features located within the Gardens of the Palace of Versailles, which is recognized as a UNESCO World Heritage Site. The fountain depicts the Greek god Apollo rising from the water in a chariot, symbolizing the sun and aligning with Louis XIV's self-identification as the "Sun King." It reflects the grandeur of Baroque garden design and the symbolic use of water as a representation of power, mythology, and celestial order (Thompson 2006).

- The Grand Canal

The Grand Canal at the Palace of Versailles was designed by André Le Nôtre in the 1670 s and constructed under the supervision of King Louis XIV. Covering an area of approximately 230,000 square meters with a depth of 2 m, the canal played both an aesthetic and functional role in the landscape design. It not only extended the visual axis of the gardens but also served as a central feature for royal boat displays and festive events. The canal exemplifies the integration of hydraulic engineering into Baroque landscape architecture, reflecting the symbolic control of nature through water [12] (Thompson, 2006).

- **Taj Lake Palace**

The Taj Lake Palace, originally named *Jag Niwas*, was built in the eighteenth century by Maharana Jagat Singh II, the former ruler of Udaipur. Situated on one of the four islands in Lake Pichola, it was initially constructed as a summer retreat for the royal family.

The palace uses water as a central architectural and symbolic component. Its impressive appearance is enhanced by the visual reflections on the lake's surface, while the internal waterways and canals contribute to the unique ambiance of the structure. The integration of water into the architectural design exemplifies traditional Indian principles, where water is viewed as a source of purity, tranquility, and harmony [10].

Today, the Taj Lake Palace is one of India's most renowned heritage hotels and a major tourist attraction, celebrated for its seamless fusion of water and architecture in a royal setting [13].

2.2 Research Method

The rationale behind selecting these palaces lies in their historical and cultural significance within their respective civilizations, despite their geographical and religious differences. However, these palaces share a common symbolic use of water. Notable examples of historical architecture that use water as both a functional and aesthetic element include the Alhambra and Generalife in Spain, and the Chehel Sotoun Palace in Iran. The Sotun Palace, the Palace of Versailles in France, and the Taj Lake Palace in India each exemplify the importance of water in architectural design and demonstrate a deep understanding of the surrounding aquatic environment's capacity to enrich life through its multifaceted symbolism.

At the Alhambra, the role of water in design is exemplified through pools and gardens that reflect the evolution of Islamic thought regarding the organization of space and the use of natural resources [2, 9]. At Versailles, the design of the royal gardens was shaped by fountains and other water features that reflected the wealth and grandeur of the French monarchy [11, 12]. The Chehel Sotoun Palace in Iran represents an important example of Persian architecture, where the use of water is evident as part of traditional Persian gardens [10]. This reveals the significance of water in Persian thought as a vital component in providing comfort and maintaining ecological balance.

The Taj Lake Palace in India provides a case study in the integration of water into the design of a royal palace. Situated in the middle of a lake, the palace creates a sense of seclusion and tranquility. Its design reflects the traditional Indian view of water as a calming and purifying element, with water being used carefully and thoughtfully [10].

The following examples of palaces illustrate these concepts.

3 Research Results

Al Hamra and Jannat Al Arif are characterized by the prominent use of circular shapes and angles. The function of these spaces is not only aesthetic but also serves a religious and intellectual purpose. The designs rely heavily on natural lighting, which serves to highlight the beauty of the space and provide balanced and variable illumination throughout the day. The dynamic movement of water in these designs demonstrates the designers' engineering expertise in using natural flows to achieve a serene visual and acoustic effect. Mosaic decorations and tiles were used in a wide variety of colors, with blue being the main distinguishing feature of the water areas. The Islamic architectural tradition was shaped by a tendency toward bold tones, many of which are exemplified by the deep blue tones of the water elements. This stylistic preference is evident in a large number of works from the period. By placing plants around the fountains, it was aimed to give the water a natural quality, thus establishing a balance between light and shadow and encouraging a sense of intimacy within the space.

The architectural style of Sotoun Palace is characterized by the use of angular forms. The palace fulfills three functions: an aesthetically pleasing space, a place of worship, and a basin designed to maintain thermal balance due to its location in an area with high temperatures. This requires an architectural design that can effectively address climatic conditions. The vegetation surrounding the basin insulates the building from the environment, provides shade, and reduces the ambient temperature inside the building. The fountains located on both sides of the basin, south and north, were designed to create a sense of movement through the dynamic movement of water. The use of dark colour tones serves to enhance the visual impact of the design, thus reflecting the harmonious integration of watery elements with the surrounding natural environment.

Versailles is a symbol of French Baroque splendor. It is renowned for its distinctive architectural design, which blends aesthetic principles with artistic motifs. The palace uses angular forms, with sharp angles contributing to a sense of solidity and structural precision. The juxtaposition of angular forms facilitates a coherent organization of spaces, defining different functions. Dynamic artistic design is characterized by the presence of figures and sculptures representing various religious and historical scenes. The figures give a sense of movement and can be considered a lively artistic expression. The use of light colors serves to strengthen the perception of spatial dimensions. The gardens are designed to be an integral part of the space, creating a sense of power and beauty through the use of careful natural designs. Representing the pinnacle of artistic and architectural excellence, the design reflects the royal power and beauty of France throughout the centuries.

Taj Lake Palace, located in Udaipur, India, has a number of features as summarized in Table 7.2, with its location in the heart of Lake Pichola. This advantageous location gives the palace a distinct advantage by offering awe-inspiring views of the lake and the surrounding Aravalli Range, thus enhancing its visual appeal. The palace's central location on the lake shore provides a high degree of privacy and tranquility.

4 Conclusion

When analyzing the selected examples, it becomes evident that natural water elements dominate the spatial experience. In the case of the Alhambra and Jannat al- 'Arif, water was transported from distant sources due to its essential functional and symbolic role. Water was used not only to serve practical needs but also as a defining element of the architecture, giving the palace its unique identity and emotional atmosphere [2].

Similarly, the Palace of Versailles and its gardens represent a hydraulic and artistic marvel. Designed by André Le Nôtre, the extensive use of pools and fountains was not only decorative but also symbolic, representing the beauty, order, and power of the monarchy. This approach to water design marked an innovative shift in landscape architecture and came to be associated with elite status and royal authority [11] (Thompson 2006).

The Alhambra and Jannah Al Arif are examples of the evolution of Islamic architecture, which showcases the emergence of new design styles distinguished by their distinctive technical features. In contrast, Western architecture is characterized by a diversity of styles that reflect the country's multifaceted history and the influence of the various civilizations it has encountered. Western architecture often prioritizes the display of power and wealth. It can be said that Islamic architecture is distinguished by its high artistic standards and complex geometry, while Western architecture is characterized by its tendency to display power and wealth through a variety of styles and the use of rich materials.

Chehel Sotoun Palace demonstrates how Persian architecture integrates water and space to meet the environmental and spiritual needs of its users. The combination of shaded vegetation, reflecting pools, and columned structures creates not only a functional response to climate but also a serene and aesthetically meaningful environment [10].

In the case of the Taj Lake Palace, the decision to build in the middle of a lake reveals the user's deep appreciation for nature and the symbolic power of water. This architectural choice emphasizes both the experiential value of water and its cultural significance, turning the palace into a statement of harmony, privacy, and visual beauty [13].

The symbolic meaning attributed to water significantly contributes to the architectural identity of historical palaces, reflecting the spiritual and cultural values of their respective societies. Water functions not only as a physical element but also as a medium that enriches both the visual and emotional experiences of users. The

Table 7.2 Features analysis of examples

Samples used in the research		Features																
		Form		Water function			Lighting		Action		Ground color		Water element				Qualities: These elements with integrated	
		Circular	Angular	Visuality	Worship	to swim	Natural	Colorful	Dynamic	Static	Light color	Dark	Pond	water channel	Waterfall	Fountain	Plants	Structure
Alhambra	The Lions fountain	✓	–	✓	✓	–	✓	–	✓	–	✓	–	–	–	–	✓	–	✓
	Patio of the Arrayanes	–	✓	✓	–	✓	✓	–	✓	✓	–	✓	✓	–	–	✓	✓	✓
	The Partial Palace	–	✓	✓	–	✓	✓	–	–	✓	–	✓	✓	–	–	✓	✓	✓
	Salad of the dead brisket Hermanas	✓	–	✓	✓	–	–	✓	✓	–	✓	–	–	–	–	✓	–	✓
Generalife	EL Patio del Cipres de la Sultana	–	✓	✓	✓	–	✓	–	✓	✓	✓	✓	✓	–	–	✓	✓	✓
	The Patio of the Acequia	✓	✓	✓	✓	–	✓	–	✓	✓	–	✓	–	✓	–	✓	✓	–
Chehel Sotoun	Palace Pool	–	✓	✓	✓	✓	✓	–	✓	✓	✓	✓	✓	–	–	✓	✓	✓
Versailles	Latona's fountain	✓	–	✓	–	–	✓	–	✓	✓	✓	–	–	–	–	✓	✓	–

(continued)

Table 7.2 (continued)

Samples used in the research		Features																
		Form		Water function			Lighting		Action		Ground color		Water element				Qualities: These elements with integrated	
		Circular	Angular	Visuality	Worship	to swim	Natural	Colorful	Dynamic	Static	Light color	Dark	Pond	water channel	Waterfall	Fountain	Plants	Structure
	fountains of the fight of the animals	✓	–	✓	–	–	✓	–	✓	–	✓	–	✓	–	–	✓	✓	–
	le bassin d'apollo	–	✓	✓	–	–	✓	–	✓	–	✓	–	–	–	–	✓	✓	–
	The Grand Canal	✓	✓	✓	✓	✓	✓	–	–	✓	–	✓	✓	–	–	–	✓	–
Lake Palace	Lake Palace	✓	✓	✓	–	✓	✓	✓	✓	✓	✓	✓	✓	–	–	✓	✓	✓

selected palaces examined in this study serve as important examples of how water can be integrated into architectural design to serve aesthetic, functional, and symbolic purposes. Their diverse contexts—spanning Islamic, European, and Indian traditions—illustrate the universal yet culturally specific significance of water in built environments. This study highlights the enduring relevance of water as a design element and offers a conceptual foundation for future projects aiming to integrate water meaningfully into architectural compositions. Ultimately, water can transform a space into a living symbol, both dynamic and deeply rooted in cultural identity.

References

1. Vílchez JMP (2007) La Alhambra y el Generalife de Granada. Artigrama 22:187–232
2. Puerta-Vílchez JM (2021) The Alhambra and the Generalife. The Eternal Landmarks of Islamic Granada. In: A companion to Islamic Granada. Brill, pp 365–406
3. Lohrer A (2017) Basics designing with water. Birkhäuser
4. Burmil S, Daniel TC, Hetherington JD (1999) Human values and perceptions of water in arid landscapes. Landsc Urban Plan 44(2–3):99–109
5. Daglio L (2014) Building with water: innovative approaches for sustainable architecture
6. Prominski M, Stokman A, Stimberg D, Voermanek H, Zeller S (2012) Fluss. Raum. Entwerfen: Planungsstrategien for urban Fließgewässer. Walter de Gruyter, p 9
7. Ishmael NI (2022) Study of the architectural elements of creativity in the design vocabulary of Islamic architecture of Alhamra palace (Kingdom of Bani El Ahmar) JES. J Eng Sci 50(2):78–100
8. Gray F (2006) Designing the seaside: architecture, society and nature. Reaction Books
9. Fernandez-Puertas A, Goodwin G, Jones O (1997) The Alhambra. Saqi Books, London, pp 110–120
10. Mar'ah F, Mawardi K, Purnomo A (2021) Seni Arsitektur Dynastic Safawi from Dinasti Mughal. Tarikh from Tsaqofah: Sejarah from Jurnal Kebudayaan Islam 6(1):41–50
11. Howse, J. (2015). Palace of versailles: home to the kings of France. Weigl Publishers
12. Vernhes JD, Saulet P, Heitzmann A (2022) The Grand Canal at Versailles: Geotechnical investigation, II. In: Geotechnical Engineering for the Preservation of Monuments and Historic Sites III. CRC Press, pp 254–268
13. Tillotson GHR (2004) The tradition of Indian architecture: continuity, controversy and change since 1850. Yale University Press

Smart Technologies in Eco-Friendly Water Infrastructure

Jyoti Kumari, Praveen Singh, Archana Singh, and Arvind Nain

Abstract Sustainable water management is changing as a result of the integration of innovative technologies into eco-friendly water infrastructure. According to this study, the Internet of Things (IoT), artificial intelligence (AI), and remote sensing are increasing the efficacy of green infrastructure solutions like rain gardens, bioswales, and permeable pavements. Real-time tracking of water flow, quality, and storage capacity is now feasible thanks to smart sensors, which enable us to use resources more effectively and reduce the likelihood of flooding. AI-driven predictive analytics goes one step further and aids in better decision-making by forecasting patterns of extreme weather and water demand. Even with all these improvements, there are still some hurdles to overcome, like the big price tag up front, worries about keeping data private, and making these new systems work with the old ones. Finding ways to deal with these issues—through government policies, funding options, and latest tech breakthroughs—will be essential if we want to see these technologies used more broadly. Our research really brings to light how these smart technologies could make our city water systems tougher, more efficient, and better for the environment in the long run.

Keywords Internet of Things · Artificial intelligence · Remote sensing · Green infrastructure · Geographic information system

---

J. Kumari (✉) · P. Singh · A. Nain
Department of Management Studies, Graphic Era Deemed to Be University, Dehradun, Uttarakhand, India
e-mail: kumaruj92@gmail.com

P. Singh
e-mail: praveensingh@geu.ac.in

A. Singh
School of Management, Graphic Era Hill University, Dehradun, Uttarakhand, India

V. Kandpal and T. Choudhury (eds.), *Blue Revolution: Sustainable Water Use and Circular Economy Innovations*, Urban Sustainability,
https://doi.org/10.1007/978-981-95-1916-3_8

1 Introduction

The process of distributing water is a vital element that links the different elements of urban life. In order to maintain the general sustainability of cities, promote economic growth, safeguard the environment, and maintain the well-being of urban populations, proper water management is essential. Life, economic development, and the health of our environment all depend on water. However, the systems that supply water globally are facing more challenges as our cities expand rapidly, more people require water, and climate change impacts us. Many water systems are outdated and unprepared to handle problems like flooding, pollution, and a lack of water, as well as increased operating costs.

It is one of the major challenges that will become the focus of research and policymakers' attention if we combine our technologies with smart water infrastructure. Along with providing information about any system, including inaccessible areas like pipes, it also aids in speedy process access. The technology automates continuous, controlled mediation processes while monitoring the water and equipment to improve operations and compliance. IoT devices (sensors), real-time monitoring systems, smart meters, AI and ML algorithms, advanced analytics, and connectivity (5G) are all components of smart water infrastructure. Together, these technologies and tools monitor water demand, predict problems with infrastructure, and detect pollutants that traditional monitoring systems would have overlooked. It is noteworthy that AI is being used in smart water infrastructure.

Examples of smart infrastructure that can increase water conservation and strengthen city resilience include Singapore's Smart Water Grid, Amsterdam's Digital Delta, and India's Atal Mission for Rejuvenation and Urban Transformation (AMRUT). However, many developing nations continue to face organizational, financial, and technological obstacles that hinder their ability to adopt these innovations, raising questions about whether these solutions can be expanded, implemented at a reasonable cost, and distributed equitably throughout society [1]. Even though a lot of studies are being done on smart water systems, we still don't know enough about how these technologies affect both how well water systems work and the environment, especially in different places and among different social and economic groups. Also, it's really important that we create some combined approaches that connect new technologies with policies, management, and get people involved to make sure we have water solutions that work for everyone and last a long time. Traditional water management and smart water infrastructure are different in two important ways: It gather a lot more information about the infrastructure in real-time or almost real-time, which enables water management teams to respond swiftly. Traditional water management systems use a variety of systems to gather data, including laboratory information management systems (LIMS), computerized maintenance management systems (CMMS), and supervisory control and data acquisition (SCADA) systems. However, data fragmentation is a problem that makes it hard to quickly grasp the entire picture. The demand for food has increased due to the substantial increase in the world's population over the last few decades.

2 Literature Review

2.1 Evolution of Smart Water Infrastructure

Climate change and the need for effective resource management have affected corporate technologies in a smart water system. Soares Ascenção et al. [2], after systematically reviewing 62 publications, discovered 100 intelligent water management system applications and concluded it was vital to implement the framework control and remediate wasted water and to conserve.

Li et al. [3] assert that the development and implementation of smart water systems depend on standardized frameworks. They provided a framework in their review that includes elements like "smartness" and "cyber wellness" as measures of the performance of the systems. Also, they argued for greater stakeholder cooperation to promote the uptake of smart water systems. Adeoti et al. [1] used the framework of developing regions to evaluate the efficacy of smart water management systems in Nigeria. They learned through their research and demonstrated that these systems improved the sustainability and resilience of our water systems by increasing monitoring and operational efficiency. Anaerobic digestion, although first introduced in practice in 1859, has become a widely used biological method of wastewater treatment and is the most ecological. Anaerobic digestion not only treats wastewater, but it also generates biogas, a renewable source of energy. Anaerobic reactors have been proven in several studies to reduce organic pollutants and produce methane while reducing the carbon footprint of wastewater treatment [4–6]. Membrane bioreactors (MBRs), which combine biological treatment and membrane filtration, are not only highly effective at removing organic pollutants and pathogens, but they also take up less physical space than conventional systems [7–9].

This comprehensive interdisciplinary review marks the first instance of the intersection of a field of digital-data technologies, smart cities, sustainable cities, and environmental sustainability, as cutting-edge forms of ICT. Thus, this review will propose new avenues from the diverse collection of studies across different disciplines, in order to contribute to the growing field of ecologically sustainable smart cities and open gateways to more understanding of coalescing models about urbanism and technological fields. There is no question that ecologically sustainable urban development will persist in this direction. So, as more attention is given to adopting data-driven technologies and smart city implementations to deal with the danger from climate change and the imperatives of environmental sustainability, given the increasing growth in the academic interest in the topic, it will be much more impactful. To assist the academic world, practitioners and policy makers to gain a better understanding of the processes at the core of its emergence, insertion, operation, prominence, seduction, and evolution it is very important to question the newly developed interests, language, developments, applications, dynamics, trends, and directions in the flourishing domain of ecologically sustainable smart cities.

2.2 Advances in Technologically Driven Smart Water Management

Blockchain, AI, and IoT technologies are transforming water infrastructure management. Intelligent water management systems that are implemented in the US and Africa, which support the application of AI and IoT in real-time monitoring and decision-making with respect to water systems, and factors of significance in developing policy frameworks that allow organizations to adopt such technology [10]. Homaei et al. [11] took a system-wide approach that fused together blockchain-based digital twins, machine learning algorithms for intrusion detection, and LoRa WAN for data collection with the aim of guaranteeing security and transparency in the management of water in rural areas. The system was scalable and economically feasible to the extent that it was able to do transactions as soon as possible.

Bartos et al. [12] illustrated how smart technology can aid in stormwater management and flood forecasting to establish a resilient, water-smart city. The mainstay of contemporary water management processes is real-time data collection and analysis. Water management is a continual, ongoing operation that involves collecting, monitoring, and evaluating water quality, quantity, and distribution through IoT (Internet of Things) sensors, machine learning, remote sensing, and other processes. These sensors enable authorities to detect leaks, optimize the treatment of waste, and quickly respond to contaminant events. Technology, such as processes for producing fresh water through desalination and purification, is rapidly advancing. For example, forward osmosis, solar desalination, and graphene filters ideally provide greater accessibility, greater efficiency concerning energy, and a cleaner water supply. Technology, such as this, is crucial for areas faced with freshwater scarcity.

Technology has to play a part in water management; it is not just an option. Given the looming water crisis, we need to use innovation to protect this valuable resource for future generations. We can accomplish effective water management while encouraging environmental stewardship by implementing sustainable practices, smart monitoring, and cutting-edge treatment techniques. Everyone has a role to play in the effort to ensure future water security, including governments, corporations, and individuals. To address the urgent global water challenges we face, cooperation, investment, and a dedication to responsible technology deployment will be essential. We can strive towards a future in which water is a symbol of sustainability and abundance rather than a source of scarcity if we have technology on our side.

2.3 Smart Water Solutions for Distribution Systems

Smart water meters provide users, water providers, and municipalities with more accurate feedback on their water consumption in real time. These improvements provide better management of water sources; more accurate billing, and a timely alerting for a dam's water levels. If the levels fall too much, the right people receive an automated alert. These common solutions for agricultural uses can work in isolation for a smart farm. Smart irrigation practices not only assist farmers in using water efficiently and minimizing waste, but sensors also tell the farmers how much moisture is really in the soil and to alter watering schedules if needed. Ultimately, monitoring water quality is also imperative to farming success and is an important contributor to increased productivity.

3 Considerations for Policy and Framework

Policies are one of the most important instruments for achieving goals in any organisation in a consistent, effective, and efficient manner. If your business is similar to mine, you have policies covering everything from the daily tasks of staff management to important matters pertaining to technology, security, and human resources. Policies are the means by which you establish and disseminate to the entire organisation the expected standards of conduct and business practices. Policies give stakeholders and staff members direction and clarity on how your company wishes to operate. They also help you to create a strong reputation for the brand by creating a culture and defining the values that support the brand.

3.1 Innovative Technology Changing the Water Game

Environmental degradation, the effects of climate change, ageing infrastructure, the rising demand for water, and antiquated methods and solutions for water management are just a few of the issues facing the water industry. Sensors, data analytics, and artificial intelligence/machine learning are examples of digital technologies that can help the water sector choose a different course by tackling some of these issues and allowing a change from depending only on the increased supply to managing water resources by balancing the supply and demand sides. Water utilities and municipalities can improve their understanding of water supply and demand patterns, spot inefficiencies and opportunities for water usage optimisation, cut down on water loss, embrace water reuse and the principles of the circular economy at all scales, and create more efficient water management plans with the aid of digital transformation. Additionally, by tracking when, where, and what kind of water is needed, these technologies can help provide real-time monitoring of water quality, allowing for quick responses to emergencies and problems. To accelerate change and meet SDG 6, we need to take advantage of new technologies and digital transformation. But to maximally leverage these technologies, we need to strengthen our infrastructure, improve capacity, and create the skills necessary to successfully manage and utilize digital tools. Furthermore, the benefits that come from digital transformation must be made available to all, including those in underserved communities.

Digital transformation presents opportunities and challenges for open and transparent management of water systems. Using the benefits that arise from digital transformation can help water utilities to improve operations and service delivery while building transparency and trust with stakeholders by navigating the challenges and embracing new technologies. Digital transformation can enable more open and transparent management of water in several ways. First, digital transformation offers real-time water consumption monitoring and data analytics to help utilities identify leaks, eliminate waste, and increase efficiency in the management of water resources. Better resource management, cost savings, and enhanced customer service are all possible outcomes of this. For instance, by employing decentralised ledgers, we can enhance the effectiveness of agricultural water markets and more effectively enforce water rights.

Only 11% of wastewater from homes and businesses is currently recycled. The presence of heavy metals, hazardous chemicals, and nutrients in industrial wastewater that necessitate specific treatment procedures is a major obstacle. Water treatment uses a staggering 4% of the world's electricity, making it a very energy-intensive process. Nanobubbles are one technological advancement that is revolutionizing treatment effectiveness. Simply put, nanobubbles transfer gases more effectively because they are incredibly small. Using 100% sustainable technology, Kran Nanobubble's (Chile) system uses nanobubbles to purify food, treat wastewater, restore contaminated environments, and boost crop yields.

3.2 Technology Integration with Sustainable Infrastructure

The power of technology to increase intelligence and efficiency is most evident when eco-friendly infrastructure is paired with it. Green roofs, bio-retention cells, and permeable pavements are examples of sustainable urban drainage systems (SUDS), which are made to naturally handle stormwater and reduce pollution and runoff. These systems can maximize water collection and reuse by responding to rainfall in real-time when smart sensors are added.

Today, rainwater harvesting is becoming more technologically advanced and isn't limited to rural or urban areas. These systems are now able to modify flow, storage, and filtration in response to evolving weather conditions and present demands thanks to IoT-enabled controls. Using weather information and soil moisture sensors, smart irrigation systems in farming provide precisely the right amount of water to crops, reducing waste and increasing growth. Both efficiency and environmental protection are greatly enhanced by these developments. A company can prepare for the future by adopting sustainability. However, choosing the incorrect place to start can drag out planning stages when efforts to promote sustainable development ought to be underway. As all of the ideas and components are integrated into overall sustainability, the teams involved will require additional assistance.

By incorporating technology into sustainable development, a company can streamline operations, gain a deeper understanding of the sustainability environment, and make more informed decisions and strategies. When people can use innovations and technological breakthroughs to guide social and environmental progress, the future is bright.

The future is not that far off. Sustainability shifts around the world have influenced innovation. It is best to begin integrating technology into business operations in order to secure a prosperous and sustainable future. Decarbonization-focused technology and services are referred to as climate tech. This includes a broad range of services aimed at lowering greenhouse gas emissions and lessening the effects of climate change. Decarbonization is approached differently by various industries. Companies can lower their carbon footprint in their offices and facilities by investing in renewable energy sources. A logistics and transportation firm might switch to electric cars. An architectural firm might focus their new designs on energy-efficient infrastructure and collaborate with suppliers and construction firms that use more environmentally friendly appliances and materials. Climate technology can be found in all industries and is applicable across value chains, guiding economies around the world toward more carbon–neutral and sustainable practices.

3.3 Harvesting Rainwater and Conserving Water

Water conservation and rainwater collection are key elements of Earthship design, which complements the sustainable and self-sufficient ethos of the environmentally friendly homes. Here's how collecting rainwater and conserving water are related. Planning for rainwater harvesting can be done more effectively in a watershed than in any other type of geographic unit because watersheds are hydrologic units that are chosen based on the flow of water in a land area. A reservoir created by building dams across streams of any order, check dams built across the streams, or dugout ponds can store direct rainfall, overland runoff, and lateral seepage flow (emerging within the watershed boundaries). By lifting at a suitable moment, water that has seeped into the soil profile can also be utilized as subsurface storage.

Among the world's biggest users of water is agriculture. Our precious water resources can be preserved, and freshwater withdrawals can be greatly reduced by implementing sustainable water treatment in this industry. Large volumes of water are needed for the operations of many industries. These sectors can drastically cut back on wastewater production and water consumption by putting in place water treatment systems that permit the recycling and reuse of water.

Another essential component of Earthship designs is the safe treatment of wastewater. We build free, permanent homes that provide you with all the essentials. Waste from gray water enters the inner plant cells, where the plants use and purify it until it is sufficiently clean to collect in a well at the end of the bowl and be pumped into the toilet tank for flushing when necessary. Forty percent of the water used in a typical household is for toilet flushing. After passing through a traditional septic tank, the toilet's sewage overflows into an outdoor plant cell lined with rubber and stocked with landscaping plants. Earthships reduce their dependency on nearby water supplies by combining rainwater collection Greywater recycling is the process of treating wastewater from showers, sinks, and water-saving devices. These actions support water management, improve the relationship with natural resources, and aid in sustainability. This non-drinking recycled water is used to flush toilets and water plants indoors and out. Rooftop Social An Earthship's wide, sloping roof is designed to channel rainfall into gutters and drains.

The collected rainfall is transferred from the pipe to sturdy tanks that are kept underground or in chimneys. Water conservation is done so that we can reuse rainwater for other household purposes and even implement water-saving techniques in industries. "Earthship" promotes water-saving behaviours, such as repairing leaks immediately, conserving or minimizing water use, and reducing water use. Earthships feature low-flow toilets, showers, and faucets which are designed to use less water while still being functional.

4 Benefits and Difficulties

Incorporating smart technology in our daily lives, we can save electricity and reduce water waste. Its efficiency is increased by low energy consumption, wastewater, etc. Apart from that, we can use the technology to save water resources. Another difficulty is finding people with the technical expertise required to maintain and operate these intelligent systems. Since digital networks are so essential to our communication, cybersecurity must also be a concern. Additionally, excluding some communities, such as those in rural areas or with lower incomes, from these advancements could worsen already-existing disparities.

Accessing foreign cultures, including their cuisine, music, films, and artwork, is now simpler than ever thanks to globalization. You can have Thai food delivered to your apartment while you watch a Bollywood film or listen to your favourite artist from the United Kingdom because of this unrestricted movement of people, goods, art, and information. Since many nations are always connected, information and technological advancements spread swiftly. Scientific advancements made in Asia can be implemented in the United States within a few days due to the rapidity of knowledge transfer. Globalization has facilitated travel to other nations, but it has also started to combine disparate societies. Some cultures were so successful around the world that other nations tried to imitate them. However, we lose the diversity of the world when cultures start to lose their unique characteristics. Conventional water distribution systems frequently face inefficiencies brought on by issues like leaks, imprecise consumption monitoring, and a dearth of real-time data. Through the use of a variety of cutting-edge technologies, such as sensors, complex data analytics, and reliable communication networks, smart water systems seek to overcome these constraints.

The political environments in Europe and the United States demonstrate that opinions on the effects of globalization differ. It is more difficult for immigrants to find employment in new countries because of many nations tightening their immigration laws. The main cause of this rise in nationalism is resentment over the idea that foreigners take jobs in the country or that businesses relocate their operations overseas to cut labour costs.

5 Prospects for the Future

Future developments are typically associated with technological advancements and progress. However, rethinking the future means doing things in a sustainable and ever-improving manner. Businesses have a great opportunity to innovate and mold business practices for sustainability as they continue to adjust to changes in the world. We can better guarantee a sustainable world for coming generations by altering the way we operate. "Technology and sustainability work together to make progress easier, track and forecast performance, simplify systems and processes, and help businesses better understand the ever-evolving and complex world of sustainability reporting and governance. Sustainability becomes a practice that is ingrained in daily life rather than an unfamiliar environment when technology is incorporated into business operations. Future developments in water infrastructure will be fueled by a more robust integration of technology, environmental concerns, and community engagement. Solar-powered sensors and pumps are among the innovations in renewable energy that will increase the sustainability of intelligent water networks. Digital twins and edge computing are two new trends that offer new opportunities for quick, location-specific decisions. To overcome financial and organizational obstacles, policy support, public–private sector partnerships, and international collaboration will be essential.

Stability in education is key, but so is the ability to create programs that can create and preserve these system edges. The development of trust and the advancement of the cause will depend on having strong frameworks for data management, privacy protection, and interoperability. As intelligent technology develops further, water management appears to have a promising future. The effectiveness of water management is set to increase with new developments in artificial intelligence, machine learning, and the Internet of Things. Improved sensor accuracy, more advanced data analytics, and improved interoperability with current infrastructure are all features of future systems. We can expect even more solutions that can develop one that can look after each and every part of the issues that are there to clear out. the innovation will help our water system to adapt to environmental changes and can also reduce a lot of wastage happening every day. These systems are able to analyse data from various sources and spot patterns and instances of water waste or deteriorating water quality by using AI and ML algorithms. The problem of water scarcity can be significantly reduced by AI-driven technology, which is crucial for locating and stopping water leaks. These tools can continuously measure the water's force and flow, promptly identifying any signs that could point to an issue. By removing the error, an AI can automate the water management system, which can improve the accuracy of smart water procedures. Water management is located near the border of the smart city, which can be integrated frequently. They can supervise the maintenance of water-related work, manage the billing system of water and space, and monitor the quality of the water. These applications promote sustainability, enhance water resource supervision, reduce incident rates, make it easier to allocate city funds more effectively, and increase transparency in water management. On the basis of these significant benefits, it is reasonable to expect a consistent increase in future funding for IoT-powered water management projects.

6 Conclusion

Effective water management in urban areas is now crucial to ensuring freshwater sustainability and security, as urbanisation has accelerated and water scarcity has emerged as a major environmental challenge. The digitisation of various economic activities is also gaining traction in the water sector, based on this rationale. This study successfully contributes to the exploration of how digital technologies have sparked efficient and sustainable water management by offering a road map of best practices for intelligent water stewardship in urban supply and distribution networks. To this end, the systematic literature review analysis that follows attempts to categorise the gathered publications based on their research scope/focus, technologies employed, and level of analysis.

It is important to note that the review analysis done here is a first attempt at reviewing the most relevant cutting-edge research on the relationship between urban water management and digitisation. A more thorough techno-economic and environmental analysis of the smart technologies (such as sensors) that are given a lot of attention in the field of urban water monitoring should therefore be the focus of future research projects. Furthermore, the analysis focused on digitisation could be expanded to include other areas that were deemed outside the current research's purview, like drinking water supply, urban wastewater treatment, or flood prevention. Future research should concentrate on creating a comprehensive framework for digital urban water management, building upon similar frameworks for smart water stewardship in agri-food systems. This will help policymakers and/or decision-makers choose the best technological interventions for effectively tracking and predicting water use in urban environments. Notably, these innovative frameworks might consider the participation of citizens in intelligent water monitoring in order to promote participatory water management. The viability and dependability of smart urban water management systems should also be protected by researching cybersecurity and other legal concerns that are indicative of the use of big data. In particular, we concentrated on cutting-edge technologies that track, regulate, and manage water levels, water usage, and water leakage, beginning at the building scale. Lastly, sustainable systems, like water harvesting systems, which are typically used to conserve and reuse water resources, were combined with innovative technologies to analyse them.

By reducing technological losses and human overuse, this thorough review made clear how crucial it is to investigate the possibilities of cutting-edge, sustainable water-saving technologies in order to maximise resource management.

References

1. Adeoti OS, Haremi R, Kandasamy J, Vigneswaran S (2025) Evaluating the effectiveness of smart water management systems in enhancing the resilience and sustainability of water infrastructure in Nigeria. Aqua Water Infrastruct Ecosyst Soc 74(2):253–266. https://doi.org/10.2166/aqua.2025.291
2. Soares Ascenção É, Melo Marinangelo F, Meschini Almeida CF, Kagan N, Dias EM (2023) Applications of Smart Water Management Systems: A Literature Review. Water 15(19):1. https://doi.org/10.3390/w15193492
3. Li J, Yang X, Sitzenfrei R (2020) Rethinking the framework of smart water system: a review. Water 12(2):1. https://doi.org/10.3390/w12020412
4. Nayeri D, Mohammadi P, Bashardoust P, Eshtiaghi N (2024) A comprehensive review on the recent development of anaerobic sludge digestions: Performance, mechanism, operational factors, and future challenges. Results in Engineering 22:102292. https://doi.org/10.1016/j.rineng.2024.102292
5. Otieno EO, Kiplimo R, Mutwiwa U (2023) Optimization of anaerobic digestion parameters for biogas production from pineapple wastes co-digested with livestock wastes. Heliyon 9(3):e14041. https://doi.org/10.1016/j.heliyon.2023.e14041
6. Yang Q, Liu H, Liu L, Yan Z, Chui C, Yang N, Wang C, Shen G, Chen Q (2025) Enhancing Methane Production in Anaerobic Digestion of Food Waste Using Co-Pyrolysis Biochar Derived from Digestate and Rice Straw. Molecules 30(8):1. https://doi.org/10.3390/molecules30081766
7. Al-Asheh S, Bagheri M, Aidan A (2021) Membrane bioreactor for wastewater treatment: A review. Case Stud Chem Environ Eng 4:100109. https://doi.org/10.1016/j.cscee.2021.100109
8. Judd S, Judd C (2011) The MBR book principles and applications of membrane bioreactors for water and wastewater treatment. Elsevier
9. Mahboubi A, Ylitervo P, Doyen W, De Wever H, Taherzadeh MJ (2016) Reverse membrane bioreactor: Introduction to a new technology for biofuel production. Biotechnol Adv 34(5):954–975. https://doi.org/10.1016/j.biotechadv.2016.05.009
10. Olatunde TM, Adelani FA, Sikhakhane ZQ (2024) A review of smart water management systems from Africa and the United States. Eng Sci Technol J 5(4):1231–1242. https://doi.org/10.51594/estj.v5i4.1014
11. Homaei M, Morales VG, Gutierrez OM, Gomez RM, Caro A (2025) Smart water security with AI and blockchain-enhanced digital twins. http://arxiv.org/abs/2504.20275
12. Bartos M, Wong B, Kerkez B (2018) Open storm: a complete framework for sensing and control of urban watersheds. Environ Sci: Water Res Technol 4(3):346–358. https://doi.org/10.1039/C7EW00374A

IoT in Water Resource Management: Sustainable Use and Circular Economy Innovations

G. R. Gnana King and J. H. Jensha Haennah

Abstract The global water crisis, driven by factors such as population growth, rapid urbanization, and climate change, necessitates innovative approaches to manage water resources sustainably. The Internet of Things (IoT) has emerged as a powerful tool to transform water management through real-time monitoring, automation, and data-driven decision-making. By enabling proactive strategies, IoT helps optimize water usage, detect leaks, improve distribution, and monitor water quality. Applications range from smart water grids and precision agriculture to flood control and wastewater management, all contributing to more efficient and sustainable water systems. While the benefits include cost savings, environmental protection, and improved public health, challenges such as high implementation costs, data privacy, and technical complexity must be addressed. Looking ahead, the integration of IoT with technologies like AI, blockchain, and 5G, along with global cooperation, holds promise for building resilient, smart water systems as part of a broader vision for sustainable development.

Keywords IoT · Water resource management · Sustainability · Real-time monitoring · Precision agriculture · Flood control · Blue revolution

1 Introduction

Growing unavailability and volatile patterns in freshwater supplies have raised the demand for fresh and sustainable water management strategies. The Internet of Things (IoT) has been the impetus for transformation in this aspect through real-time information, automation, and improved decision-making mechanisms. This

G. R. G. King · J. H. J. Haennah (✉)
Sahrdaya College of Engineering and Technology, Kodakara, India
e-mail: jenshahaennah@gmail.com

G. R. G. King
e-mail: kings.326@gmail.com

V. Kandpal and T. Choudhury (eds.), *Blue Revolution: Sustainable Water Use and Circular Economy Innovations*, Urban Sustainability,
https://doi.org/10.1007/978-981-95-1916-3_9

chapter gives an overview of how IoT is transforming the water resource management paradigm in accordance with the concepts of the circular economy.

IoT infrastructure includes networked sensors and actuators and data analysis platforms which monitor variables like water flow, quality, consumption trend, leakages, and reservoir levels at all times. IoT systems have continuous data gathering and communication capabilities to allow responsive and dynamic water management within agriculture, industries, and urban areas.

In agriculture, the smart irrigation systems employ IoT sensors to track soil moisture and weather, conserving water and wasting less. In urban areas, homes and utilities employ smart meters and leak detection sensors to measure usage in real-time, saving non-revenue water and facilitating conservation. Industrial facilities apply IoT to wastewater recycling and reuse according to the principles of the circular economy, reducing freshwater dependence.

The advantages of IoT-based water management include greater efficiency, lower operating expenses, better water quality, and better regulatory compliance. Moreover, IoT's predictive capacity can foresee equipment breakdown, flooding, or water shortage so intervention can be initiated in time and long-term strategies can be developed.

However, challenges such as high implementation costs, data security, interoperability, and lack of technical skills hinder mass adoption. Scalable infrastructure, integration with existing systems, and data privacy are some of the most critical issues.

Despite such limitations, the prospects for IoT in water resource management are bright. Artificial intelligence, edge computing, and blockchain technologies will be in a position to further augment the capabilities of IoT, making it possible for systems to become yet smarter, safer, and efficient. As water becomes an even more precious asset, IoT is going to become the driving force behind a Blue Revolution, where sustainable water usage, circular economy business models, and technology-based solutions converge to address global water challenges.

2 The Use of IoT in Water Resource Management

2.1 Real-Time Monitoring and Data Collection

The key to effective water resource management is precise, timely, and complete data. Conventional practices of monitoring water assets, for example, through manual inspections or sporadic sampling, are normally slow, error-prone, and not fine-grained enough for the water issues of the modern era. IoT breaks this environment by making real-time monitoring and data gathering possible along the entire water chain.

IoT sensors like smart sensors, flow meters, pressure sensors, and drones are strategically deployed across touchpoints such as reservoirs, groundwater wells, rivers, treatment plants, pipelines, and end-user connections. These IoT sensors continuously monitor parameters such as water levels, turbidity, pH, chemical composition, flow rate, and even biological contaminants. For instance, smart buoys can monitor lake and river conditions in real-time and detect changes due to pollution or weather conditions.

Water management information collected by such sensors is transmitted wirelessly based on protocols such as LoRaWAN, NB-IoT, or 5G to cloud infrastructure or data center clusters. Such a constant stream of high-resolution data develops a "digital twin" of the water network, enabling real-time monitoring and assessment of system performance.

The return on investment is enormous. Regulators can detect anomalies like leaks, pipe bursts, contamination events, or pressure loss—issues that may otherwise go unnoticed for days or weeks using traditional methods. In municipalities, smart water meters can measure domestic and commercial usage in real-time, encouraging conservation and enabling differential pricing based on rates of consumption. And drones' sensors are able to image inaccessible lands, track watershed conditions, and estimate the effects of land use shift. Solar IoT sensors in the countryside or outside the grid provide a low-key way to continuous monitoring, overcoming the infrastructure deficiency. This information revolution enables anticipation. Rather than responding to crisis situations, water utilities and authorities can track patterns, forecast concerns, and eliminate issues, reinforcing water systems to become stronger, clearer, and answerable.

2.2 *Predictive Analytics and Decision Support*

While real-time data capture forms the essence of IoT systems, their complete potential is realized through predictive analytics and decision support. With vast volumes of data available, water management agencies can now move beyond descriptive analytics (what is happening) to predictive (what will happen) and prescriptive (what should be done) analytics.

Predictive analytics leverages machine learning (ML), artificial intelligence (AI), and statistical modeling to detect patterns, predict trends, and offer early warnings. For example, historical consumption behavior, weather conditions, seasons, and socio-economic data can be utilized by AI models to predict future water demand with great accuracy. Utilities can subsequently fine-tune water supply, capacity development planning, and over-extraction or shortage avoidance.

One of its principal applications is leak detection. The algorithms can compare pressure and flow data to search for slight fluctuations that indicate leaks, even before visibility or the extensive destruction. It saves water, saves repair costs, and avoids infrastructure damage. AI systems also determine which leaks should be repaired, on the basis of parameters such as location, amount of water lost, and repair costs—thus allowing more informed resource allocation.

In agriculture, predictive analytics helps to create the most effective irrigation schedules. With analysis of projections for the climate, soil conditions, type of plant, and evapotranspiration rates, systems generate adaptive watering schedules that achieve maximum production without waste. Farmers are no longer working with stationary calendars; they determine based on science-driven suggestions specific to their fields.

Decision Support Systems (DSS) which are based on IoT platforms summarize such analytics in simple dashboards and mobile apps and present actionable information to various stakeholders—policymakers, farmers, utility operators, and emergency response officials. As an illustration, a water treatment plant manager will be alerted of increased turbidity, along with suggested actions for minimizing risk. A city planner will be able to plan water-conserving housing projects using forecasts of demand.

These systems permit scenario analysis, too. If a drought is more prolonged than anticipated? If a key industrial customer doubles their water consumption? DSS tools make it possible for stakeholders to build such scenarios and analyze the effect of various interventions, allowing long-term resilience planning.

The combination of GIS (Geographic Information Systems) with IoT data provides another level of decision-making. Space visualization of information—leakage hotspots, pollution points, or over-exploited aquifers—makes information more transparent and allows effective interventions.

In effect, predictive analytics and decision support turn IoT data into foresight, making it possible to shift from reactive to proactive water management.

2.3 Automation and Control

One of the wonderful things that IoT makes possible is automation—the ability to execute processes and respond to systems based on real-time data with minimal human intervention. Automation translates to higher efficiency, reliability, and consistency, lowering human error and cost of operations.

IoT automation pervades deeply into agriculture. Intelligent irrigation systems, for instance, utilize soil moisture sensors, temperature sensors, and weather APIs to decide exactly how much water will be needed by a crop and when. They can be automated to turn off or on irrigation pumps, valves, or sprinklers based on numerous conditions. This saves water and enhances the health and yield of crops. Farmers also are able to remotely monitor and control their systems via smartphones, without necessarily employing on-farm personnel.

Automation ensures system balance in urban water supply systems. Valves and pumps can regulate flow rates depending on demand levels and reservoir conditions to ensure balanced distribution and avoid pressure imbalance. For example, when the demand for water in a district falls during the evening, the system automatically reduces the supply to conserve energy and water. Similarly, during peak time, the system can redistribute flow to high-demand zones to ensure service levels.

Sewage plants and water treatment plants are also automated. Sensors continuously monitor pH, dissolved oxygen, chemical levels, and microbial contamination. According to this information, actuators can real-time adjust chemical dosing, filtration rates, or aeration levels to ensure water quality requirements while optimizing resource utilization.

In flooded areas, smart drainage systems will open or close gates based on rain amounts and river levels. In times of emergency, automated alerts and response protocols could initiate pumping devices or disconnect affected components to prevent contamination or destruction. One of the new uses is the development of smart water grids, like smart electrical grids. Smart water grids employ a mixture of IoT sensors, control systems, and analytics to dynamically manage water delivery. Smart water grids facilitate decentralized sources like rainwater harvesting, greywater reuse, and local groundwater to be connected, in line with circular economy philosophy.

Remote control and automation are particularly beneficial in geographically isolated or inaccessible areas, where human oversight is not convenient. Automation reduces the workload, enhances safety, and allows 24/7 operation.

But to be "trustworthy," computer systems must have strong cybersecurity measures. They need to be designed with safeguards and calibrated and maintained regularly. AI control systems should also incorporate transparency features to allow decisions to be audited and verified. In brief, the IoT contribution to water resource management is an integrated platform of real-time monitoring, predictive analysis, and automation. All three work together to optimize the utilization of water, minimize loss, and pave the way for sustainable interventions. In the backdrop of rising water pressures globally, IoT becomes an important support base in driving the Blue Revolution, pushing more intelligent, more circular, and more inclusive water management systems.

3 Applications of IoT in Water Resource Management

3.1 Intelligent Water Grids

Smart water grids are a paradigm shift in the control of urban and regional water supply networks. In the past, water distribution has been a passive and predominantly linear process with restricted real-time awareness. Smart water grids, driven by IoT technologies, turn the system into a smart, responsive, and effective network.

Smart water grid refers to an organized system of digital flow meters, actuators, sensors, smart valves, and communication modules that collaborate in an integrative way for controlling and monitoring the entire process of water distribution. All these systems provide real-time information on key parameters such as water quality, pressure, and flow rate. By being capable of detecting anomalies such as pipeline leaks, pressure loss, and contamination events in real-time, smart water grids enable real-time intervention, thereby reducing water loss and preventing service disruptions. Arguably the most significant advantage of smart water grids is loss reduction and leakage detection. For most cities, vast quantities of treated water are lost along the way to end-users through eroding infrastructure and hidden leaks. IoT sensors deployed throughout the piping system can identify the exact spot where leaks occur so that maintenance teams can get there as quickly as possible. This kind of maintenance culture, which spares saves the life of the pipeline while largely reducing non-revenue water (NRW).

Second, smart water grids facilitate demand-side management. By providing customers with information about real-time actual consumption via smartphone apps or intelligent displays, smart water grids enable households and businesses to monitor consumption, set usage targets, and alert to anomalous usage behavior. This openness facilitates water savings and behavior modification, making consumers active participants in water sustainability.

Smart water grids with state-of-the-art technology are also integrated with predictive analytics systems, allowing the utility firms to forecast water demand, plan pumping, and control reservoir levels. This data-based approach lowers energy use and cost, making the water supply system more efficient.

Overall, smart water grids are the urban water management future—smart, green, and highly compatible with the tenets of the circular economy.

3.2 Precision Agriculture

Agriculture is the biggest user of freshwater in the world and is responsible for almost 70% of water withdrawals. In most countries, irrigational inefficiencies and water overuse have resulted in soil degradation, lower water tables, and decreased agricultural yields. Precision irrigation management based on science-led, data-driven practices on the grounds of IoT-based precision agriculture is transforming this reality.

Precision agriculture is defined as the use of IoT solutions to monitor, analyze, and manage soil and crop variability. The heart of the system is the soil moisture sensors that continuously check the water level in the soil. Strategically located throughout fields, the sensors are positioned to detect localized conditions, providing accurate and real-time information that is used for irrigation calendars. These sensors are complemented by automated weather stations and crop health monitors. Rainfall, temperature, humidity, wind speed, and solar radiation are measured by weather stations—principal variables that influence evapotranspiration rates and irrigation demands. Crop health is monitored using sensors and drone-mounted multispectral imagery, detecting stress conditions such as nutrient deficiencies, pest infestations, or water shortages.

This information is processed through cloud-based IoT platforms, and machine learning algorithms generate precise recommendations for irrigation. The platforms also facilitate automatic control of irrigation systems to guarantee the supply of water to the targeted crop, at the required level, and during the appropriate season. Such a high level of optimization significantly enhances water-use efficiency, enhances crop yields, and minimizes detrimental impacts on the environment, like leaching of nutrients and runoff.

Its benefits extend further than water conservation. By ensuring that the soil has appropriate moisture, precision agriculture obviates too much use of fertilizers and pesticides, reduces pumping energy consumption, and contributes to climate resilience by enhancing crop viability during droughts or uncharacteristic weather conditions. Furthermore, IoT agriculture solutions are also becoming cheaper and scalable, with solar-powered, GSM-based sensors suitable for smallholder farmers in developing countries. As these technologies mature further, they will be the driving force to make agriculture a water-efficient, sustainable enterprise.

3.3 Water Quality Monitoring

Water quality is a pillar of public health, environmental balance, and sustainable development. Traditional water quality monitoring techniques, while effective, are typically intermittent and manual, resulting in gaps in measurements and response delay in case of contamination. IoT-based water quality monitoring addresses this with its real-time and continuous monitoring. Smart water monitoring systems use a variety of sensors to monitor parameters such as pH, turbidity, electrical conductivity, dissolved oxygen, temperature, and some pollutants such as heavy metals, nitrates, or microbial pathogens. Sensors are placed at various points—water sources, treatment plants, distribution networks, and even customer taps—providing an end-to-end water quality view.

Such data received by sensors is transferred to the cloud or centralized databases via a wireless data communications protocol. Real-time analysis of data checks over this information and detects changes from tolerance limits the moment they are detected, signaling cases of potential contamination events at a glance. The immediate drop in dissolved oxygen and the immediate increase in turbidity, for example, might both indicate the onset of pollution, with immediate remedy. IoT systems enable early warning and rapid response, which are necessary in preventing waterborne diseases, conserving ecosystems, and adhering to regulations. For industrial effluents or urban stormwater runoff, IoT systems give confidence that the pollutants are identified before they are released into natural water bodies, protecting aquatic organisms and downstream consumers. One of the greatest benefits of IoT systems is that they are remotely controllable. In remote or inaccessible locations, such as mountains or distant rivers, solar-powered IoT devices can function independently, sending data to monitoring centers miles away. This improves monitoring in rural or disaster zones where regular monitoring is not feasible.

In addition, the integration of water quality data with geospatial mapping allows for monitoring of pollution sources, charting contamination spread, and identifying areas of intervention. Such data is indispensable for city planning, infrastructure planning, and public policy design. Lastly, IoT-enabled monitoring of water quality enhances accountability, responsiveness, and environmental responsibility in water management.

3.4 Flood Management

Floods are among the most frequent and destructive natural disasters, causing massive losses to life, property, and infrastructure. As climate change amplifies the variability of rainfall and urban growth raises impermeable surfaces, the threat of flooding is growing globally. IoT plays a transformative role in building effective, real-time flood management systems.

IoT-enabled flood monitoring is accomplished by establishing an array of hydrological and meteorological sensors to gather rainfall intensity, river and stream stage, soil water, and groundwater recharge data. Sensors yield real-time, high-resolution data that are critical for capturing and anticipating flood behavior.

For example, rain sensors and water level sensors placed in rivers and stormwater drains measure increasing water levels, while soil moisture sensors aid in estimating levels of saturation and the likelihood of surface runoff. Data from such sensors is sent to command centers and processed with predictive model tools.

Machine learning models and hydrological models analyze this data to predict the timing, severity, and likelihood of floods. Automatic systems have the capability to issue early warning alerts through SMS, sirens, apps, or community radio as soon as given thresholds are crossed. These alerts give critical lead time for evacuation, mobilization of resources, and protection of infrastructure.

Aside from predicting, IoT systems support real-time decision-making during flood events. Sluice gate control, drainage pump control, and diversion channel control can be automated to divert water flow and mitigate the effects on urban settlements. Dynamic management of smart reservoirs can optimize flood risk against storage requirements for water, avoiding downstream flooding while storing water for dry seasons.

After the event, IoT systems facilitate damage and recovery requirement evaluation. Flood extents are mapped using drones and sensor data, water contamination is measured, clean-up operations are directed, and timely and informed disaster management is facilitated.

With the fusion of IoT with GIS, satellite imagery, and emergency systems, cities and locations can develop intelligent flood resilience frameworks. These systems not only save lives and property but also minimize the economic loss and long-term recovery period after floods.

4 Benefits of IoT in Water Resource Management

4.1 Increased Efficiency

Perhaps the most direct and powerful advantage of bringing IoT into water resource management is increased operational efficiency. Traditional water management infrastructure is usually founded on manual checking, routine visits, and repair-based maintenance, which are all time-intensive, open to delays and mistake. IoT improves this process by allowing continuous real-time monitoring and smart automation.

Flow meters, pressure meters, and smart meters are IoT sensors that provide an endless stream of data with facts regarding every detail of the water distribution and consumption process. Policy influencers and utility operators can use the data to decide inefficiencies like pressure imbalance, leaks, and abnormal consumption levels almost in real time. For example, in the event of an unexpected loss of pressure that happens in a pipe, the system will alert maintenance personnel in good time, thus preventing unnecessary bursting or loss of water.

Smart irrigation technology in farming illustrates this efficiency even more. Plant needs, weather forecast, and moisture level in the soil are utilized to provide the exact amount of water at the right time without overwatering or underwatering. Precision irrigation not only conserves water but also yields healthy crops and higher crop yields.

Also, IoT provides predictive maintenance, where equipment performance history is utilized to forecast failure ahead of time. This prevents downtime, prolongs asset longevity, and results in a more stable water supply. Use of decision-support systems based on AI and machine learning also increases ease of operation, with the ability to reconfigure water distribution in real time depending on consumption patterns, thereby optimizing overall network efficiency.

4.2 Cost Savings

IoT in water management means virtually cost-cutting across the board. These come through various channels—energy saving, non-revenue water reduction, reduced manual labor, and improved asset management.

Non-revenue water (NRW) is treated water that is lost along the way to consumers, typically due to leaks, theft, or metering failures. Intelligent water networks with IoT sensors can detect and pinpoint leaks in real-time, enabling faster repair and minimizing water loss. This reduces costs directly and increases the amount of billable water.

Another key benefit is energy conservation. Millions of units of energy are used pumping, treating, and conveying water. IoT-driven optimization of the processes—i.e., streamlining pump speeds or operating them when off-peak energy times are available—can reduce energy expenses. For example, smart irrigation controls regulate water supply based on current needs and therefore prevent pumping beyond requirements and unnecessary power consumption.

In addition, IoT predictive maintenance prolongs infrastructure life and reduces the frequency and severity of failures. This saves on maintenance costs and prevents costly unplanned repairs. Farmers and utilities are more favored with improved stock management, minimal service disruption, and improved planning capability, all of which save them money. In addition, with automated reporting and data gathering, IoT eliminates the requirement for on-site visits and manual intervention, releasing human resources for more cerebral pursuits. The cumulative impact of all these efficiencies results in substantial cost savings and improved return on investment on water management infrastructure.

4.3 Environmental Sustainability

Environmental sustainability is a shared goal in the age of climate change and limited natural resources, and IoT plays an important part in guaranteeing that this goal is achieved in water management. In promoting the wisdom of utilizing water and minimizing environmental footprint, IoT inspires the spirit of the circular economy and sustainable development.

IoT networks help in achieving sustainability by reducing wastage of water through real-time monitoring and control. Unnecessary use of water and leakages can be easily detected and eliminated, avoiding unnecessary withdrawal from nature. Freshwater conservation is very important, particularly in drought-prone and population-pressure areas.

In farming, IoT-based precision irrigation significantly reduces river and aquifer withdrawals. Through the supply of water where and when needed, the systems avert soil salinization, overdraft, and leaching of nutrients—programs that ruin soil structure and pollute downstream water bodies.

Additionally, IoT allows for the prevention of pollution. Pollutants can be detected in real-time by sensors installed in water bodies and wastewater treatment plants, allowing for immediate response before the contaminants diffuse. This preserves aquatic life, promotes biodiversity, and ensures safety for downstream consumers. Furthermore, smart water systems promote efficient use of energy, which in turn reduces greenhouse gas emissions from water treatment and pumping. Interdependent water and energy, maximizing one tends to maximize the other.

By enabling an evidence-based approach to policymaking for the environment and water, IoT facilitates long-term sustainability. With correct environmental information in real-time, there is improved regulation, evidence-based policymaking, and public awareness—key components of sustainable management of water resources.

4.4 *Improved Public Health*

Water is central to human health, and IoT technology further supports public health protection through ensuring water safety, reliability, and quality. Routine and sometimes reaction-based water quality monitoring methods have a low ability for quick response to contamination. IoT redefines this convention by virtue of continuous real-time monitoring of water quality parameters.

Intelligent sensors can detect pH, turbidity, dissolved oxygen, heavy metals, and microbial pollution and alert immediately when water quality crosses safety limits. This detection provides the utilities with a quick response to post public advisories, close contaminated sources, or reroute clean water, before waterborne disease has a chance to spread.

In the third world or disaster scenarios in which centralized water distribution might get disrupted, portable IoT-based testing kits and sensors can monitor drinking water sources in real-time. These types of systems are especially important in humanitarian environments, where there is limited access to clean water and public health risks are high. IoT also plays an important role in the treatment of wastewater. Through monitoring the quantity and quality of sewage, these systems ensure that untreated sewage does not find its way into natural watercourses, reducing the spread of pathogens and poisonous chemicals. IoT allows sophisticated treatment plants to regulate ideal processing conditions in order to shield the environment as well as humans. In addition to acute contamination detection, IoT facilitates long-term epidemiological research and health planning. Linking water quality information to healthcare data can show disease outbreak patterns, which public health officials can use in planning more effective prevention and education plans.

Finally, enhanced transparency and openness to water quality information enhance citizen trust and participation. Citizens have the potential to be enabled to more effectively observe the safety of their water system, aid conservation, and ensure improved infrastructure—a healthier, smarter society in any case.

5 Challenges and Barriers

5.1 *High Initial Costs*

One of the main obstacles to the widespread use of IoT in water resource management is the enormous amount of initial capital involved. A solid IoT system entails a lot of hardware procurement and installation, such as sensors, smart meters, gateways, and communication modules. The equipment must be strategically deployed throughout water infrastructure networks—distribution networks, treatment plants, reservoirs, agricultural fields, and natural bodies of water—to gather a full complement of real-time data. Besides hardware, there is a need for high-end data analytics platforms, cloud storage, and cybersecurity solutions, all of which add to the capital expenditure.

Integrating such aspects into existing water management systems tends to involve extensive upgrades or retrofitting, adding to the costs.

For most water utilities, especially in low- and middle-income nations, such spending can be daunting. Organizations in the public sector usually run on lean budgets, and justifying significant expenditures in new technology can be daunting, particularly where the return on investment is in the long-term and indirect sense. In addition, funding mechanisms for IoT projects are still being developed. Unlike conventional infrastructure projects that might be funded through familiar institutions, IoT deployments might demand new funding institutions like public–private collaborations, performance-based deals, or donor-funded schemes.

To bridge cost-related hurdles, the stakeholders will have to seek cost-sharing models, modular and scalable solutions, and policy incentives like tax incentives or subsidies. Having unambiguous cost–benefit analyses and potential pilot success can also lend weight to the potential for wider implementation.

5.2 Data Privacy and Security

The application of IoT in water management leads to the generation of humongous amounts of sensitive information, such as trends in water usage, weaknesses of infrastructure, environmental observations, and user actions. Though valuable for water systems optimization, the data also poses substantial security and privacy threats.

These consist of cyber assaults in the guise of hacking, data robbery, and denial-of-service attacks, which can disrupt essential infrastructure, resulting in general service disruptions or exposing information that feeds decision-making processes. For public utilities, a cyber-attack can result in a contaminated water supply, unauthorized access to control systems, or even manipulation of billing systems.

Privacy is also inherent in the case, particularly if consumption data is associated with individual homes or businesses. In the wrong hands, that data could be used for surveillance, discrimination, or profit. In order to avoid such attacks, solid cybersecurity architectures should be incorporated at the design phase of IoT installations. This involves the application of encryption, secure communication protocols, multi-factor authentication, and routine vulnerability scanning. Water utilities should also have firm data governance policies that lay out who can access data, how it will be stored, and how long it will be kept.

Compliance with international data protection law, including the General Data Protection Regulation (GDPR), is most critical for the confidence of the public. Building cyber defense procedures' capabilities and ongoing monitoring of the threat landscape will also give some resilience.

5.3 Interoperability and Standardization

The IoT ecosystem is characterized by high variability between devices, manufacturers, standards, and data types.

While these enable innovations, this level of variation creates enormous interoperability and standardization challenges. Merging different systems and devices onto a common platform for managing water in proportion as a whole becomes difficult and expensive if the standards do not converge. For example, a meter from company A might not be compatible with an analytics platform from company B, causing inefficiencies or expensive custom integration options. This type of fragmentation makes sharing data, interagency coordination, and scaling up or replicating successful methods elsewhere more difficult.

Interoperability also applies to support and maintenance. Utilities can find themselves stuck in proprietary stacks, more and more reliant on particular vendors and less able to make adjustments to support new technologies. To address these issues, open-source standards and industry-standard protocols must be created and implemented in a hurry. Organizations such as the International Telecommunication Union (ITU), the Institute of Electrical and Electronics Engineers (IEEE), and the Open Geospatial Consortium (OGC) are currently attempting to standardize data models, security methods, and communication protocols for IoT.

Close collaboration between governments, technology providers, and standardization organizations has to be established in order to enable harmonization. Building national or regional IoT architectures and schemes for regulation can also increase interoperability, limit vendor lock-in, and speed up digital transformation in the water industry.

5.4 Technical Expertise

Installation and operations of IoT devices for water resource management involve an interdisciplinary competence spanning hardware design, application programming, data analysis, and computer security. Specifically, these high-technology products cannot be easily installed and operated effectively by water companies or agronomic businesses, with some exceptions. Knowledge gaps are especially wide in rural or resource-poor environments, where little IT infrastructure exists and few technical experts.

Even in urban settings, the rapid pace of development of IoT technologies necessitates ongoing training and reskilling to maintain compatibility with new tools and methods. Insufficiently skilled professionals yield suboptimal system designs, inefficient use of data collected, and increased risk of cyber-attacks. Furthermore, without proper expertise, organizations might not gain useful insights from data analytics platforms due to reduced return on investment in IoT. To fill this lacuna, coordinated efforts must be made to create local capacity via education, training programs, and knowledge transfer systems. Collaborative efforts with the academic community, vocational training facilities, and development agencies can serve to develop an IoT-capable workforce.

Moreover, intuitive interfaces, automation software, and managed service models can reduce the technical hurdle to adoption. Vendors and system integrators will also need to invest in end-to-end customer support and post-sales service to guide organizations through the transition process. Ultimately, developing technical skills is not solely about technology—it is about creating an ecosystem that fosters innovation, resilience, and sustainability in water resource management. Policy-makers and governments need to appreciate the strategic value of developing digital skills and incorporate this as a top priority in national water and digital transformation plans.

6 Future Prospects

6.1 Synergy with Next-Generation Technologies

Combining IoT with other advanced technologies is likely to transform water resource management in deep ways. Artificial Intelligence (AI), blockchain, and 5G are among the most promising enablers that are likely to enhance the impact of IoT further.

Artificial Intelligence (AI): AI can sort through vast quantities of data from IoT sensors and convert raw information into useful intelligence. For example, AI programs can analyze usage patterns for water, identify abnormalities that show evidence of leaks, and make extremely accurate forecasts for peak demand. Machine learning routines can learn over time based on historical records, such that dynamic, adaptive, and self-adjusting water management is achievable. AI can also be utilized to make decisions in real-time, such as modifying valve operations or modifying water flow based on predictive data.

Blockchain Technology: The transparency and integrity of data are critical to water management, particularly in the multi-stakeholder and regulatory contexts. Blockchain offers a decentralized ledger to record every data point and transaction with time stamps and cryptographic hashes. This enables water consumption data, quality data, and billing information to be traceable and tamper-proof. It can also offer backend support for the installation of smart contracts for water trading or conservation incentive programs, which can build trust between parties.

5G Networks: 5G brings low latency, high capacity, and bandwidth to enable a tremendous number of devices to be connected simultaneously. This matters for large-scale IoT deployment in high-density cities or vast agricultural fields. 5G supports real-time data transmission from dispersed sensors, allowing for instantaneous monitoring and control. It also enhances the responsiveness and reliability of autonomous systems such as drones and robotic water quality monitors.

The convergence of these technologies will not only increase the capability of IoT systems but also hasten the shift towards data-driven and adaptive water management practices. Pilot projects and public–private partnerships will be critical in proving feasibility and iterating on integrated solutions.

6.2 Smart Cities and Water Management

The smart city vision imagines urban cities that are effective, robust, and sustainable, with technology infused in each layer of infrastructure and service provision. Water management is a foundation of the vision, and IoT is the common tissue that weaves it together with other city systems. In smart cities, Internet of Things-enabled water networks exchange information seamlessly with energy networks, transportation systems, and waste treatment facilities. For instance, a smart city may change irrigation schedules in open spaces dynamically depending on weather predictions and the availability of water supply in order to avoid wastage. Interconnected dashboards may provide a common view for all utility services to the managers of the city so that they can design an integrated reaction to occurrences like pipe bursts or flooding. Smart meters in residential and commercial buildings enable real-time measurement of water usage, enabling customers to make decisions and detect leaks in time. Water conservation can be promoted by municipalities through the use of dynamic pricing and gamification schemes based on consumption patterns.

Moreover, IoT enables water resource fair delivery in urban areas. Big data can trace out underserved regions, guide infrastructure investment, and render water services inclusive and resilient. With climate variability, smart water networks can enhance resilience by supporting adaptive measures such as groundwater storage, rainwater collection, and decentralized treatment.

The more cities turn to smart technologies, the lower the costs of IoT water solutions will be, driving further innovation. Open data standards and interoperable platforms will be key to integration among systems and stakeholders.

6.3 Global Cooperation and Knowledge Sharing

The water crisis is a cross-boundary one that impacts ecosystems, economies, and human health across the world. It can be solved only through collective effort based on the joint wisdom and experience of nations, institutions, and communities. IoT is the key to facilitating such international cooperation and sharing of knowledge.

With IoT, nations can have transboundary monitoring networks for their common water bodies like rivers, lakes, and aquifers. Real-time flow rate information, pollution levels, and usage patterns can shape joint management decisions and encourage cooperation. It is particularly important for areas with a water conflict or a low-water resource history. IoT-based global data platforms may also be used as case studies, best practices, and innovations. Consider, for example, a precision irrigation pilot that has succeeded in Israel and can be used to advise on similar initiatives in India or sub-Saharan Africa. Open-source software and data sets may share technology with the masses, and low-income countries can sidestep the usual infrastructure hurdles.

Global institutions like the United Nations, World Bank, and UNESCO can leverage IoT data to track progress toward the attainment of Sustainable Development Goals (SDGs), particularly Goal 6: Clean Water and Sanitation. Standardized indicators of IoT systems can enhance the accuracy and comparability of reporting across nations. Research networks of collaboration and online forums can put researchers, engineers, policy-makers, and community leaders in touch. Webinars, open conferences, and e-training modules can spread information rapidly, enhancing capacity in the most vulnerable areas. Low-cost IoT kits, which make citizen science projects possible, also enable local communities to participate in monitoring and safeguarding water assets.

However, international cooperation also demands respect for ethical and legal aspects, including data sovereignty, intellectual property, and access equity. Governance mechanisms must be established to ensure that data sharing is reciprocal and responds to the rights of participants in an equivalent manner. The future of the Internet of Things in water management is not tech but is forging wiser, more sustainable, and more inclusive futures. As the new technologies converge, smart cities are empowered, and nations collaborate, IoT can be an effective driver in making water security in the twenty-first century and beyond a reality.

7 Conclusion

The Internet of Things (IoT) is a revolutionary force in the field of water resource management, redefining the rules governing how water is monitored, distributed, and conserved. As global water scarcity continues to rise, as populations swell, and as climate-related issues, IoT is the solution because it allows for a data-driven, automatic, and very efficient way of using and dispensing water. IoT infrastructure enables real-time monitoring by networked smart devices and sensors that offer valuable information on system performance, usage, quality, and water flow. Such streams of data enable water authorities and utility companies to identify inefficiencies, leakages, and contamination in real time.

Predictive analytics based on IoT platforms enable proactive prediction and maintenance to provide a more assured and uninterrupted supply of water. The use of IoT in precision farming, intelligent water networks, and water quality monitoring has led to tremendous benefits. Water conservation and enhanced crop yields are made possible through optimized irrigation based on real-time needs as per the soil moisture and weather forecasts. Urban residents are aided by smart meters and dynamic pricing, promoting proper use. IoT-based water quality sensors also promote public health by enabling early warning systems for contamination threat detection and enabling corrective action to be taken in a timely manner. Nevertheless, various issues need to be tackled in order to realize the entire capability of IoT in this area. Exorbitantly high costs of deployment are a major hurdle, particularly for less developed countries. The difficulty in sustaining cybersecurity alongside data privacy within massive-scale IoT systems cannot be underestimated. Moreover, non-cross-platform compatibility and the non-availability of IoT installation experts are compelling concerns that work against large-scale adoption. Governments, the private sector, and civil society need to join forces to create enabling regulatory environments, invest in capacity development initiatives, and advocate for open standards for device interoperability for development to occur. Public–private partnerships and international cooperation can also break cost barriers.

In the coming times, the combination of IoT with emerging technologies like AI, blockchain, and 5G will further revolutionize water resource management. These synergies will facilitate having more responsive, resilient, and scalable solutions tailored to different sets of environmental and social conditions. Smart cities will increasingly depend on IoT in an attempt to balance cross-pressures on water and adjust to climate risks. In fact, IoT is not merely an enabler but the pivot itself of the Blue Revolution—its revolutionary revolution towards sustainable, equitable, and smart water management. By appropriating its capabilities and surmounting the hurdles in its present path, IoT can assume the central role for ensuring long-term water security as well as developing the agenda for a circular economy.

References

Arshad B, Ogie R, Barthelemy J, Pradhan B, Verstaevel N, Perez P (2019) Computer vision and IoT-based sensors in flood monitoring and mapping: a systematic review. Sensors 19(22):5012. https://doi.org/10.3390/s19225012

Ashish T, Arush J, Arun KS, Pushpa C, Krishna KM, Prem CV (2022) Chapter 9—The internet of things in agriculture for sustainable rural development. https://doi.org/10.1016/B978-0-12-823694-9.00012-8

Ashton K (2009) That 'internet of things' thing. RFID J 22(7):97–114

Food and Agriculture Organization (2017) Water for sustainable food and agriculture: a report produced for the G20 Presidency of Germany. FAO

Hossain MS, Muhammad G (2016) Cloud-assisted industrial internet of things (IIoT)—Enabled framework for health monitoring. Comput Netw 101:192–202. https://doi.org/10.1016/j.comnet.2016.01.009

Jan F, Min-Allah N, Düştegör D (2021) IoT based smart water quality monitoring: recent techniques, trends and challenges for domestic applications. Water 13(13):1729. https://doi.org/10.3390/w13131729

Helmbrecht J, Pastor J, Moya C (2017). Smart solution to improve water-energy nexus for water supply systems. https://doi.org/10.1016/j.proeng.2017.03.215

García-Martín JP, Torralba A, Hidalgo-Fort E, Daza D, González R (2023) Carvajal,IoT solution for smart water distribution networks based on a low-power wireless network, combined at the device-level: a case study. Internet of Things. https://doi.org/10.1016/j.iot.2023.100746

Krishnan SR, Nallakaruppan MK, Chengoden R, Koppu S, Iyapparaja M, Sadhasivam J, Sethuraman S (2022) Smart water resource management using artificial intelligence—A review. Sustainability 14(20):13384. https://doi.org/10.3390/su142013384

Lee I, Lee K (2015) The Internet of Things (IoT): applications, investments, and challenges for enterprises. Bus Horiz 58(4):431–440. https://doi.org/10.1016/j.bushor.2015.03.008

Singh M, Ahmed S (2021) IoT based smart water management systems: systematic review, Vol. 46. Part 11. https://doi.org/10.1016/j.matpr.2020.08.588

Pranav H, Prasad G, Rohan S (2025) IoT-based smart irrigation system for efficient water use in agriculture. J Clean Prod 321:128981.https://www.ijprems.com/uploadedfiles/paper//issue_3_march_2025/39370/final/fin_ijprems1743335187.pdf

Satilmisoglu TK, Sermet Y, Kurt M, Demir I (2024) Blockchain opportunities for water resources management: a comprehensive review. Sustainability 16(6):2403. https://doi.org/10.3390/su16062403

Syed TA, Khan MY, Jan S, Albouq S, Alqahtany SS, Naqash MT (2024) Integrating digital twins and artificial intelligence multi-modal transformers into water resource management: overview and advanced predictive framework. AI 5(4):1977–2017. https://doi.org/10.3390/ai5040098

Turskis Z, Šniokienė V (2024) IoT-driven transformation of circular economy efficiency: an overview. Math Comput Appl 29(4):49. https://doi.org/10.3390/mca29040049

Lakshmikantha V, Hiriyannagowda A, Manjunath A, Patted A, Basavaiah J. Anthony A (2021) IoT based smart water quality monitoring system. https://doi.org/10.1016/j.gltp.2021.08.062

World Bank (2020) Water security and the global water agenda. World Bank Publications

Blockchain-Based Solutions for Water Resource Management and Tracking Using Smart Contracts and IPFS

Suseta Datta, Rajdeep Roy, Sourav Banerjee, and Utpal Biswas

Abstract Water resource management is a critical issue globally, requiring efficient tracking and accountability. Traditional methods of managing water usage often face challenges related to transparency, data integrity, and resource allocation. Blockchain, along with smart contracts and decentralized storage systems like IPFS (InterPlanetary File System), presents innovative solutions to address these challenges. This chapter explores the use of Blockchain for transparent and immutable water resource management by securely recording water usage data. Smart contracts can automate compliance with water policies, enforce penalties for violations, and incentivize sustainable water practices. Water data can be stored on a decentralized storage system known as IPFS to ensure the integrity and accessibility of information while preventing data tampering. By integrating these technologies, real-time tracking and verification of water usage across industries and regions becomes possible, encouraging accountability and enhancing water conservation efforts. Blockchain-based solutions can significantly improve water resource management practices by providing a secure, transparent, and automated framework to ensure sustainability and better allocation of water resources. This chapter discusses the potential benefits, challenges, and practical applications of using Blockchain, smart contracts, and IPFS in the context of water sustainability. Finally, the chapter concludes with some future directions.

S. Datta
Department of Computer Application, Narula Institute of Technology, Kolkata, India
e-mail: suseta.datta@nit.ac.in

R. Roy (✉) · U. Biswas
Department of Computer Science and Engineering, University of Kalyani, Kalyani, India
e-mail: rrajdeep2@gmail.com

U. Biswas
e-mail: utpalbiswas@klyuniv.ac.in

S. Banerjee
Department of Computer Science and Engineering, Alipurduar Government Engineering and Management College, Alipurduar, India
e-mail: mr.sourav.banerjee@ieee.org

V. Kandpal and T. Choudhury (eds.), *Blue Revolution: Sustainable Water Use and Circular Economy Innovations*, Urban Sustainability,
https://doi.org/10.1007/978-981-95-1916-3_10

Keywords Blockchain · IPFS · Smart contracts · Sustainability · Water resource management · Water tracking

1 Introduction

One of the most important resources is water among all the resources on the planet [24], taking part in an imperative role in agriculture, industry, and most importantly in daily life. So far, rising in population, climate fluctuation, and unsustainable consumption patterns have revolved efficient water management into a global challenge. In the past few years, the focus on sustainable water management applications has built up, operated by mounting pressure on freshwater systems [1]. Regardless of the progress in optimizing water utilization, subsisting monitoring and tracking techniques continue to combat matters such as limited transparency, susceptibility to fraudulence, and complications in imposition. At the core of these difficulties is the deficiency of a secure, transparent, and automated framework to inspect water usage over different sectors. To take care of these situations, decentralized technologies are appearing as a transformative solution for revamping water management. With their potential to impart real-time tracking, data security, and operational efficiency, decentralized systems detain eminent assurance. Blockchain technology [50], put together with smart contracts [29] and decentralized storage tools like the InterPlanetary File System (IPFS) [49], offers practical answers to these acute issues. Blockchain can reliably record invariable water usage data, while smart contracts can compel sustainability regulations instinctively. In the meantime, IPFS makes certain decentralized, transparent, and effortlessly attainable storage of water-related information. When integrated, these upheavals have the capability to build a more explicable, efficient, and sustainable system for handling global water resources.

1.1 Current Works in Water Resource Management

Over the past few years, water resource management [14] has accomplished remarkable progress, with multiple initiatives and technologies coming out to improve water conservation and efficiency, specifically in regions scuffling with water shortage. Revolutions such as advanced irrigation methods, water recycling systems, and smart metering devices have been launched to identify these situations [4]. In spite of these developments, many of these solutions face hindrances like unreliable data, limited transparency, and weak enforcement technologies. Standard water monitoring techniques, for instance, frequently pivot on manual data collection and centralized databases, assembling them susceptible to inaccuracies, data tampering, and reporting delays. In addition, the lack of coordination and integration amongst different water-using sectors often results in unacceptable allocation and management of resources. Even though diverse technologies have been initiated to enhance water use efficiency, the convoluted nature of water systems and the incrimination of multiple stakeholders make carrying out fair and consistent management a complicated task. Developments in IoT, artificial intelligence, and machine learning have definitely advanced data gathering and analysis in this domain, so far these systems frequently remain centralized, uncovering data to potential manipulation and inaccuracies. As a result, there is an increasing demand for solutions that can deliver transparent, reliable, and tamper-resistant records of water consumption.

1.2 Challenges in Water Resource Management

In defiance of remarkable advancements in water management, the sector still combats various notable challenges [15]. A prime issue lies in the lack of transparency adjacent to how water resources are observed and dealt out. Centralized systems typically fail to offer real-time perceptions into water usage, making it complex for stakeholders to accurately evaluate the effectiveness of management strategies. These systems are also exposed to manipulation and fraudulence, mainly when data reporting relies on manual processes. Another preserving problem is the ineffective allocation of resources. Standard water distribution models are often encountered to optimize usage over various sectors, with agriculture being a foremost example. Water use in farming is customarily wasteful due to the lack of accurate measurement and control technologies. Adding to these troubles, climate change continues to augment water shortages, causing inconsistent weather conditions and lengthening shortages that put further strain on already inadequate supplies. Ultimately, securing regulations and clinching compliance residues a considerable hindrance within conventional systems. Water management policies are frequently passed over because there are no automated methods to track and carry out them. Without definitive enforcement tools, both organizations and individuals can circumvent consequences for consuming water or splitting environmental standards [21].

1.3 Decentralized Systems for Addressing Water Management Issues

The requirement for higher transparency, data integrity, and efficiency in water resource management can be identified through the use of decentralized systems. Standard centralized databases frequently face issues associated with data security, accessibility, and transparency. By moving towards a decentralized model, water usage data can be recorded and stored in a distributed approach, ensuring that all stakeholders have access to real-time, explicit information. Decentralized systems offer a remarkable advantage over standard centralized models by providing greater data security [42]. Since Blockchain is an immutable ledger, once data is recorded, it cannot be tampered with or amended. This makes it particularly convenient for tracking water resources, as all transactions (water usage, allocation, etc.) can be upheld and validated by all parties involved, clinching trust and accountability. Moreover, decentralized systems can entitle real-time observation and management of water resources. Using IoT sensors and smart meters, water usage can be tracked constantly, and data can be stored on a decentralized network, ensuring its integrity and availability to all admissible stakeholders [33]. This type of system also eases automation through smart contracts, which can be used to activate actions such as penalties for imprudent water use or benefits for conservation efforts, clinching compliance with water sustainability goals.

1.4 Blockchain, Smart Contracts, and IPFS for Water Resource Management

The collaboration of Blockchain, Smart Contracts, and IPFS accords sturdy solutions to the ongoing issues in water resource management [28]. Blockchain permits transparent, tamper-proof, and secure recording of water usage data. Its decentralized ledger makes sure that once data is entered, it remains immutable and even more convenient, giving all stakeholders a definitive and verifiable record of water consumption. Smart contracts augment this system by automating the imposition of water regulations. These self-driven agreements, presided over predefined rules, can automatically implement actions such as applying penalties for overexploitation or granting rewards for conservation efforts on the basis of actual usage data. By abolishing the requirement for intermediaries, smart contracts efficient administrative processes and intensify the effectiveness of policy enforcement [3]. In addition, IPFS offers a decentralized procedure for storing essential water-related details, inclusive of consumption logs, compliance records, and quality assessments. In comparison to centralized cloud storage, IPFS distributes data over multiple nodes, upgrading security and making the system more resistant to data manipulation. This decentralized storage model clinches the integrity and availability of critical information, permitting all participants' approach to accurate, verified data.

1.5 Organization of the Chapter

The chapter is organized as follows: Sect. 2 instigates the latest works and challenges in water resource management. It traverses subsisting methods for water tracking and addresses key issues such as data transparency, resource allocation, and compliance imposition. Section 3 searches into the problems and incompetencies in current water management systems, pointing up the requirement for solutions that offer higher transparency, security, and automation. Section 4 talks through decentralized systems, inclusive of Blockchain, and how they identify many of the issues identified in the foregoing sections. The role of decentralized technology in enhancing water management is thoroughly examined. Section 5 provides an in-depth analysis of how Blockchain, Smart Contracts, and IPFS work together to offer inventive solutions for managing water resources efficiently and sustainably. Real-world examples and use cases are considered in this section. Section 6 culminates the chapter by encapsulating the major points discussed and delimiting future directions for the incorporation of Blockchain and decentralized technologies in water resource management.

2 Current Works and Challenges in Water Resource Management

Water resource management remains a notable global concern, as increasing water scarcity and inefficient usage continue to affect key sectors such as agriculture, industry, and household consumption. Effectively managing and observing water resources entails addressing a range of persistent issues, such as sparse data access, lack of transparency, inefficient allocation, and feeble implementation of regulations. Although several techniques have been implemented to enhance water management, many still fail to provide real-time, precise, and applicable discernments. This section searches into the latest systems and practices in place to manage water resources, addressing the vital challenges and analytic gaps that must be labelled to enhance overall efficiency.

2.1 Traditional Water Resource Management Systems

Standard water resource management systems have long served as the premise for looking after global water supply and demand. These systems mainly depend on centralized databases to manage information related to water usage. Although, an essential impediment is their incapacity to offer real-time observing, along with constricted access to data for stakeholders. In many occurrences, water usage is documented manually, which not only causes reporting delays and imprecisions but also opens the access to potential fraudulence or tampering with the data [39]. Another major deficiency is the splitted nature of water management over different sectors. Areas like agriculture, urban development, industry, and domestic consumption are frequently governed by dissimilar bodies, resulting in poor coordination and inefficient resource distribution [11]. A fragmented strategy to water management frequently results in allocation decisions made without a holistic understanding, leading to shortages in critical sectors. In addition, many conventional systems do not incorporate sophisticated tools like predictive analytics, which are crucial for forecasting demand and optimizing resource distribution. While some standard frameworks have begun adopting technologies such as smart meters and IoT sensors, their effectiveness is often hindered by poor integration and the absence of standardized protocols. Furthermore, centralized systems remain susceptible to data breaches and manipulation. As a result, there is an urgent requirement for more advanced solutions that provide real-time data access and support automated processes to address these inefficiencies more efficiently.

2.2 Issues with Data Transparency and Integrity

A major impediment in managing water resources effectively is the absence of transparency and consistency in data management. Many existing systems face problems related to data manipulation, where records are altered to conceal inefficiencies conservation successes. This undermines trust amongst stakeholders and complicates decision-making at both governmental and institutional levels. Besides, with most water usage data housed in centralized systems, it becomes extremely susceptible to cyber threats, tampering, and unauthorized access. Limited availability and access to accurate water data only augment the situation. In many parts of the world, especially in developing nations, water-related data [13] is not voluntarily available to the public. This lack of receptiveness averts local communities, Non-governmental organization (NGOs), and other stakeholders from actively taking part in monitoring and decision-making processes regarding water management. Additionally, the absence of standardized formats for recording and sharing water data creates hindrances to cross-regional collaboration and the exchange of best practices. As well, water quality data [44] is frequently either not collected or retained from the public, preventing communities and industries from gaining condemnatory perceptions into the health of their water resources. This paucity of visibility can lead to long-term environmental degradation and public health concerns. Improving transparency within water resource management systems would highly strengthen the eminence of water-related information and empower better, more sustainable decision-making for equitable resource distribution.

2.3 Challenges in Resource Allocation and Efficiency

Constructive water resource distribution [19] continues to be a significant challenge, distinctly in areas facing expanding water scarcity. Recurrently, water is given out without an inclusive understanding of the overall demand over various sectors, leading to excessive use in some areas while others are left inadequately serviced. A common example is agriculture, where former irrigation practices and the absence of real-time consumption data contribute to excessive water use. In the meantime, urban centers may suffer from water scarcity even when remaining resources exist close to underutilized industrial zones. An utmost contributing factor to this polarity is the lack of precise data regarding water availability and consumption needs. Governments and private organizations often make allocation decisions based on incomplete information, resulting in inefficient use of resources. In addition, political and economic impacts can lead to needful management and uneven water distribution, frequently adverse to susceptible populations. In some cases, water is averted to sectors with least influence on public welfare, like industries using water for non-essential resolutions. Over misallotment, the inefficiency of the latest water management systems further inhibits conservation efforts. Many agricultural operations still depend on antiquated irrigation technologies, which consume large quantities of water in spite of more efficient unconventional being available. Likewise, many water treatment amenities lose a considerable amount of water due to former processes. To improve water distribution and maximize conservation, integrating advanced technologies like IoT sensors [32], real-time analytics, and predictive modeling into water management systems is significant.

2.4 Regulatory and Compliance Enforcement Challenges

Making certain water management strategies and regulations remains a significant obstacle within the sector [40]. Many latest frameworks deficit robust monitoring and enforcement tools, making it challenging to ensure compliance to water conservation laws and standards. Standard strategies usually depend on manual inspections and self-reporting, which can result in errors, delays, and a general lack of accountability. In a lot of developing nations, infirm regulatory structures and limited resources further impede the government's ability to compel water management effectively. Factors like corruption, political instability, and deficient enforcement infrastructure contribute to indigent implementation, leading to imprudent extraction and wasteful use of water resources. Non-compliance is specifically extensive in agriculture and manufacturing, where water-intensive activities often are monitored without proper supervision. The lack of automated systems to monitor usage and enforce regulations makes it difficult to endorse water management policies consistently. Moreover, with the evolution of smart contracts, there is an opportunity to automate enforcement processes. Smart contracts can launch automatic penalty and reward mechanisms based on real-time water usage data, promoting higher compliance with sustainability goals in a transparent and reliable manner [5].

2.5 *Integration of Technological Innovations in Water Management*

Technological progress holds enormous contracts for transforming water management practices. Even though many countries and organizations have started implementing smart water meters, IoT sensors, and data analytics tools, these upheavals have yet to be fully made up into existing water management structures. The imploded nature of current systems and the lack of interoperability between technologies present major hindrances to achieving maximum efficiency. While some areas have launched smart water grids for real-time monitoring, the lack of standardized protocols continues to limit their scalability and complete effectiveness. A further significant impediment is the adoption of these new technologies by several stakeholders. High initial costs associated with deploying advanced water tracking systems, along with the expense and effort required for training, often make them inaccessible to developing regions and small businesses. In addition, the failure to develop a coherent strategy for incorporating emerging technologies such as Blockchain, smart contracts, and AI-powered analytics—into water management frameworks [34] keeps systems disunited and inefficient. Identifying these issues will require communal initiatives lined up at developing standardized platforms that can consistently incorporate new technologies into existing infrastructures. Blockchain and decentralized technologies, for instance, offer a sturdy foundation for securing, making transparent, and universally sharing water data, ultimately driving greater innovation and efficiency over the sector [37].

3 Problems and Inefficiencies in Current Water Management Systems

Existing water management systems are perforated with challenges that obstruct their ability to effectively set about water scarcity, revamp resource allocation, and upgrade sustainability. These deficiencies are especially distinct in regions facing limited water availability or relying on former management practices. An insufficiency of transparency, security, and automation often leads to unproductive observation, mismanagement, and significant resource wastage. This section explores the cardinal inefficiencies within the system and points up the critical need for innovative strategies to address them. Prioritizing real-time data acquisition, deploying secure monitoring systems, and embracing automated compliance tools can greatly enhance the transparency and effectiveness of water management. Moreover, integrating advanced technologies like Blockchain and smart sensors presents viable solutions to these issues, encouraging more sustainable and efficient water usage practices.

3.1 Lack of Real-Time Data and Monitoring

One of the vital challenges in contemporary water management lies in the deficiency of real-time data and continuous surveillance. Many latest systems continue to rely on outdated or manual data collection methods, resulting in delayed reporting and potential inaccuracies in data analysis. The lack of up-to-date information impedes the ability of water authorities to promptly identify issues such as supply shortages, contamination, or system inefficiencies. As water networks become more complex and demand continues to grow, real-time monitoring has become essential. In agriculture, for instance, water usage is frequently recorded retrospectively, making it arduous to track consumption patterns or make timely adjustments to irrigation strategies. Likewise, in urban settings, leaks within water distribution systems can remain undetected for long periods, resulting in major water losses. Traditional monitoring lacks the capacity to deliver instant insights, causing significant inefficiencies and wastage. The paucity of real-time information also obstructs swift and informed decision-making. Authorities may not have access to major data regarding water reserves, quality, or distribution patterns, detaining critical interventions. The incorporation of smart meters, IoT sensors [20], and advanced data analytics could transfigure real-time water tracking [23]. These technologies offer explicit and immediate information, accrediting water managers to base their decisions on current data, inflate distribution efficiency, and hold up conservation initiatives.

3.2 Data Integrity and Security Risks

Making certain data integrity and security remains a prime obstacle in latest water management systems. Frequently, water consumption data is put away in centralized databases that are susceptible to manipulation, cyber threats, or unauthorized access. In the deprivation of robust security measures, data can be tampered with, forged, or even erased, leading to unreliable information that erodes sound decision-making. This, in turn, compromises the effectiveness of water management strategies and can result in indigent policy choices and inefficient resource distribution. The threat of data violations adds another layer of concern. Many water management platforms operate on former infrastructure that lacks modern security features, revealing both consumer personal data and delicate industrial water usage information to potential risks. As digital technologies become more unified into water systems, the importance of conserving this data has become even more pressing [7]. Moreover, limited transparency in data reporting makes the situation worse. Major stakeholders such as regulators, communities, and businesses often cannot access the critical water usage data needed to ensure accountability and avert mismanagement. Appearing technologies like Blockchain [18] present a promising solution by creating immutable, transparent records of water data, giving all parties access to the same reliable information. Recognized for its decentralized design, Blockchain also strengthens data security by eliminating single points of failure, diminishing the perils of manipulation and cyberattacks.

3.3 Inadequate Resource Allocation and Mismanagement

Numerous existing water management systems face significant issues with ineffective resource distribution. Water is frequently improperly allocated to agriculture and industry sectors, while urban populations and marginalized communities endure from shortages [6]. This imbalance is largely due to former or inaccurate data on water consumption, imprecise predicting models, and insufficient coordination amongst water users. In several cases, water allocation decisions are made based on authentic usage patterns rather than current demands or environmental needs. As a result, regions may experience over-allocation during deficit conditions or insufficient distribution during times of affluence. Water wastage [10] is also common when industries receive more water than necessary or use it inefficiently, without considering long-term sustainability. Besides, in many countries, water pricing structures do not truly reflect the actual value of water. Subsidized rates frequently encourage excessive consumption, as neither individuals nor industries are given strong incentives to use water responsibly. To identify these inefficiencies, advanced data analytics and predicting tools are essential for accurately divining water needs and guiding fairer allocation strategies. By integrating automated systems capable of real-time water consumption tracking and environmental monitoring, more effective and sustainable water distribution practices can be attained.

3.4 Insufficient Enforcement of Regulations and Policies

Making certain water management policies and regulations continues to be a major challenge over different regions. Even though governments and water agencies have established rules to promote sustainable water usage, these regulations are regularly implemented inconsistently or insufficiently. A notable contributing factor is the absence of robust monitoring systems competent of effectively tracking compliance. Customarily, enforcement varies broadly between sectors and locations, leading to often violations and unnecessary water wastage. In the agricultural sector, for instance, irrigation practices are repeatedly left unchecked, resulting in hacking of water resources. Likewise, industries that discharge untreated wastewater into rivers or lakes frequently avoid significant penalties due to inadequate oversight and monitoring of their waste disposal practices. These challenges often peduncle from limited financial and human resources, poor governance, or corruption within the water management system. Inventive technologies like Blockchain and smart contracts offer viable solutions to identify these problems. Smart contracts can automate the implementation of water regulations by instantly imposing penalties for violations or offering incentives for environmentally responsible behaviour. Blockchain technology [9] makes certain that all activities are securely documented and fully traceable, thereby boosting transparency and accountability. By incorporating these technologies into existing water management frameworks, governments and institutions can greatly enhance their capacity to enforce and monitor compliance more effectively.

3.5 High Operational Costs and Resource Inefficiencies

Operational expenses associated with water management often place a heavy burden on both governments and industries [12]. These costs encompass infrastructure upkeep, data collection, system monitoring, and regulatory enforcement. Many conventional water management models rely heavily on manual labour, including physical meter readings, data input, and on-site inspections, which drive up operational expenditures. In addition, outdated and inefficient water distribution networks further escalate costs. Water losses from leaks and deteriorating infrastructure are especially prevalent in older urban centre and rural regions. The financial demands of repairing and maintaining aging systems can be substantial, and inefficiencies in water treatment facilities often lead to wasted water and energy, compounding overall expenses. Furthermore, the absence of real-time data and analytics frequently results in poor resource allocation [25], where surplus water is misdirected while other areas experience shortages. The adoption of advanced technologies—such as IoT-enabled sensors and predictive analytics—offers a solution to reduce these operational costs. By enabling real-time monitoring and automated detection of leaks or inefficiencies, smart systems help minimize waste and lower maintenance expenses. These technologies also enhance the performance of water treatment plants by optimizing energy and resource consumption.

4 Decentralized Systems and Their Role in Enhancing Water Management

Decentralized technologies like Blockchain are emerging as significant breakthroughs in the field of water resource management. As previously discussed, traditional centralized systems often face challenges such as inefficiency, weak data security, and limited transparency. Decentralized solutions address these issues by enhancing transparency, strengthening security, enabling automation, and ensuring improved regulatory adherence. Blockchain, in particular, safeguards data integrity and allows real-time access to information, reducing the likelihood of human error and minimizing the risk of fraud or mismanagement. Furthermore, smart contracts can streamline essential processes and enforce regulations automatically, eliminating the need for third-party intervention and increasing operational efficiency. These decentralized frameworks also encourage broader involvement by empowering local communities and stakeholders to participate. This section explores how Blockchain and other decentralized innovations can effectively overcome the shortcomings of current water management systems, supporting improved transparency, accountability, and more strategic decision-making.

4.1 Blockchain for Transparent and Immutable Water Data

A major hindrance in water management today is the deficiency of transparency and the vulnerability of data integrity. Standard systems typically depend on centralized databases for storing and handling information, making them susceptible to manipulation, unauthorized alterations, or even data loss. Blockchain technology offers a potent solution by providing a decentralized, immutable ledger that enables secure and transparent tracking of water usage. Through Blockchain, water consumption records can be securely documented, ensuring that all stakeholders have access to reliable and tamper-proof data [46]. The decentralized nature of Blockchain creates a key advantage of adding transparency in the system. Every transaction is recorded over multiple nodes within the network, making it nearly impossible to modify or delete any piece of information once entered. This covenants the authenticity and reliability of water usage data over various levels whether it's individuals, industries, or government bodies. For instance, Blockchain can ease real-time tracking of water consumption, delegating users to observe their usage patterns and their environmental impact [30]. Industries, nevertheless, can improve their sustainability efforts by publicly sharing verified water usage data on the Blockchain, building up their corporate accountability. As well, Blockchain supports instant access to data for water managers, regulatory authorities, and the wider public. This open visibility into water distribution systems promotes greater accountability and enables more effective, data-centric decisions in managing water resources.

4.2 Smart Contracts for Automated Water Management and Compliance

The integration of smart contracts into decentralized systems instigates greater automation and coherence to water management operations. Smart contracts are self-driven agreements with their conditions encoded directly into software, permitting automatic enforcement without the implication of third parties. Applied to water management, smart contracts can rationalize compliance with regulations, monitor water consumption, and upgrade sustainable usage practices. For occurrence, in the case of irrigation management, smart contracts can compel water usage limits instinctively, responding to intended factors such as weather patterns or current water supply levels. If certain thresholds are crossed, the smart contract can instantly inflict penalties for immoderation or offer incentives for conservation, lessening the requirement for manual oversight and diminishing the chances of human error or delays in enforcement [36]. In addition, smart contracts empower peer-to-peer exchanges within water markets [17], where individuals and businesses can securely and transparently deal with water rights. This capability is specifically valuable in areas facing water scarcity, as it upgrades more efficient and impartial distribution of water resources through market-oriented mechanisms. By immersing rules directly into the system and automating imposition, smart contracts remarkably boost the effectiveness and reliability of water management frameworks.

4.3 Decentralized Autonomous Organizations (DAOs) for Community-Based Water Management

Decentralized Autonomous Organizations (DAOs) [45] present a transformative model for reimagining water management, particularly in community-based projects. DAOs function as decentralized governance systems where all stakeholders share in the decision-making process, eliminating the need for central authorities or intermediaries. This model enables local communities, water users, and other participants to engage directly in the management of water resources and have a meaningful impact on policy and planning. By leveraging DAOs, communities can address local water management challenges through open, transparent, and decentralized collaboration [27]. DAOs allow members to collectively make important decisions regarding water allocation, pricing, and distribution, conducted by both community input and data-driven analysis. Blockchain technology ensures that every decision and vote is permanently recorded, creating an open and empiric ledger of all community actions and agreements. Additionally, DAOs can uplift responsible water use by implementing incentive systems. Community members could earn rewards for preserving water or face penalties for overuse, all governed automatically by the DAO's smart contracts. Because controlling power rests with those directly affected, DAOs stimulate stronger community engagement, accountability, and a profound sense of ownership over local water management applications.

4.4 Distributed Ledger Technology for Water Quality Monitoring and Enforcement

Behind tracking water usage, observing water quality is another major facet where decentralized technologies can make a significant impact. Standard water quality monitoring systems typically rely on centralized sensors and reporting mechanisms, which are often susceptible to inaccuracies, data tampering, or information loss. Distributed Ledger Technology (DLT) [35], the foundation of Blockchain, offers a solution by creating a transparent, secure, and tamper-proof record of water quality data congregated from multiple sources, such as sensors installed in water bodies or treatment facilities. Integrating Distributed Ledger Technology (DLT) with real-time water quality monitoring enables sensor data to be securely stored on a Blockchain, guaranteeing that the information remains tamper-proof and easily verifiable. This level of transparency facilitates the quick detection of irregularities or regulatory breaches, boosting public confidence in water safety and ensuring that industries are held responsible for discharging wastewater into natural bodies of water. Moreover, smart contracts implemented within Blockchain networks can automate actions based on water quality data. For example, if water quality falls below established standards, a smart contract can instantly notify relevant authorities or activate appropriate treatment processes [16]. Such automation reduces the risk of human error and enables rapid response, thereby enhancing the efficiency and dependability of water quality management systems.

4.5 Enhancing Water Resource Allocation and Distribution with Blockchain

Blockchain technology introduces a groundbreaking method for improving the allocation and distribution of water resources. Traditional water management approaches usually rely on manual processes or centralized systems, which often operate without real-time insights or transparency. This leads to inefficiencies, misallocation, and excessive wastage of water. By exploiting Blockchain, water consumption and distribution can be monitored in real time, providing transparent and accessible data over vital sectors such as agriculture, industry, and urban utilities [26]. A Blockchain-executing system can facilitate the flawless tracking and exchanging of water rights, ensuring that resources are directed where they are most required and used most efficiently. Smart contracts can be programmed to automatically endorse allocation rules, pledging that water distribution aligns accurately with sector-specific and regional requirements. In areas scuffling with acute water scarcity, Blockchain can support the evolution of decentralized water markets, enabling industries and individuals to buy and sell water rights in a fair, open, and efficient environment. Through decentralization, Blockchain [41] improves the transparency, equity, and responsiveness of water resource management, making sure that users pay a fair price and receive a proper share based on the latest demand and availability.

5 Blockchain, Smart Contracts, and IPFS for Efficient and Sustainable Water Resource Management

This section examines the integration of Blockchain, Smart Contracts, and the IPFS in addressing the expanding need for efficient and sustainable water resource management. While each technology offers distinct advantages, their integration yields a powerful solution to the challenges considered in previous sections. Blockchain offers secure and transparent data tracking, smart contracts, accredit automation and execution of water management rules, and IPFS allows for decentralized storage of major data and documents. Together, these technologies offer a complete approach that can corroborate how water resources are managed more efficiently, with a focus on sustainability, transparency, and accountability.

5.1 Blockchain and Smart Contracts for Automating Water Resource Allocation

Blockchain technology incorporated with smart contracts can revolutionize the automation of water resource allocation and distribution, offering a substantial upgrade across standard methods [22]. Water allocation typically involves several steps, starting from gathering data on available resources to determining how much water should be distributed to sectors like agriculture, industry, and urban areas. When these processes are managed manually, they often suffer from inefficiencies, human errors, and a lack of transparency, resulting in flawed distribution. Smart contracts offer a way to automate these tasks by embedding predefined conditions within their code, which can be triggered by real-time data such as weather forecasts, reservoir capacities, and sector-specific demands. When these conditions are met, the smart contract executes the distribution of water based on established guidelines. For example, if a drought is forecasted in a certain region, the smart contract could automatically restrict water usage for non-essential purposes, prioritizing vital sectors like agriculture. Exploiting Blockchain technology [9] ensures that each water distribution transaction is securely logged on an immutable, tamper-resistant ledger. This inherent transparency allows all stakeholders to audit and validate the fairness and effectiveness of distribution decisions. By integrating Blockchain and smart contracts into the process, water management becomes more streamlined, transparent, and adaptable, minimizing human error and reducing the risk of resource misallocation.

5.2 IPFS for Secure and Decentralized Storage of Water Data

IPFS presents an inventive approach to securely and transparently storing and sharing water management data. Standard storage systems frequently depend on centralized servers, making them susceptible to data breaches, loss, and unauthorized alterations. Such susceptibilities trade-off the coherence of essential water-related information, inclusive of consumption records, treatment activities, and environmental monitoring data. By embracing IPFS, water management systems can hold a decentralized, distributed storage model that improves both security and pliability. On the IPFS network, data is divided into smaller fragments and spread over multiple nodes, ensuring continued accessibility and protection even if certain nodes fail or are come to terms. This distributed architecture eradicates single points of failure, guaranteeing that condemning water data remains consistently secure, available, and tamper-resistant. Additionally, IPFS entitles the creation of transparent, verifiable records that any network participant can access. Stakeholders can simply obtain information on water usage, treatment protocols, and quality assessments, promoting greater transparency and accountability throughout the water management process [31]. This level of openness is important for ensuring adhesion to sustainability goals and holding industries and government bodies responsible for their water management practices.

5.3 Real-World Use Case: Blockchain in Agricultural Water Management

Agriculture reckons for a substantial portion of global water consumption, making it an important sector for improving water resource efficiency. Blockchain technology integrated with smart contracts offers a powerful solution for remodelling agricultural water management, stimulating sustainability, and diminishing wastage [47]. Along Blockchain, distinct farm water usage can be precisely tracked and recorded on a tamper-proof ledger. Smart contracts can be depicted to automate irrigation schedules by utilizing real-time weather data and information on water availability. With Blockchain's immutable archival, water consumption can be audited at any time, confirming farms observe with administrative standards and that water distribution remnants fair and efficient. Besides, Blockchain can transfigure the management of water rights in agriculture. Farmers could easily buy, sell, or allocate water rights amongst themselves based on immediate needs, using a transparent and decentralized system. In the event of droughts or water scarcities, smart contracts could automatically impede water usage and grade distribution to critical crops required for food security. This adaptable, real-time strategy to water allocation helps maximize efficiency, lessen waste, and support the sustainable usage of water resources in farming.

5.4 *Blockchain for Transparency in Water Quality Monitoring*

Blockchain technology can notably inflate water quality monitoring systems by creating a transparent, immutable, and auditable record of water quality data. Standard monitoring systems frequently depend on centralized organizations to converge, test, and report water quality, leaving room for errors, manipulation, or false reporting. By making use of Blockchain [2], water quality data can be securely logged in real-time on a decentralized ledger, making certain a constant and reliable record of water conditions over different locations and water bodies. For example, sensors placed in rivers, lakes, or reservoirs could automatically feed data into the Blockchain, pledging that the information is protected from tampering and remains intact. This imparts a reliable record for regulatory authorities, environmental groups, and the public. In addition, smart contracts can be immersed to automatically compel water quality standards. If water quality drops below an established threshold, a smart contract can set off alerts or initiate corrective actions, like activating treatment systems or notifying the responsible authorities. This real-time monitoring and response system ensures that water stays safe for consumption and use, effectively preventing contamination and safeguarding public health.

5.5 *Scaling Blockchain and IPFS for Global Water Management Systems*

Although Blockchain and IPFS offer innovative solutions for localized water management, scaling these technologies to support global water systems poses significant challenges. Expanding their application on a worldwide level requires addressing issues such as data interoperability, network congestion, and the need for consistent regulatory frameworks across different regions. With proper optimization, Blockchain can be adapted to handle the vast amounts of data generated by global water systems. Solutions like layer-2 scaling [38] and sharding [48] are currently being explored to enhance Blockchain's performance, allowing for faster transaction speeds and reduced network congestion. Likewise, IPFS, with its decentralized and distributed architecture, is competent of storing large volumes of water related data on a global scale, but it must be incorporated with other data systems to clinch smooth interoperability between diverse platforms. By endorsing global Blockchain networks based on decentralized models, nations can collaborate more effectively, sharing data about water resources and usage. This would surface the way for a global framework for monitoring and managing water use. As more regions implement these technologies, incorporating Blockchain and IPFS can lead to a unified [43], worldwide technique for improving water sustainability, minimizing inefficiencies, and enhancing resource allocation over the borders.

6 Conclusion and Future Work

In this culminated section, we outline the vital substance from the chapter, showing how Blockchain, smart contracts, and IPFS dispense inventive solutions to the challenges in water resource management. Throughout the chapter, the barriers of conventional water management systems have been examined, and the potential of decentralized technologies to enhance efficiency, transparency, and sustainability has been highlighted. From automating water distribution through smart contracts to firmly storing water data with IPFS, these technologies lay the preliminaries for more efficient and accountable water usage over the sectors named agriculture, industry, and urban management. Blockchain espousal can ensure that water related data is kept in an immutable, transparent format, while smart contracts can automate the application of water distribution and quality regulations. IPFS offers a scalable, decentralized method for storing large amounts of water data, conserving its integrity and accessibility. At once, these technologies generate a robust framework that enables more effective water resource management, offering stakeholders accurate, real-time data to produce better decisions. Concentrating on the near future, the global integration of Blockchain and decentralized technologies in water management is on the scope. Challenges around scalability and system interoperability must still be identified, but as these technologies unfold, a rise in practical implementations is expected, particularly in regions facing water scarcity and mismanagement. The integration of these technologies into smart water grids and automated systems holds the potential to transform the management of this pivotal resource, making water usage more efficient, equitable, and sustainable for future generations.

Acknowledgements R.R expresses sincere gratitude to Mr. Ravi Jagannathan (CEO), Mr. Mohit Sethi (ED and CTO), Mr. Karthik Mohan (VP-Product and Strategic Programs), and the development team at KrypC Technologies Private Limited, address: 606, 1st Main Rd, IAS Colony, 06th Sector, HSR Layout, Bengaluru, Karnataka, India, PIN- 560102, for their generous support and contributions toward the preparation of this book chapter.

Disclosure of Interests No competing interests

References

1. Ahuja S (2017) Overview: sustaining water, the worlds's most crucial resource. Chemistry and Water 1–22. https://doi.org/10.1016/B978-0-12-809330-6.00001-5
2. Alharbi N, Althagafi A, Alshomrani O, Almotiry A, Alhazmi S (2021) A blockchain based secure IoT solution for water quality management. In 2021 international congress of advanced technology and engineering (ICOTEN). IEEE, pp 1–8, July 2021. https://doi.org/10.1109/ICOTEN52080.2021.9493474
3. Al-Zoubi A, Aldmour M, Khoury A, Al-Thaher D (2024) Blockchain of things for securing and managing water 4.0 applications. Int J Online & Biomed Eng, 20(11). https://doi.org/10.3991/ijoe.v20i11.50277

4. Amir A, Fauzi R, Arifin Y (2022) Smart water meter for automatic meter reading. In IOP Conference series: materials science and engineering (Vol. 1212, No. 1, p. 012042). IOP Publishing. https://doi.org/10.1088/1757-899X/1212/1/012042
5. Arafat SM (2024) A study of blockchain applications in the water sector. https://doi.org/10.13140/RG.2.2.33809.44645
6. Bai Y, Wang L, Hu R, Li D (2024) Equal allocation, demand priority or negotiated allocation? How to allocate water resources in the Tigris and Euphrates River Basin efficiently. Heliyon, 10(11). https://doi.org/10.1016/j.heliyon.2024.e31458
7. Balta DD, Kaç SB, Balta M, Oğur NB, Eken S (2025) Cybersecurity-aware log management system for critical water infrastructures. Appl Soft Comput 169:112613. https://doi.org/10.1016/j.asoc.2024.112613
8. Banerjee S, Das D, Biswas M, Biswas U (2020) Study and survey on blockchain privacy and security issues. In Cross-industry use of blockchain technology and opportunities for the future. IGI Global, pp 80–102. https://doi.org/10.4018/978-1-7998-3632-2.ch005
9. Banerjee B, Sen U, Pramanik S, Banerjee A (2025) Smart water governance with blockchain technology: blockchain water management. In Convergence of blockchain, internet of everything, and federated learning for security. IGI Global Scientific Publishing, pp 311–344. https://doi.org/10.4018/979-8-3373-1424-2.ch011
10. Bani R (2011) Wastewater management. In Waste water-Evaluation and management. IntechOpen
11. Belay M, Bewket W (2013) Traditional irrigation and water management practices in highland Ethiopia: case study in Dangila woreda. Irrig Drain 62(4):435–448. https://doi.org/10.1002/ird.1748
12. Blokus-Dziula A, Dziula P, Kamedulski B, Michalak P (2023) Operation and maintenance cost of water management systems: Analysis and optimization. Water 15(17):3053. https://doi.org/10.3390/w15173053
13. Cantor A, Kiparsky M, Hubbard SS, Kennedy R, Pecharroman LC, Guivetchi K, Darling G, McCready C, Bales R (2021) Making a water data system responsive to information needs of decision makers. Frontiers in Climate 3:761444. https://doi.org/10.3389/fclim.2021.761444
14. Cosgrove WJ, Loucks DP (2015) Water management: current and future challenges and research directions. Water Resour Res 51(6):4823–4839. https://doi.org/10.1002/2014WR016869
15. Dinar A (2024) Challenges to water resource management: the role of economic and modeling approaches. Water 16(4):610. https://doi.org/10.3390/w16040610
16. Dogo EM, Salami AF, Nwulu NI, Aigbavboa CO (2019) Blockchain and internet of things-based technologies for intelligent water management system. Artificial Intelligence in IoT 129–150. https://doi.org/10.1007/978-3-030-04110-6_7
17. Donaghy K, Li Z, Rezwan FU (2024) Water ledger: a blockchain-based smart water resource management platform. In International symposium on distributed ledger technology. Singapore: Springer Nature Singapore, pp 119–124, November 2024. https://doi.org/10.1007/978-981-96-4442-1_8
18. El-Moustaqim K, Mabrouki J, Hmouni D (2025) How to ensure data integrity in water monitoring networks using IoT and blockchain technology. In AI-Enhanced solutions for sustainable cybersecurity. IGI Global Scientific Publishing., pp 77–96. https://doi.org/10.4018/979-8-3693-6371-3.ch004
19. Esenarro D, Lescano J, Chalco B, Tapia N, Vilchez Cairo J, Vargas Beltran C, Arriola C, Ruiz Reyes R (2024) Spatial, functional, and constructive analysis of the water resource at the archaeological Center of Tipon, Cusco, Peru, 2023. Heritage 7(12):6629–6656. https://doi.org/10.3390/heritage7120307
20. Geetha S, Gouthami SJSW (2016) Internet of things enabled real time water quality monitoring system. Smart Water 2:1–19. https://doi.org/10.1186/s40713-017-0005-y
21. Gourbesville P (2008) Challenges for integrated water resources management. Physics and Chemistry of the Earth, Parts a/b/c 33(5):284–289. https://doi.org/10.1016/j.pce.2008.02.002

22. Huseynov F, Mitchell J (2024) Blockchain for environmental peacebuilding: application in water management. Digit Policy, Regul GovAnce 26(1):55–71. https://doi.org/10.1108/DPRG-06-2023-0080
23. Iancu G, Ciolofan SN, Drăgoicea M (2024) Real-time IoT architecture for water management in smart cities. Discov Appl Sci 6(4):191. https://doi.org/10.1007/s42452-024-05855-9
24. Kılıç Z (2020) The importance of water and conscious use of water. Int J Hydrol 4(5):239–241. https://doi.org/10.15406/ijh.2020.04.00250
25. Krishnan SR, Nallakaruppan MK, Chengoden R, Koppu S, Iyapparaja M, Sadhasivam J, Sethuraman S (2022) Smart water resource management using Artificial Intelligence-A review. Sustainability 14(20):13384. https://doi.org/10.3390/su142013384
26. Liu Y, Shang C (2022) Application of blockchain technology in agricultural water rights trade management. Sustainability 14(12):7017. https://doi.org/10.3390/su14127017
27. Miller B (2021) Application of blockchain capabilities to the management of water rights and water markets. Information Systems Water and Catchment Group Department of Environment Land Water and Planning Victoria, Australia: Victoria, Australia
28. Mughal MH, Shaikh ZA, Ali K, Ali S, Hassan S (2022) IPFS and blockchain based reliability and availability improvement for integrated Rivers' streamflow data. IEEE Access 10:61101–61123. https://doi.org/10.1109/ACCESS.2022.3178728
29. Muralidharan S, Ko H (2019, January). An InterPlanetary file system (IPFS) based IoT framework. In 2019 IEEE international conference on consumer electronics (ICCE). IEEE, pp 1–2, January 2019. https://doi.org/10.1109/ICCE.2019.8662002
30. Naqash MT, Syed TA, Alqahtani SS, Siddiqui MS, Alzahrani A, Nauman M (2023) A blockchain based framework for efficient water management and leakage detection in urban areas. Urban Science 7(4):99. https://doi.org/10.3390/urbansci7040099
31. Nododile T, Nyirenda C (2025) A hybrid blockchain-IPFS solution for secure and scalable data collection and storage for smart water meters. arXiv preprint arXiv:2502.03427
32. Pagano A, Garlisi D, Tinnirello I, Giuliano F, Garbo G, Falco M, Cuomo F (2025) A survey on massive IoT for water distribution systems: challenges, simulation tools, and guidelines for large-scale deployment. Ad Hoc Netw 168:103714. https://doi.org/10.1016/j.adhoc.2024.103714
33. Pincheira M, Vecchio M, Giaffreda R, Kanhere SS (2020). Exploiting constrained IoT devices in a trustless blockchain-based water management system. In 2020 IEEE International conference on blockchain and cryptocurrency (ICBC) (pp 1–7). IEEE, May 2020. https://doi.org/10.1109/ICBC48266.2020.9169404
34. Rokaya M, Hemdan DI, Alajmani SH, Alyami RY, Elmarhomy G, Hashim H, Atlam ES (2025) Optimized AI and IoT-Driven framework for intelligent water resource management. IEEE Access. https://doi.org/10.1109/ACCESS.2025.3572067
35. Romero Ugarte JL (2018) Distributed ledger technology (DLT): introduction. Banco De Espana Article 19:18
36. Sarantakos T, Amaxilatis D, Pagano A, Garlisi D, Taloma RL, Cattai T, Chatzigiannakis I, Vythoulka V, Zaroliagis C (2024) A tool to facilitate the design of smart contracts in smart water distribution networks. In 2024 IFIP Networking Conference (IFIP Networking). IEEE, pp 708–713, June 2024. https://doi.org/10.23919/IFIPNetworking62109.2024.10619818
37. Satilmisoglu TK, Sermet Y, Kurt M, Demir I (2024) Blockchain opportunities for water resources management: a comprehensive review. Sustainability 16(6):2403. https://doi.org/10.3390/su16062403
38. Sguanci C, Spatafora R, Vergani AM (2021) Layer 2 blockchain scaling: A survey. arXiv preprint arXiv:2107.10881
39. Sharma A, Ji S (2024) Linkages between Traditional Water Systems (TWS) and Sustainable Development Goals (SDGs): A case of Govardhan. India. Social Sciences & Humanities Open 9:100816. https://doi.org/10.1016/j.ssaho.2024.100816
40. Singh BJ, Chakraborty A, Sehgal R (2023) A systematic review of industrial wastewater management: evaluating challenges and enablers. J Environ Manage 348:119230. https://doi.org/10.1016/j.jenvman.2023.119230

41. Singh BJ, Singh R, Rout C (2024) Next-generation water resource management: leveraging blockchain and IoT. Introduction to Sustainable Solution Techniques in Civil and Environmental Engineering Science, 107
42. Sundaresan S, Kumar KS, Kumar TA, Ashok V, Julie EG (2021) Blockchain architecture for intelligent water management system in smart cities. In Blockchain for smart cities. Elsevier, pp 57–80. https://doi.org/10.1016/B978-0-12-824446-3.00006-5
43. Thakkar DS, Thakker A, Gupta R, Jadav NK, Tanwar S, Pau G, Sharma G, Alqahtani F, Tolba A (2024) Blockchain-Orchestrated intelligent water treatment plant profiling framework to enhance human life expectancy. IEEE Access 12:49151–49166. https://doi.org/10.1109/ACCESS.2024.3384607
44. Uddin MG, Nash S, Olbert AI (2021) A review of water quality index models and their use for assessing surface water quality. Ecol Ind 122:107218. https://doi.org/10.1016/j.ecolind.2020.107218
45. Wang S, Ding W, Li J, Yuan Y, Ouyang L, Wang FY (2019) Decentralized autonomous organizations: concept, model, and applications. IEEE Transactions on Computational Social Systems 6(5):870–878. https://doi.org/10.1109/TCSS.2019.2938190
46. Xia W, Chen X, Song C (2022) A framework of blockchain technology in intelligent water management. Front Environ Sci 10:909606. https://doi.org/10.3389/fenvs.2022.909606
47. Xiong H, Dalhaus T, Wang P, Huang J (2020) Blockchain technology for agriculture: applications and rationale. Frontiers in Blockchain, 3, 7. https://doi.org/10.3389/fbloc.2020.00007
48. Yu G, Wang X, Yu K, Ni W, Zhang JA, Liu RP (2020) Survey: Sharding in blockchains. IEEE Access 8:14155–14181. https://doi.org/10.1109/ACCESS.2020.2965147
49. Zheng Z, Xie S, Dai H, Chen X, Wang H (2017) An overview of blockchain technology: Architecture, consensus, and future trends. In 2017 IEEE international congress on big data (BigData congress). Ieee, pp 557–564, June 2017. https://doi.org/10.1109/BigDataCongress.2017.85
50. Zheng Z, Xie S, Dai HN, Chen W, Chen X, Weng J, Imran M (2020) An overview on smart contracts: challenges, advances and platforms. Futur Gener Comput Syst 105:475–491. https://doi.org/10.1016/j.future.2019.12.019

Smart Urban Management: Innovations, Policies and Future Directions for Sustainable Cities

Priyanka Gupta, Manu Sharma, Kapil Ahalawat, and Sudhanshu Joshi

Abstract Existing issues in urban water management have worsened with increased urbanization coupled with increasing climate change impacts. The application of new innovations based on emerging technologies and data processing methods allows cities to develop innovative solutions to environmental issues. New water infrastructure depicted revolutionary changes with new digital technologies, including Internet of Things (IoT), Artificial Intelligence (AI), Digital Twins, and Edge Computing. The analysis comprises policy mechanisms for facilitating appropriate technology deployment in addition to governance and outreach options for the people. The chapter also analyzes the integration of smart water grids and predictive analytics technology for real-time observation, in addition to leak detection, quality management, and demand forecasting functions. The chapter also highlights that circular economy models need to go hand in hand with the inclusivity of stakeholders in all projects. Lastly, the chapter offers a strategic model that assists municipal planners, along with utility management staff and research professionals, in coming up with water systems that enable smart cities and resilience standards for climate.

Keywords Smart cities · Urban water management · Artificial Intelligence (AI) · Water sustainability · Climate resilience · Smart infrastructure

P. Gupta (✉) · M. Sharma · K. Ahalawat
Graphic Era Deemed to Be University, Dehradun, India
e-mail: Priyankagupta.mgt@geu.ac.in

M. Sharma
e-mail: Manu.sharma@geu.ac.in

K. Ahalawat
e-mail: Kapilahalawat.mgt@geu.ac.in

S. Joshi
Doon University, Dehradun, India
e-mail: sudhanshujoshi@doonuniversity.ac.in

V. Kandpal and T. Choudhury (eds.), *Blue Revolution: Sustainable Water Use and Circular Economy Innovations*, Urban Sustainability,
https://doi.org/10.1007/978-981-95-1916-3_11

1 Introduction

Water resource management receives the most significant impacts from infrastructure development. The management of urban water systems presents complex operations for sustainable solutions because age-deteriorated pipelines need to support expanding water needs within unpredictable weather patterns alongside environmental contamination [23]. Cities currently face a challenging water management situation because their aging infrastructure must meet expanding demand within uncertain climate conditions and environmental pollution [28]. The concept of smart cities has come into the limelight, offering a technological framework in which Internet of Things (IoT), Artificial Intelligence (AI), and Analytics are integrated to tackle problems of the urban environment in real time. The structure of smart water systems includes sensor networks and automatic control systems with predictive analysis components, which enhance services while lowering water losses while maintaining system operations [17]. Through the combination of Digital Innovation, policy platforms, and citizen engagement, smart cities are able to develop strong, equitable, and circular water systems. The chapter creates a strategic framework for city planners and policymakers regarding the connection between urbanization and urban water governance with digital technologies.

Technical solutions alone are insufficient to manage urban water systems because they need to unite elements of social order with economic requirements and institutional mechanisms. Several governance deficits appear alongside inadequate utility funding, as well as insufficient education about water conservation, in developing world urban areas. The design of efficient smart water solutions has dual requirements for operational efficiency improvement and equal treatment of all parties and environmental preservation in the long run. Achieving desired outcomes from inclusive transformation demands that institutions use digital solutions that both support governmental reforms and incorporate public engagement policies.

Smart water systems introduce a fundamental redesign of urban planning practices which entails improved water resources management. Smart water governance employs real-time monitoring along with participatory planning and predictive analytics as main elements to develop city resilience. The methods underpin early warning, facilitate demand-responsive supply networks, and inform decision-makers with fine-grained, scalable insights. For this reason, the deployment of smart technologies should be understood as being integral to a larger systemic shift one that ties in with the values of sustainable urban development, climate resilience, and the circular economy. This chapter will seek to unravel these interdependencies and provide strategic advice on how to harness digital transformation to resolve the urban water problem on equitable and forward-looking terms.

1.1 Overview of Urbanization Trends and Water Management Challenges

The rate of urbanization surpasses all previous levels recorded in human history. According to the United Nations population prediction, by 2050, almost 70% of worldwide residents will live in cities, starting from the current 55%. The massive migration of population creates unbearable stress on urban infrastructure because water systems form the backbone of public health, along with economic productivity and environmental sustainability. The current water infrastructure is advanced in years, and the distribution is unbalanced, while leaks persist and contamination rises because of climate-related water scarcity [8]. The lack of control during urban growth leads to insufficient sanitation and inadequate stormwater management that heighten pollution risks in crowded population centres [6].

The effects of climate change intensify the amount of stress water management systems encounter in urban areas. The occurrence of severe weather disasters has increased, which causes significant disturbances when sectors try to meet supply and demand requirements. Higher temperatures lead to shrinking water resources, but unpredictable rainfall patterns increase water evaporative losses, which then raise operational costs alongside pumping and water treatment expenses.

Separate governance units that manage Urban areas fail to coordinate proper water management since their operations stay independent from one another. Different municipal units that handle water supply, stormwater services, and sanitation cannot coordinate decisions because they operate independently. Various sectors in water management operate independently, which creates challenges for planners to construct sustainable plans and blocks the development of new smart water systems [3]. Rapid urbanization causes several linked water problems in cities because of changing social conditions and environmental patterns that demand novel innovative approaches combined with holistic methods.

1.2 Introduction to Smart City Concepts and Their Relevance to Water Management

Urban development follows the Smart City model, which combines high-end information and communication technologies (ICTs) to optimize urban service performance and enhance its quality. Water pressure and flow measurements provided by IoT sensors enable immediate detection of leaks as well as environmental contaminations within the water system. Artificial intelligence algorithms predict water usage patterns to optimize water processing and the city planners use data analytics tools for long-term water resources assessments [5]. Tourism controllers in Singapore and Copenhagen demonstrate leading performance. The "NEWater" system, which recycles wastewater for national water needs, receives optimizations through AI-based analytics in Singapore alongside IoT sensing in Copenhagen to cut down

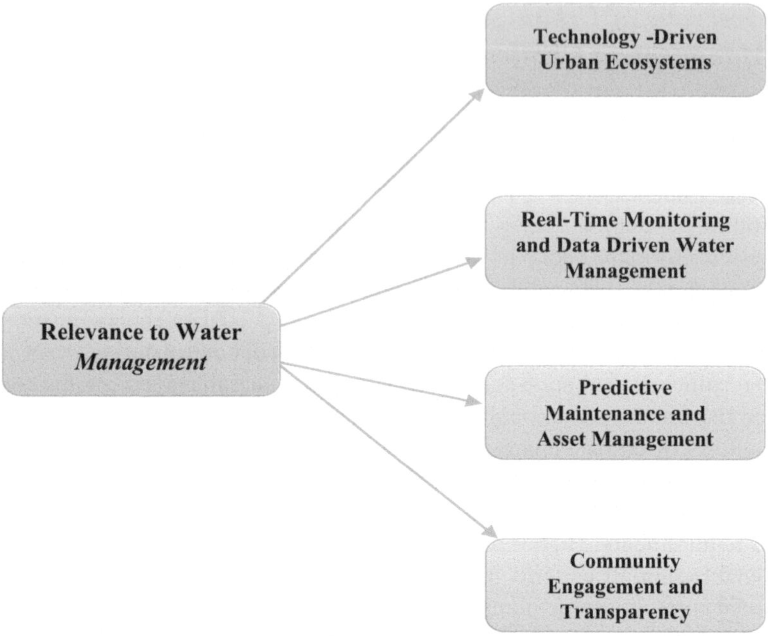

Fig. 1 Smart cities and its relevance to water management

water loss rates and boost leak monitoring [14]. Smart cities and its relevance to water management as shown in Fig. 1:

a. **Smart Cities as a Technology-Driven Urban Ecosystems:** Both infrastructure and public utilities services in urban systems integrate key technologies, which include Internet of Things (IoT), Artificial Intelligence (AI), and use cloud computing and Geographic Information Systems (GIS). Digital advancement takes precedence in water supply and treatment along with distribution systems [1].
b. **Real-Time Monitoring and Data Driven Water Management:** Alerts regarding water system leakages can be delivered in real time because of the gathered information. Through the capabilities utilities accomplish lower water loss rates while improving customer services together with system upgrade decisions made thanks to evidence-based reasoning [22].
c. **Predictive Maintenance and Asset Management:** The combination of historical and sensor data enables predictive models to produce failure predictions for treatment units and pumps and pipelines. Cities with aging water infrastructure should focus on this matter since it aids their operations [18].
d. **Community Engagement and Transparency:** Activity participants utilize the technology to both preserve water resources and join collective decision platforms that permit group-based choices. The active involvement of people allows

them to form trust relationships as they collectively handle the maintenance of sustainable water practices [10].

1.3 Objective of the Chapter

The chapter assesses how smart technology transforms urban water management systems throughout this chapter. The primary goals are:

a. The research aims to evaluate core difficulties in water resource management because of fast urban growth while studying smart city development.
b. To examine the use of IoT, AI, and data analytics in promoting water efficiency, sustainability, and resilience.
c. To investigate the policy frameworks and models of governance that enable the inclusion of smart water technologies.
d. Developing a framework that allows stakeholders and community members to participate in order to make smart water systems operational effectively.

2 Technological Innovations in Urban Water Management

2.1 Applications of IoT in Real Time Water Monitoring and Leak Detection

The Internet of Things (IoT) is aiding the modernization of water infrastructure in a revolutionary way by enabling real-time monitoring, control, and automation of urban water systems. The IoT devices, consisting of smart meters, flow sensors, and pressure monitors, help utilities track real-time consumption of water and pipeline status and water quality elements, which allows immediate responses to anomalies, including leaks, contamination, and low-pressure events [11].

The Internet of Things (IoT) ensures real-time dynamic management of water infrastructure by continually monitoring urban water networks, since it facilitates live system surveillance and automated control of systems. Smart meters, along with flow sensors and pressure monitors, gather continuous water measurement data that allows utilities to identify potential problems like pipe leaks, water contamination, and pressure issues [26]. Leak detection is the top useful application that IoT has to offer in urban water infrastructure. The Barcelona intelligent water project utilizes a dispersed IoT sensor system to detect water distribution problems in real time, which results in the reduction of both economic costs and water waste [25]. The use of Internet of Things devices allows continual measurement of essential water quality indicators, including pH, together with turbidity and conductivity.

2.2 *Utilization of AI and Data Analytics for Predictive Maintenance and Demand Forecasting*

Artificial Intelligence (AI) and Big Data Analytics are transforming water management by facilitating predictive maintenance and demand forecasting two capabilities critical to effective and resilient urban water systems. Predictive maintenance employs machine learning algorithms to analyze sensor data and historical logs to predict equipment failures and infrastructure problems before they happen [7]. This not only minimizes downtime but also maximizes the life of critical infrastructure and reduces operational expenses. AI models use massive datasets that range from weather conditions, consumer history, and demographic data to forecast short- and long-term water demand trends in demand forecasting. This allows utilities to schedule resource distribution, streamline pumping routines, and avoid overconsumption during peak times [13] as shown in Fig. 2.

a. **Predictive Maintenance Using AI**

Water system monitoring happens through predictive maintenance which uses Artificial Intelligence (AI) to find early signs of equipment breakdown in advance. Using predictive maintenance enables water utility operators to formulate proper scheduling of maintenance and reduces breakdown costs while making infrastructure last longer [30]. AI-based predictive systems operating in London and Tokyo have employed the technology to identify leaks and fix pipeline vulnerabilities before they occur thus minimizing non-revenue water.

b. **Demand Forecasting with AI**

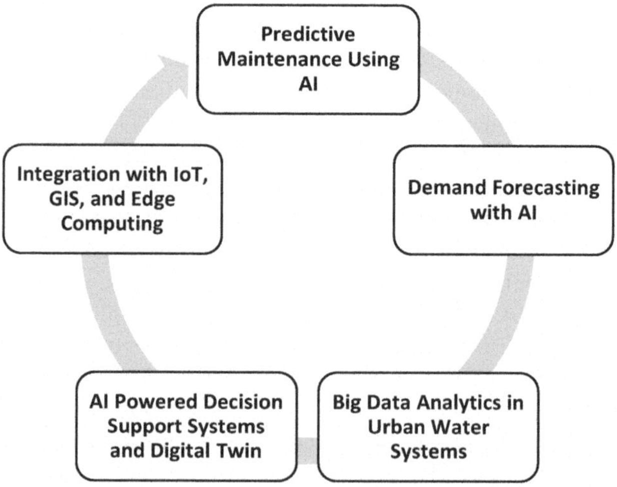

Fig. 2 Use of AI and data analytics for predictive maintenance and demand forecasting

The use of artificial intelligence stands crucial in predicting water demand conditions. The correct prediction of demand enables water managers to optimize both production, storage and supply operations. Such predictions allow organizations to plan better while avoiding excessive consumption when demands rise. The Singapore Public Utilities Board utilizes AI-based demand forecasting systems through PUB to control water supply and minimize operational wastages especially during droughts [20].

c. **Big Data Analytics in Urban Water Systems**

The analytics platforms collect diverse datasets from cloud computing facilities to analyze massive information which discovers usage patterns and detects anomalous behavior and enables strategic decision-making. Utility companies leverage big data analytics to construct vital performance indicators known as KPIs while tracking network performance and transmitting real-time alerts to their staff members [12].

d. **AI Powered Decision Support Systems and Digital Twin**

Decision support systems platforms work with digital twin modeling of infrastructure to allow engineers and city planners assessment of different scenarios through prediction modeling while testing solutions before actual software deployment [21].

e. **Integration with IoT, GIS, and Edge Computing**

When AI system joins forces with other technologies it creates an urban water system which provides versatile solutions to fluctuating urban problems [2].

2.3 Role of Artificial Intelligence in Water Quality Management

AI plays a transformative role in water quality management by facilitating real-time monitoring, predictive analysis, and automated decision making. AI systems, based on machine learning algorithms and sensor fusion, are able to identify pollutants, forecast trends in pollution, and streamline treatment processes, making water resource management more efficient and proactive, as shown in Fig. 3.

a. **AI for Real-Time Water Quality Monitoring and Contamination Detection:** The central role of Artificial Intelligence for immediate water quality assessment

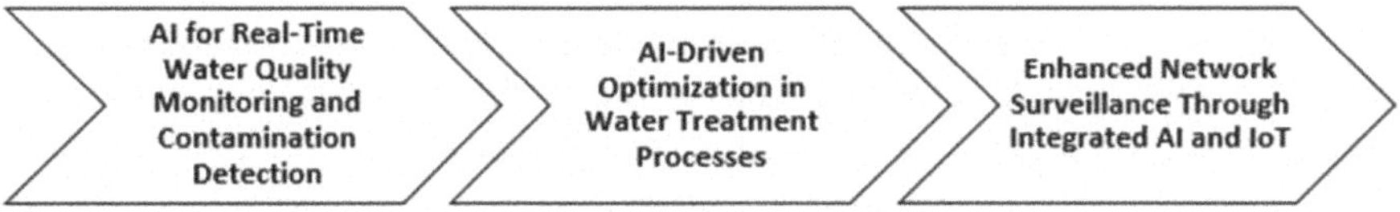

Fig. 3 Role of artificial intelligence in water quality management

as well as contamination detection operates throughout urban environments. Transportation between water authorities enables them to spot water contamination hazards when they monitor combined pH systems with turbidity operations and temperature, together with conductivity measurements. This capability enhances waterborne outbreak management because it defends the population health of densely populated urban zones.

b. **AI-Driven Optimization in Water Treatment Processes:** AI supports functions in water treatment facilities beyond detection because it drives operational optimization systems. The combination of machine learning processes enables automatic operations of water plant variables such as chemical dosing (including chlorine and coagulants) alongside filtration speed and disinfection cycle duration by using real-time water quality measurements.
c. **Enhanced Network Surveillance Through Integrated AI and IoT:** The intelligent surveillance framework forms when AI joins forces with IoT to cover the entire water infrastructure. The integrated system enables steady water safety outcomes and predictions of water contamination risks that originate from industrial outflows or pipeline defects and environment influences. Location-based understanding allows utility managers to execute practical targeted actions and achieve more efficient results through this solution.

3 Policy and Governance Framework

3.1 Analysis of Existing Policies Supporting Smart Water Management

The effectiveness of smart water management systems relies significantly on the existence of strong policy and governance structures that facilitate innovation, efficiency, and accountability. In most nations, national and local governments have established policy measures to embed digital technologies into water infrastructure [3]. For example, the European Union's Water Framework Directive (WFD) inspires member states to adopt digital tools for monitoring water resources and stakeholder participation. Likewise, India's Smart Cities Mission encourages smart water meters, leak detection systems, and automated controls as part of its urban infrastructure development strategy [29].

Though active attempts have been made various existing policies continue to operate in a fragmented manner as well as in a reactive state instead of being adaptive. The current infrastructure stands poor in terms of establishing standardized interoperability as well as common real-time data exchange methods and financial support for utility adoption of intelligent systems. The successful implementation of smart water technologies faces significant challenges in developing countries because they lack technical capabilities and their legal structures need modernization and also have insufficient collaborative partnerships between public entities and private organizations [19].

3.2 Recommendations for Developing Regulatory Frameworks to Facilitate Technology Adoption

A flexible regulatory approach must support new developments to support Innovative practices as shown in Fig. 4.

a. **Enables seamless integration through unified data standards:** The primary need for modern water management systems relies on establishing integrated operating procedures for technological platforms. Local regulations must require standard data structures alongside network protocols which enable all smart devices alongside software frameworks to share information easily. When city-wide systems lack standard protocols they tend to split into uncoordinated

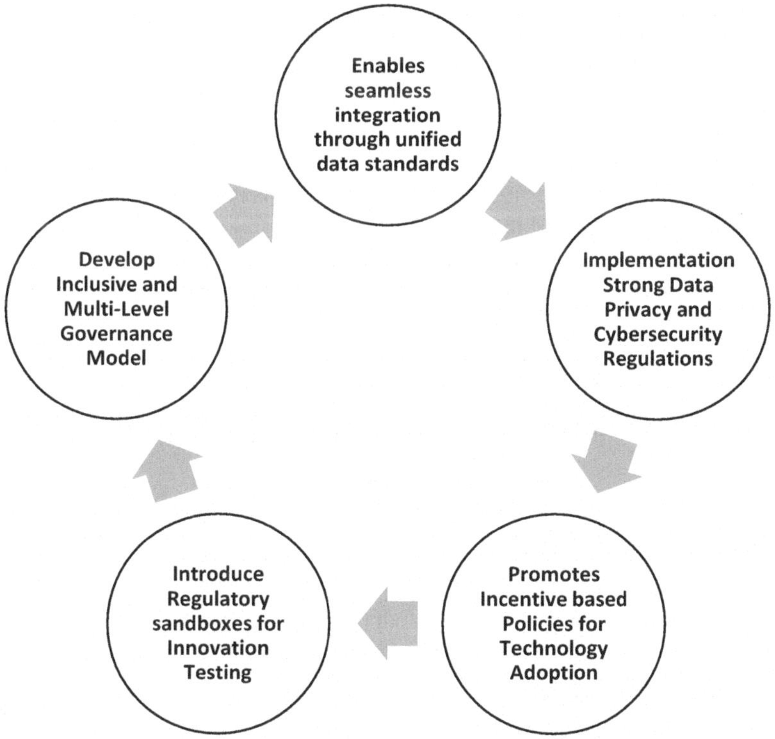

Fig. 4 Recommendations for developing regulatory frameworks to facilitate technology adoption

sections that reduce their operational capability. Open data policies support both transparency initiatives and third-party system integration which establishes a collaborative framework between smart utilities [22].

b. **Implementation Strong Data Privacy and Cybersecurity Regulations:** Residential and commercial users provide sensitive data to operational smart water technologies. Governments need to pass complete data privacy and cybersecurity laws that shield users from information breach and improper usage. The GDPR serves as an international standard that companies can adopt to develop regional privacy standards [16].
c. **Promotes Incentive based Policies for Technology Adoption**: In order to promote the uptake of smart technologies, regulatory systems ought to incorporate fiscal incentives in the form of tax credits, rebates, or utility and municipal subsidies to roll out smart water solutions. Such incentives lower the cost of high upfront investment and spur innovation. Performance-based financing models, in which utility companies are incentivized to minimize water loss or boost efficiency, can prompt substantive improvements [24].
d. **Introduce Regulatory sandboxes for Innovation Testing:** New technologies require testing environments which authorities can establish through regulatory sandboxes to test technologies under minimal regulatory requirements. The approach enables innovators along with startups and utility providers to evaluate smart water solutions through testing alongside collecting proof and engineering their technologies before implementing at scale [27].
e. **Develop Inclusive and Multi-Level Governance Model:** Policy development needs full participation from national to local authorities together with private sector members and citizens in order to succeed. The governance system built through collaboration ensures that policy initiatives match the issues people face in their communities and their current requirements. Frameworks need to make space for stakeholders to engage throughout planning stages as well as implementation phases and evaluation steps for developing trust and accountability [9].

4 Community Engagement and Stakeholder Collaboration

4.1 Strategies for Involving Citizens and Stakeholders in Smart Water Initiatives

The success of smart water initiatives depends heavily on citizen participation because their involvement brings together inclusive decision-making with enhanced transparency, as shown in Fig. 5.

a. **Develop Participatory Digital Platforms:** The development of approachable digital systems enables citizens to monitor and report water system issues and provide feedback by using real-time solutions. Through their platforms, these

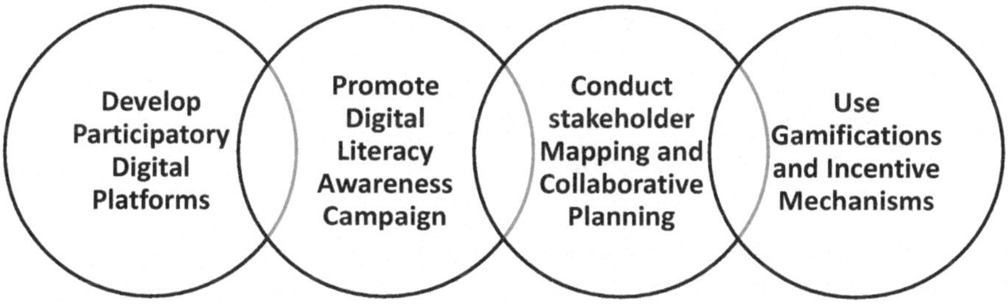

Fig. 5 Strategies for involving citizens and stakeholders in smart water initiatives

platforms link users with utility providers so they can enhance both transparency and prompt action. The cities of Amsterdam and Melbourne operate open-data dashboards which show water consumption patterns and let community members watch their service quality [19]. Such community engagement practices establish public trust while developing an environment where accountability remains constant in urban water administration.

b. **Promote Digital Literacy Awareness Campaign:** In order to ensure inclusive participation from citizens in smart water systems, investment in education and awareness is the key. Most people might not be aware of technologies like IoT, AI, or data analytics. Organizing community workshops, school outreach, and public drives will enhance digital literacy and environmental awareness. Informing the public of the advantages of smart water systems and sustainable water utilization aids in closing the knowledge gap and encouraging more substantive involvement.
c. **Conduct stakeholder Mapping and Collaborative Planning:** The foundation of inclusive water management emerges from the successful involvement of stakeholders that include municipal governments with NGOs and utility companies and academia as well as vulnerable community members. By using stakeholder mapping planners receive valuable information about participated groups' responsibilities their interests and their specific concerns. The participatory governance method implements smart water policies that maintain both fairness and exact fulfillment to regional circumstances.
d. **Use Gamifications and Incentive Mechanisms:** The combination of tracking systems and rewards for best water savers among individuals or neighborhoods through digital applications produces behavioral change through certification programs and public acknowledgements or discount benefits. The creative approaches utilize interactive strategies to provide enjoyable ways to engage people which ultimately enhances their commitment toward long-term water sustainability.

5 Challenges and Future Directions

5.1 Discussion on Integration, Data Management and Scalability Challenges

While smart water technologies have great potential, their adoption is usually delayed by several technical and institutional issues. Integration with legacy systems is among the biggest hurdles most urban water systems have existing legacy infrastructures that do not interface well with new IoT systems or AI platforms. Limited interoperability and common standards tend to result in stand-alone systems, which are costly and make data consolidation across departments difficult [2].

Data management is also a major challenge. Smart water networks produce huge amounts of real-time data, which need to be stored, processed, and analyzed effectively. Most water utilities, particularly in the developing world, do not have the infrastructure or capacity to deal with such sophisticated data environments. Moreover, issues regarding data security, privacy, and ownership are still open. Cyberattacks on water networks can undermine not only operational effectiveness but also public safety [2].

Scalability is another urgent concern. Pilot projects might be successful in individual neighborhoods, but scaling up smart solutions to the city or regional level is frequently hindered by budgetary limitations, the absence of qualified staff, and uneven regulatory support. The upfront cost of sensors, digital platforms, and training staff can discourage small cities from implementing smart water systems at scale. In addition, the urban-peri-urban or rural digital divide can result in unequal access to these sophisticated service.

5.2 Exploration of Future Research Directions, and Emerging Technologies in Smart Water Management

a. **Advancements and Application of Digital Twin:** The implementation of digital twins enables utility organizations to maximize their pumping operations while also conducting digital disaster response simulations and optimizing their maintenance procedures. Digital twins act as an advanced support system for cities which steers them away from passive toward predictive water management methods
b. **Integration of Edge Computing for Real-Time Processing:** The adoption of edge computing technology by future smart water systems aims to overcome two major processing challenges that exist during cloud-based centralized computing operations. The location of data processing operations at or near sensor devices or local gateways provides utilities enhanced real-time monitoring capabilities together with fast response times. The application of edge computing produces

powerful benefits for leak detection together with pressure monitoring across areas featuring minimal network reliability. The combination of IoT and AI technology will establish responsive automated systems.

c. **Adoption of Blockchain for Secure Water Data Transactions:** Blockchain technology enables transparent contracts during both public–private partnerships and water trading system applications. The researchers recommend further investigations of blockchain technology to establish its role in strengthening data security and trust-based monitoring while creating decentralized control systems for water management frameworks.
d. **Citizen Centric Innovation and Gamification:** There is an increasing desire for creating citizen-centered smart water apps that foster civic engagement. Mobile apps that offer rewards for conserving water through gamification, virtual assistants enabled by AI to answer user queries, and participatory sensing apps that allow citizens to monitor their local water quality are some examples of new trends. The tools both expand user knowledge and enable users to actively change their behaviors which keeps them engaged in the process of water management. The advancement of accessible research tools demands improvements in their usability as well as inclusion elements.
e. **AI Driven Forecasting for Climate Resilience:** Another essential area of research is the application of AI for climate-resilient water management. Future research should aim to investigate the ability of AI models to combine climate information (e.g., variability in rainfall, drought regime) to predict water supply and demand with greater accuracy. The models can inform investment in infrastructure, enable early warning systems, and enable cities to become climate-resilient against water stress.

6 Strengths and Weakness of the Chapter

6.1 Strengths

a. **Interdisciplinary Approach to Urban Water Challenges:** The chapter stands out because it presents real-world views through a combination of environmental science and urban planning and public policy and information technology knowledge The integrative approach of uniting multiple viewpoints enables this chapter to establish a comprehensive understanding which benefits different readers from the academic field to local government social workers.
b. **Alignment with Global Sustainability Goals (SDGs):** The research content demonstrates strong equivalency with the United Nations Sustainable Development Goals (SDGs) through the Combination of SDG 6 (Clean Water and Sanitation) SDG 11 (Sustainable Cities and Communities) and SDG 13 (Climate Action).

c. **Policy and Governance Integration:** Technology alone does not produce outcomes effectively because it requires supportive configurations of regulations alongside financial motivators and coordinated engagement of stakeholders. The article directs particular relevance to public sector leaders because its governance-centered approach.
d. **Practical Recommendations for Implementation:** Practical Recommendations for Implementation represent a strong feature since they present use cases throughout the chapter. Steps to deploy the programs are included together with professional guidelines which are paired with decision-making frameworks and strategies that guide actual implementation activities and processes. This resource incorporates helpful recommendations that develop an efficient instrument to assist personnel who handle smart cities and utility networks
e. **Emphasis on Equity and Inclusion:** The section promotes both equal technology access as well as equal service availability to ensure vulnerable population groups can use smart water systems.

7 Conclusion

The chapter demonstrated how latest technologies facilitate real-time monitoring and forecasted servicing and smart decisions from end to end water delivery systems. The chapter emphasizes the critical connection between smart water solutions and circular economy and climate resilience principles and United Nations Sustainable Development Goals (SDGs). The progress in smart water technology will revolutionize urban operations of critical water infrastructure by bringing revolutionary thinking towards its utility. Smart technology innovations create key solutions that assist in managing both the rapid urban population increase alongside environmental change effects.

a. **Smart Technologies are reshaping Urban Water Infrastructure:** Smart Technologies such as IoT, AI, Digital Twins, and edge computing are transforming the way water infrastructure in cities is managed. These various technologies enable real-time monitoring, preventive maintenance, as well as autonomous decision-making, thereby enhancing service delivery, curbing wastage of resources, and providing increased system resilience, Through data exploitation, cities are shifting from reactive to proactive systems.
b. **Governance and Policy are essential Enablers:** Technological success necessitates complete policy-based architectures as well as adequate governance mechanisms. The discussion highlights basic standards required for data collaboration and data protection security between organizations and systemic policies should promote system interoperability between sectors.
c. **Community Engagement Builds Trust and Adoption:** Correct handling of the water resources is not just simple technical deployment as it involves societal participation. For the chapter citizens should be made an integral part of the

system using digital dashboards, cell phones and gaming instruments. Participants in monitoring decision-making enhance community people's confidence and facilitate easier accountability which fosters enhanced utilization behavior of the community members.

d. **Real Time Monitoring and Predictive Analytics Drive Efficiency:** By shifting from crisis management to strategic planning, organizations achieve enhanced performance while extending the lifespan of assets which assures better service in emergency situations like floods or droughts.

8 Summary of Key Insights

The chapter analyzed shifts in intelligent water management systems that operate within towns with quick population expansion and amid weather-related stressors. The effectiveness of these modern implementations depends on strong policy frameworks as well as appropriate community interaction together with approaches that can grow according to need.

a. The current water infrastructure system does not match present operational standards because urban growth continues at a fast rate amidst climatic changes. Modern urban water systems require effective management of escalated resource consumption and decreasing source availability together with extreme environmental changes from climate change.
b. **Intelligent Technologies Are Revolutionizing Water Infrastructure:** The technology provides two essential benefits that optimize operational performance and manage complete data along with decreasing water consumption
c. **Real-Time Monitoring and Predictive Analytics Enhance Efficiency**: Through real-time data utilities can locate leaks as well as stop failures and enhance their service delivery in advance before issues happen.
d. **Strong policy frameworks and governance structures are necessary enablers:** To achieve successful implementation all frameworks need to consist of standards for data flow, security measures, privacy preservation and government transparency. Technology solutions experience difficulty in reaching mass adoption through time because institutions need legislative backing combined with operational readiness.
e. **Community Involvement Is the Key to Adoption and Trust:**.A smart water system's effectiveness stems from its features that focus on citizens at the forefront to achieve adoption and trust from the population. When people actively participate in monitoring and decision-making systems their behavior changes more effectively since their presence increases transparency and inspires all sections of the population to follow suit. Mobile applications and dashboard interfaces integrate gamification platforms to let users participate in governing and achieve water conservation which builds trust in public utility organizations.

f. **Integration, Cost, and Legacy Infrastructure are the Main Obstacles:** The main deterrents towards efficient integration and cost-effective smart water system implementation stem from the combination of old infrastructure with departmental disconnects in aging cities. High capital outlays and scanty technical staff members and the resistance to change represent vital barriers to implementation.
g. **Ethical, Privacy, and Security Issues Need to Be Resolved:** Rapid resolution of ethical, security and privacy matters is vital prior to adoption of smart technologies. Operations within systems must lead to augmented data volume while comprehensive controls must tackle personal and business data collection and safeguards. Distorted data perception or exploitation by firms makes individuals feel intruded thereby fostering public dissatisfaction that renders privacy management indispensable in order to control data abuse and ethical oversight paramount for resolving such matters.

8.1 Recommendations for Practitioner and Policymaker

Practitioners together with policymakers should create scalable pilot programs through modular technologies and build strong digital infrastructure cybersecurity standards along with inclusive regulations which combine AI ethics standards and data protection and stakeholder engagement to enable equitable smart water services as shown in Fig. 6.

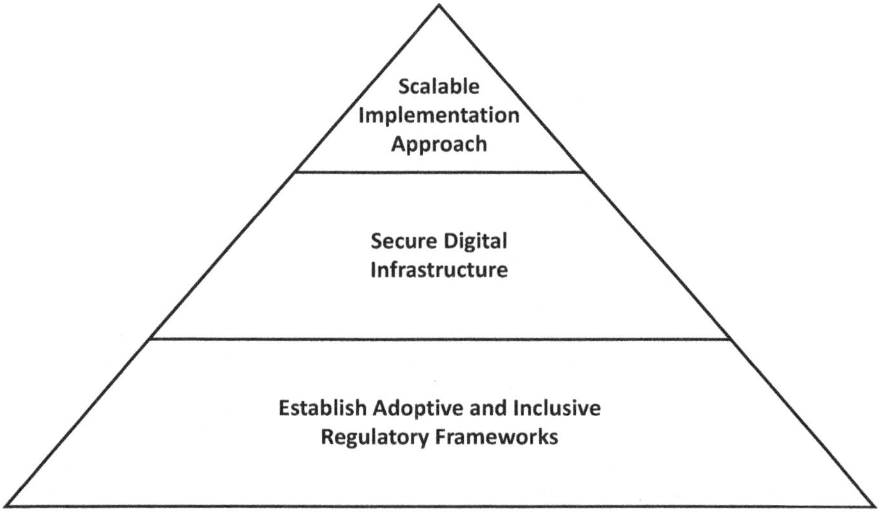

Fig. 6 Recommendations for practitioner and policymaker

a. **Scalable Implementation Approach:** Research demands the execution of pilot programs across selected regions to implement testing areas for assessing new technologies and resolving unexpected complications. Building flexible and cost-effective urban regional expansion systems depends on modular interoperable technology deployment, making it a top priority factor.
b. **Secure Digital Infrastructure:** Public officials need to create standard principles and security standards that defend digital networks from unauthorized entry while protecting sensitive data.
c. **Establish Adoptive and Inclusive Regulatory Frameworks:** The necessary policies need to combine data protection measures with ethical standards for AI systems and guidelines for involving stakeholders in decision processes. Smart water services should offer protection to all groups regardless of their vulnerability status

8.2 Final Thoughts on the Role of Smart Technologies in Shaping the Future of Urban Water Management

Urban water management of the future is inextricably tied to the uptake and adaptation of smart technologies. With urban populations growing ever larger and with pressures from climate change growing ever more intense, the old ways of managing water are no longer adequate, reactive, and unsustainable. Smart technologies fuelled by technologies like Internet of Things (IoT), Artificial Intelligence (AI), digital twins, blockchain, and edge computing provide a proactive, smart, and data-driven approach to managing water as a valuable and limited resource. The water infrastructure benefits from IoT sensors which deliver automatic and ongoing information about flow rates alongside pressure measurements and quality status and consumption rates. Through digital twin technology which creates virtual replicas of physical systems the city improves its capability to test potential responses in simulations before implementing actual changes without endangering operational systems. The enhanced public trust brings together shared responsibilities which leads citizens to bring about behavior modifications. The water ecosystem welcomes citizen participation through Gamification and AI chatbots and smart billing systems among other example tools.

Any digital water system transformation without proper security protocols might cause existing social differences to become worse and raise doubts about monitoring practices and data management. Smart technological innovations provide a substantial capacity to transform modern urban water management operations in the twenty-first century.

References

Adedeji KB, Ponnle AA, Abu-Mahfouz AM, Kurien AM (2022) Towards digitalization of water supply systems for sustainable smart city development—Water 4.0. Appl Sci 12(18):9174

Ahmad K, Maabreh M, Ghaly M, Khan K, Qadir J, Al-Fuqaha A (2022) Developing future human-centered smart cities: Critical analysis of smart city security, Data management, and Ethical challenges. Comput Sci Rev 43:100452

Aivazidou E, Banias G, Lampridi M, Vasileiadis G, Anagnostis A, Papageorgiou E, Bochtis D (2021) Smart technologies for sustainable water management: An urban analysis. Sustainability 13(24):13940

Ali AS, Abdelmoez MN, Heshmat M, Ibrahim K (2022) A solution for water management and leakage detection problems using IoTs based approach. Internet 18:100504

Asmal I, Syarif E, Amin S, Walenna MA (2022) The impact of the environment and people's attitudes on greywater management in slum coastal settlements. Civ Eng J 8(12):2734–2748

Ayvaz S, Alpay K (2021) Predictive maintenance system for production lines in manufacturing: a machine learning approach using IoT data in real-time. Expert Syst Appl 173:114598

Becher O, Pant R, Verschuur J, Mandal A, Paltan H, Lawless M, Raven E, Hall J (2023) A multi-hazard risk framework to stress-test water supply systems to climate-related disruptions. Earth's Futur 11(1), e2022EF002946

Bouramdane AA (2023) Optimal water management strategies: paving the way for sustainability in smart cities. Smart Cities 6(5):2849–2882

Han X, Boota MW, Soomro SEH, Ali S, Soomro SGH, Soomro NEH, Tayyab M et al. (2024) Water strategies and management: current paths to sustainable water use. Appl Water Sci 14(7):154

Kanyama MN, Shava FB, Gamundani AM, Hartmann A (2024) Anomalies identification in Smart Water Metering Networks: fostering improved water efficiency. Phys Chem Earth, Parts A/B/C 134:103592

Karina R, Siti-Nabiha AK, Jurnali T (2025) Big data integration in performance management and control: a socio-technical perspective. J Account Organ Chang

Kavya M, Mathew A, Shekar PR (2023) Short term water demand forecast modelling using artificial intelligence for smart water management. Sustain Cities Soc 95:104610

Khosravi A, Ashkpour M (2025) Machine learning and digital innovation for managing and monitoring water resources. In Emerging trends and technologies in water management and conservation. IGI Global Scientific Publishing, pp 241–284

Khozani ZS, Banadkooki FB, Ehteram M, Ahmed AN, El-Shafie A (2022) Combining autoregressive integrated moving average with Long Short-Term Memory neural network and optimisation algorithms for predicting ground water level. J Clean Prod 348:131224

Koech R, Cardell-Oliver R, Syme G (2021) Smart water metering: adoption, regulatory and social considerations. Australas J Water Resour 25(2):173–182

Krishnan SR, Nallakaruppan MK, Chengoden R, Koppu S, Iyapparaja M, Sadhasivam J, Sethuraman S (2022) Smart water resource management using Artificial Intelligence—A review. Sustainability 14(20):13384

Le Gat Y, Curt C, Werey C, Caillaud K, Rulleau B, Taillandier F (2023) Water infrastructure asset management: State of the art and emerging research themes. Structure and Infrastructure Engineering, 1–24

Liu Q, Yang L, Yang M (2021) Digitalisation for water sustainability: barriers to implementing circular economy in smart water management. Sustainability 13(21):11868

Masud MM, Shamem ASM, Saif ANM, Bari MF, Mostafa R (2025) The role of artificial intelligence in sustainable water management in Asia: a systematic literature review with bibliographic network visualization. Int J Energy Water Resour 9(1):247–265

Mazzetto S (2024) A review of urban digital twins integration, challenges, and future directions in smart city development. Sustainability 16(19):8337

Mounce SR (2021) Data science trends and opportunities for smart water utilities. ICT for smart water systems: measurements and data science, 1–26

Nieuwenhuis E, Cuppen E, Langeveld J, de Bruijn H (2021) Towards the integrated management of urban water systems: Conceptualizing integration and its uncertainties. J Clean Prod 280:124977

Oluokun OA, Akinsooto O, Ogundipe OB, Ikemba S (2025) Strategic policy implementation for enhanced energy efficiency in commercial buildings through Energy Performance Certificates (EPCS)

Ramírez-Moreno MA, Keshtkar S, Padilla-Reyes DA, Ramos-López E, García-Martínez M, Hernández-Luna MC, Lozoya-Santos JDJ et al. (2021). Sensors for sustainable smart cities: a review. Appl Sci 11(17):8198

Rousso BZ, Lambert M, Gong J (2023) Smart water networks: a systematic review of applications using high-frequency pressure and acoustic sensors in real water distribution systems. J Clean Prod 410:137193

Schaffers H, Korakis T, Pham C, Rahim A, Jara A, Serrano M (2022) European challenges for experimental facilities. In Building the Future Internet through FIRE. River Publishers, pp 3–42

Siddique I (2022) Sustainable water management in urban environments. Chem Res J 7(4):95–101

Sushma N, Suresh HN, Mohana LJ, Kumar KS (2024) Experimental investigation on wireless integrated smart system for energy and water resource management in Indian smart cities. Results Eng 23:102687

Yazdi M (2024) Maintenance strategies and optimization techniques. In Advances in computational mathematics for industrial system reliability and maintainability. Cham, Springer Nature Switzerland, pp 43–58

Exploring Emerging Contaminants: Their Sources, Industrial Role, and New Treatment Technologies

Mamilla Vijaya Kumar

Abstract Technological, agricultural, and medical advancements have enhanced human life and introduced environmental challenges due to emerging contaminants, such as pharmaceuticals and personal care products. Often inadequately monitored, these substances accumulate in water bodies from sewage treatment plant effluents, posing significant risks to human health and wildlife. Many contaminants, such as endocrine-disrupting chemicals, are often excluded from municipal testing, and conventional treatment facilities struggle to eliminate them. As a result, current waste treatment systems are inadequate and pose risks to human health and the environment. The effectiveness of existing wastewater systems remains uncertain, allowing these contaminants to infiltrate aquatic environments. This paper details the relevant emerging contaminants and their sources, including industrial activity, and their far-reaching effects on the environment. It also reveals novel effective mitigation impact strategies on the environment implemented to address the emerging contaminants.

Keywords Emerging contaminants · Pharmaceuticals · Sewage treatment plant · Aquatic environments · Endocrine-disrupting chemicals

1 Introduction

The uncontrolled living conditions of society have tremendously deteriorated the quality of air, water and soil [1]. This has resulted in more critical challenges of concern, such as toxic wastes, pesticide usage, and emission of greenhouse gases [2]. Contrary to the aims of social welfare, imprisonment, Pharmaceuticals, household wastes and industrial discharged wastewater are examples of various pollutants that circulate negatively through water [3]. Due to economic expansion and an ever-growing population, the consumption of fresh water has tremendously increased, contributing to the pollution of water bodies [4]. The emerging contaminants are in a

M. V. Kumar (✉)
Department of Civil Engineering, KSRM College of Engineering, Kadapa, India
e-mail: civilvijay1@gmail.com

V. Kandpal and T. Choudhury (eds.), *Blue Revolution: Sustainable Water Use and Circular Economy Innovations*, Urban Sustainability,
https://doi.org/10.1007/978-981-95-1916-3_12

problematic situation because of the tendency to accumulate into plants and species making them a great source of danger to the humans [5]. Most critical challenges facing humanity emerges from the depletion of water resources and incessant pollution of freshwater sources which are vital to human health and disrupt the balance in the ecosystem [5, 6].

The aforementioned changes result from fundamental problems like the generation of harmful waste, the use of pesticides, and greenhouse gas emissions [2]. Pharmaceuticals, domestic products, and industrial waste contribute towards water pollution, which negatively influences the water cycle [3]. Economic progress, combined with population surge, necessitates the availability of freshwater, further causing the deterioration of water bodies [4]. What are regarded as environmental pollutants have serious consequences because of their ability to accumulate in the tissues of living organisms, greatly putting human life at risk [5]. Some of the most demanding issues include the water supply scarcity and problem of pollution in water resources, which is fatal to peoples' health and disrupts the balance in the ecosystem [5, 6].

Emerging contaminants (EMs) comprise overlooked chemicals and microorganisms that evade routine monitoring yet potentially jeopardize human and ecological health. Prominent categories include pharmaceuticals, personal care additives, plasticizers, and endocrine disruptors [7, 8]. Unlike regulated contaminants like heavy metals, pollutants such as EMs often lack adequate oversight. These pollutants primarily come from industrial activities and wastewater discharges, and current treatment processes may not effectively eliminate them, leaving their long-term effects largely unknown [3]. Despite their importance in modern life [9], many of these compounds are released unmetabolized into sewage systems, posing challenges for future facility design and raising concerns about the lack of regulations [10] (Fig. 1).

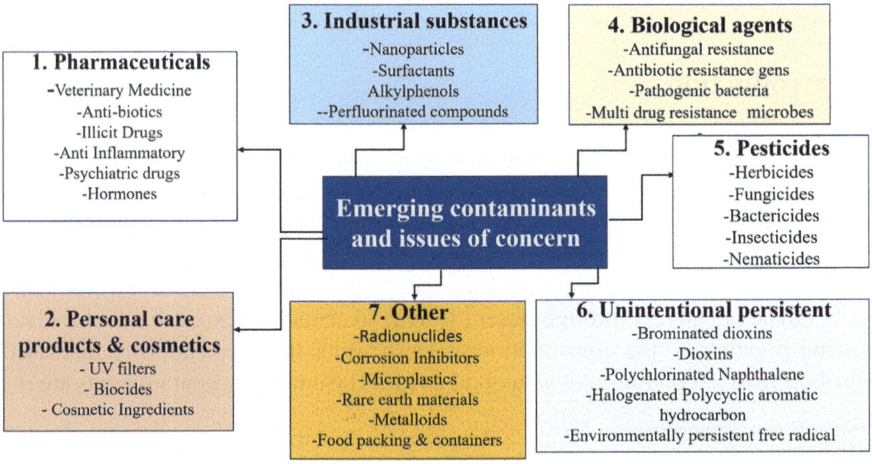

Fig. 1 Evolving spectrum of emerging contaminants: a strategic classification [11]

2 Sources of Emerging Contaminants

Mitigating the risks of emerging contaminants demands a deep dive into their nature, pathways of release, and far-reaching effects on ecosystems [12]. Domestic, industrial, and agricultural wastewater discharges bring these contaminants into water bodies, endangering ecological balance. These substances seep into water, sediment, soil, and organisms, impacting even the most remote regions and ecosystems, from tiny organisms to the top of the food chain, such as polar bears [13, 14]. Unlike traditional pollutants, emerging contaminants remain in sediments and are bioaccumulated by benthic organisms, thus moving up the food web [15]. The concentration of these contaminants in fish tissues reflects the health status of the water body and indicates risks to human health [16] (Fig. 2).

Emerging contaminants' transport and occurrence are affected by a range of factors, including environmental conditions and geographical location [18]. Their attributes, as defined by water solubility, temperature, and organic content, are crucial for evaluating the contamination's environmental lifespan. A majority of these contaminants stem from treated wastewater point sources [19, 20] and are mostly present in urban and agricultural settings. Diffuse sources, such as stormwater runoff, tend to contribute a lower load of pollutants. Other key contributors include [21]. Pharmaceutical discharge and landfill leachate pose risks of groundwater contamination [22]. Continuous effluent discharge from wastewater treatment plants further aggravates the contamination [23], severely impacting populations around the contaminated water sources [24].

Fig. 2 A schematic illuminates the intricate nexus of emerging contaminants, tracing their origin, utilization, and release into the environment [17]

3 Industrial Contributions to Emerging Contaminants

Untreated industrial discharges pose new and undiscovered contaminants into water bodies by liberating synthetic chemicals and residues. These pollutants escape treatment technologies and represent concealed dangers to ecosystems and human well-being, necessitating their active monitoring.

3.1 Pharmaceutical, Chemical, and Personal Care Industries

In search of effective medications, the pharmaceutical industry poses serious threats to the ecosystem through the discharge of drug contaminants [25, 26]. Lotions and cosmetics, among other products, further worsen this pollution on a daily basis. The presence of drug contaminants ranging from cosmetics to foods lead to genotoxic, mutagenic, and ecotoxicological damages on natural water bodies, which disrupt the ecosystem, lead to the extinction of many species and threaten biodiversity [27]. Many pharmaceutical pollutants resist degradation, persisting in water [28]. Their existence is impacted by socio-economic conditions, the availability of health care services, and even weather changes [29]. With greater consumption, these pollutant streams from medicine and hospital waste, industry waste, and agricultural runoff increasingly contaminate wastewater [30]. The flowchart demonstrates the adverse impacts of the pharmaceutical sector on the environment and public health by detailing their showing the industry's footprint created on emerging contaminants, showing what the economic impact of their contributions through each, and the environmental and health impact emerging on new pharmaceuticals (Fig. 3).

3.2 Textile Industry

The textile sector releases a range of destructive pollutants such as synthetic dyes, microplastics, heavy metals, endocrine disruptors, PFAS, and volatile organic compounds during the production processes. EDCs and synthetic dyes can contaminate water bodies and affect the ecosystem [31], and microfibers shed during washing contribute to microplastic pollution [32]. Dyes contain toxic heavy metals, and VOCs contribute to the pollution of the air [33]. Moreover, PFAS are known to persist in the environment and are associated with various health problems that result in the permanent contamination of water and soil [34].

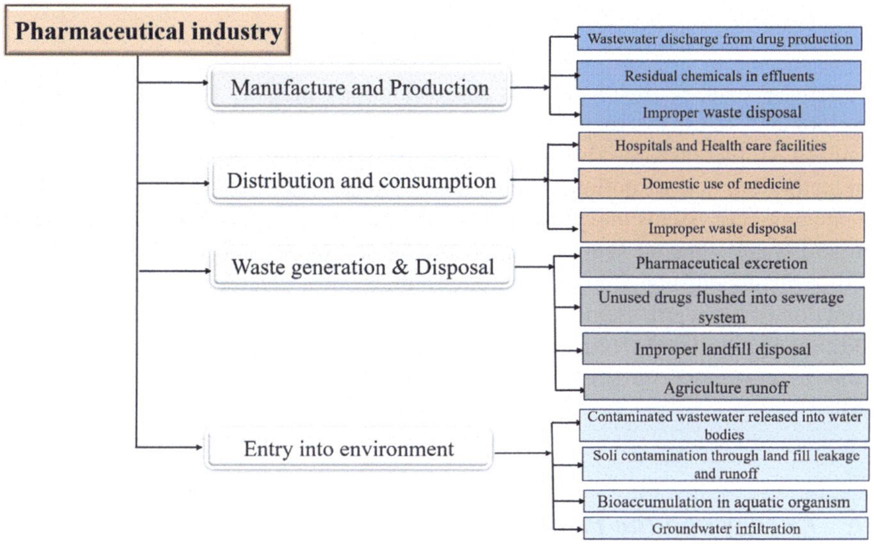

Fig. 3 Environmental trajectory of pharmaceutical emerging contaminants

3.3 Plastic and Packing Industries

Toxic emissions from the packaging and plastic industries infiltrate ecosystems, endangering both environmental balance and human well-being. Microplastics accumulate in water and the food chain, while bottled water contains an average of 325 particles per litre. Bisphenols (BPA, BPS, BPF) leach from packaging into food, acting as endocrine disruptors linked to reproductive disorders. Phthalates, used as plasticizers, also disrupt hormonal functions. PFAS in grease-resistant packaging are linked to serious health issues, including cancer, and contribute to water contamination. VOCs from plastic manufacturing cause air pollution and respiratory problems, and heavy metals in pigments can leach into the environment, leading to neurological damage [35].

3.4 Agriculture and Agrochemical Industries

Agriculture and agrochemical industries contribute to environmental pollution through emerging contaminants (ECs) from pesticides, fertilizers, and veterinary products, posing risks to human and wildlife health. Pesticides, used for crop protection, can contaminate water and soil, harming non-target organisms and accumulating in food chains; some are linked to endocrine disruption and cancer [36]. Veterinary antibiotics and hormones from livestock runoff contribute to antibiotic resistance and hormonal disruption in wildlife.

3.5 Electronic and Tech Industry

The rapid growth of the electronics and technology industry has driven economic development, but also created environmental challenges from emerging contaminants (ECs). Increased production of consumer electronics, data centres, and semiconductors has led to pollutants like heavy metals, PFAS, and e-waste residues, threatening ecosystems and human health. Toxic metals from e-waste contaminate soil and water, causing neurological disorders and higher cancer risks [37]. PFAS used in manufacturing persist in the environment near electronics plants [38]. The extraction of rare earth elements generates toxic waste harmful to biodiversity [39]. Additionally, outdated electronics contribute to significant e-waste, often processed in unsafe recycling centres, exposing workers and communities to harmful emissions [40].

3.6 Energy and the Petrochemical Industry

In the energy and petrochemical industries, contaminants from extraction, refining, and waste disposal processes pose significant risks. Per and Perfluoroalkyl substances (PFAS), found in firefighting foams, are highly persistent and can contaminate water and soil [41]. Microplastics and nanoplastics from plastic degradation affect marine and terrestrial ecosystems [42]. Lead and mercury, released during oil extraction, infiltrate aquatic ecosystems and bioaccumulate in humans, posing long-term health risks [43]. Volatile organic compounds (VOCs) from refining contribute to air pollution and respiratory issues [44]. Hydraulic fracturing fluids and wastewater from oil and gas extraction contain harmful chemicals, leading to groundwater and water contamination [45, 46].

4 Novel Treatment Techniques

4.1 Advanced Oxidation Processes

Eradicating emerging contaminants from wastewater stands as a pivotal challenge in the evolution of water treatment technologies [47]. Like traditional treatment processes, advanced oxidation technology is capable of eliminating damage-causing resistant organic pollutants, something AOT can do in a more environmentally friendly way [48]. AOPs, which came into being through the efforts of William H. Glaze in 1987, use hydroxyl radicals (•OH), which help in purifying water through the effective curing of its contaminants. Even though AOPs deal with both organic and inorganic substances, the disinfection aspect suffers a lot from efficiency due to radicals' short lifespan [49]. AOPs serve different emerging disorders and can be oxidized efficiently in the same way as their forerunners through multi-step reduction into other fission products like ketones and aldehydes [50]. The advent of Advanced Oxidation Processes in the field of AOPs enabled the treatment of wastewater by initiating powerful reactions that lead to the degradation of pollutants and their byproducts into far less harmful substances [51]. The field has been revolutionized by photocatalytic oxidation, Fenton reaction, electrochemical oxidation, and oxidized sulfate radical-based AOPs for their ability to use reactive species to disassemble molecular structures of suspended materials [52, 53]. AOPs have already set their standards clear, with their unmatched speed of oxidation, they will require low disposable resources and mark a landmark change to clean water technologies. Integrating AOPs with membrane systems further enhances contaminant removal and energy efficiency, leading to more sustainable wastewater treatment solutions [54].

AOPs are divided into homogeneous and heterogeneous categories. Homogeneous processes operate within a single phase, leveraging ozone and hydrogen peroxide oxidants to unleash hydroxyl radicals for effective contaminant breakdown. Common methods include ozone-based processes (ozone alone, ozone with UV light, and ozone plus H_2O_2), wet peroxide oxidation using hydrogen peroxide, and various Fenton processes (traditional Fenton, sono-Fenton, photo-Fenton, and others). Heterogeneous processes involve solid catalysts, such as metals on carbon, or semiconductors like titanium dioxide (TiO_2). Notable examples include heterogeneous photocatalysis activated by UV light, catalytic wet peroxide oxidation, and catalytic ozonation—each supercharging reactive species production for advanced mineralization [55] (Fig. 4).

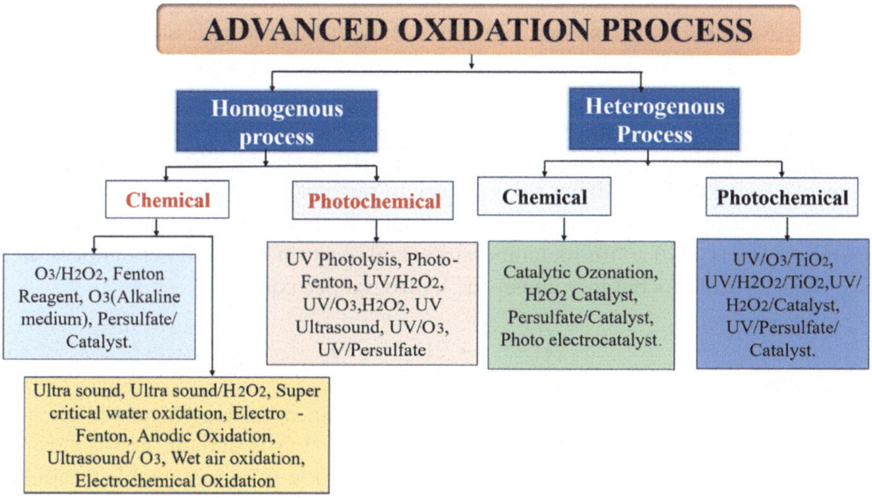

Fig. 4 Classification of AOPs

Photo-Fenton AOP

Fenton and Photo-Fenton processes stand as cutting-edge solutions, harnessing powerful oxidation to drive superior pollutant breakdown in water treatment [56, 57]. These cost-effective methods use hydrogen peroxide and ferrous ions, known as Fenton's Reagent, to oxidize various contaminants, including phenols, formaldehyde, BTEX, and complex pesticide and dye wastes [58]. Fenton processes effectively degrade pollutants such as aromatic amines, dyes, pesticides, surfactants, and explosives in industrial wastewater [59]. They utilize homogeneous ferric ions (Fe^{2+}) to catalyze hydrogen peroxide photolysis in acidic solutions, generating hydroxyl radicals (•OH) through the Haber–Weiss mechanism [60].

$$Fe^{2+} + H_2O_2 \rightarrow Fe^{3+} + OH^- + HO^{\cdot} \quad (1)$$

The Fenton reaction harnesses Fe^{2+} ions as a catalytic powerhouse, driving the rapid breakdown of pollutants for efficient water purification. The Photo-Fenton method enhances this process by incorporating UV radiation, which increases the decomposition of H_2O_2 and improves efficiency. Both methods are straightforward, require minimal specialized equipment, and provide environmental benefits through the regeneration of Fe^{2+} ions.

$$Fe^{3+} + H_2O_2 \rightarrow Fe^{2+} + H^+ + HO^{\cdot}_2 \quad (2)$$

The Fenton process effectively treats wastewater by breaking down organic pollutants below 15 mg/L into CO_2 and water at room temperature using simple reagents [61]. However, challenges include high costs of hydrogen peroxide and stricter water quality regulations, as well as the need for specific acidic conditions to maintain stable radicals [62]. Both the Fenton and Photo-Fenton methods generate sludge from iron oxides and suffer from inefficient hydrogen peroxide use and rapid Fe^{2+} consumption, leading to secondary iron pollution [61, 63]. The Photo-Fenton process enhances efficiency by utilizing solar energy to regenerate Fe (II) species.

The Fenton process is an oxidation method effective at pH 3, where efficiency depends on the H_2O_2/Fe^{2+} ratio, temperature, and reaction time. Higher temperatures can enhance reactions but may lower the hydrogen peroxide.

$$2H_2O_2 \rightarrow 2H_2O + O_2 \quad (3)$$

Furthermore, using excess hydrogen peroxide can decrease process efficiency. This system captivates with its affordability, leveraging abundant, non-toxic iron and easily manageable, eco-friendly hydrogen peroxide for a sustainable treatment approach. The photo-Fenton process improves wastewater treatment efficiency but struggles with high organic matter and turbidity. Traditional Fenton remains viable for high-COD wastewater. Both methods are less effective in high-ion waters, and precise pH control is critical to prevent excessive sludge buildup [64].

UV/H_2O_2 AOP

Ultraviolet (UV)-driven advanced oxidation processes (AOPs) employ the synergistic action of UV radiation and potent oxidants such as H_2O_2, O_3, and Cl_2 to effectively degrade recalcitrant organic pollutants in wastewater treatment systems [65]. This novel procedure uses low- and medium-pressure mercury lamps for water disinfection [66, 67] because it is portable and easy to use. Moreover, it poses no risk due to mercury's toxicity. However, the pollutant removal efficiency is lower than that achieved by other conventional UV treatment systems [68]. The synergism of OH bond destruction makes UV light reactive, resulting in the breakdown of organic pollutants when used with H_2O_2. Because of this, it is useful in the pharmaceutical industry [69] as well as in textiles [70].

Ozone breaks down into atomic oxygen and OH bond radicals when acting under UV ($\lambda < 300$ nm) with hydrogen peroxide present. Strong oxidative reactions also take place at 254 nm [71].

$$2O_3 + H_2O_2 \rightarrow 2HO^{\cdot} + 3\,O_2$$

Maintain a pH of 7–8 for best results in oxidation with ozone. Also, increase the partial pressures of the ozone and maintain low initial pollutant concentrations [72]. However, the energy demands of UV sources and ozone generators limit their use in large-scale operations [73].

Hydrogen peroxide (H_2O_2) is a naturally occurring metabolite that effortlessly breaks down into oxygen and water, ensuring an eco-friendly decomposition. It can also be produced when sunlight interacts with water. Industrially, H_2O_2 is primarily produced using the anthraquinone process developed by I.G. Farbenindustrie AG in 1940 [74]. When exposed to UV radiation (250–400 nm), H_2O_2 generates hydroxyl radicals (•OH) [75].

$$H_2O_2 + h\nu \rightarrow 2HO^{\cdot}$$

Proper dosing is essential in UV-based AOPs for safety and effectiveness [72]. Higher H_2O_2 concentrations enhance pollutant degradation but may affect economic feasibility [63].

Hydrogen peroxide (H_2O_2) can decompose into oxygen and water and can be produced through the interaction of sunlight with water. It undergoes industrial manufacturing using the anthraquinone process [74]. H_2O_2 forms hydroxyl radical (•OH) with UV radiation (250–400 nm) [75]. In UV Advanced Oxidation Processes, proper dosing is necessary [72]. Although high H_2O_2 concentration can enhance pollutant removal, it may also decrease the economic feasibility of the process [63].

The photolysis of hypochlorous acid (HOCl) and hypochlorite ions (ClO^-) into reactive radicals such as Cl• and HO• occurs when chlorine-containing water is subjected to ultraviolet (UV) irradiation. These radicals also react with chlorine and chloride ions to yield secondary oxidants ClO• and $ClOH^{\bullet -}$, which are more effective in degrading persistent organic pollutants [76]. The UV/chlorine advanced oxidation process has been effective in removing pesticides, pharmaceuticals, and other emerging contaminants [77–79]. Other by-products formed, however, chlorate and perchlorate and organochlorine compounds can form 10–30% of the total by-products. Tri and halo methanes (THM and NDMA formed can be effectively post-treatment reduced to below detection level [77].

When compared with conventional methods of treatment, UV-driven AOPs have shown better performance compared to Contaminant shred, decomposition precision, hazardous waste production and safe disposal of water. Issues still exist with regard to high energy requirements, expensive chemicals, and shallow penetration of UV light. To improve the effectiveness of these systems, reactor design, residence time, and dosage should be optimised [58].

Photocatalysis AOP

Without being consumed, a photocatalyst makes use of light energy for reaction acceleration [80]. Both photochemical and photocatalytic degradation are crucial for wastewater treatment, as they result in the complete mineralization of organic compounds under relatively low temperature and pressure conditions [81]. This process relies on the use of metal oxides as well as other oxide nanoparticles, which convert light energy into chemical energy [82, 83]. Chemical reactions can commence with a combination of light and a catalyst to molecules adsorbed on the catalyst surface. The light-activated catalyst can decompose these molecules, thereby allowing the products to desorb and subsequently use the catalyst to enhance some more reactions. The application of light to certain semiconductors generates UV or Visible light, which leads to the production of electron–hole pairs that undergo redox reactions, which lead to the formation of OH Hydroxide and O^2 Oxygen radicals which are needed for the degradation of contaminants [84, 85] (Fig. 5).

Photocatalysis is categorized into homogeneous and heterogeneous types, defined by the phase and dispersion characteristics of the photocatalyst within the reaction medium [87]. Heterogeneous photocatalysts, such as TiO_2, effectively remove refractory organics without auxiliary chemicals and are easier to maintain. Heterostructure photocatalysts enhance electron–hole pair separation, improving charge carrier lifetimes [88, 89]. Enhancing infrared (IR) photocatalytic efficacy for environmental remediation and hydrogen production [90]. Rutile and Anatase forms of TiO_2, while simultaneously noting obstacles regarding separation and agglomeration [91]. Photocatalytic activity can be improved by coupling TiO_2 with semiconductors like SnO_2 [92]. The primary difference between homogeneous and heterogeneous photocatalysis is the physical state of the catalysts: homogeneous photocatalysis employs soluble molecular catalysts, generally built from transition metals, while heterogeneous photocatalysis utilises solid semiconductors. Heterogeneous photocatalysis is preferred for the degradation of organic pollutants due to simple catalyst separation and effective regeneration once light irradiation initiates the process [93].

Titanium dioxide (TiO_2) is extensively studied for its remarkable stability, environmental safety, superior photocatalytic performance, and affordability [94]. In

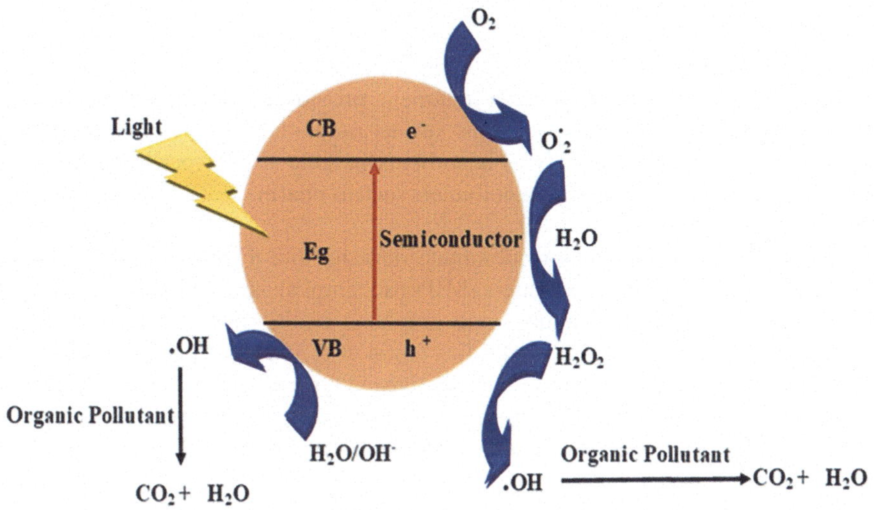

Fig. 5 The TiO_2 reaction mechanism [86]

Advanced Oxidation Processes (AOPs), Titanium Dioxide (TiO_2) along with Zinc Oxide (ZnO) acts as a crucial catalyst due to its elevated efficiency with pollutant degradation [95, 96]. Photocatalysis in sunlight is increased further due to ZnO's ability to capture a broader spectrum of solar energy [97]. Striking photocatalytic capabilities are exhibited by TiO_2; however, its economical use becomes a reality when used in combination with beads or films, which improve recyclability [98]. The search for alternatives that can be activated by visible light is preferred due to the attained restriction of UV and quantum efficiency dependence. While VLA materials can be effective in generating oxidative species, they lack broad-spectrum efficacy in purifying water [99]. Additional refinement of parameters such as pH level, catalyst amount, and light exposure duration can greatly improve photocatalytic efficiency. Overall, photocatalysis presents an eco-friendly, cost-efficient method for refining wastewater, yet the formation depletion of hydroxyl radicals by high-bandgap catalysts disrupts the process's effectiveness [100, 101].

The destruction of organic pollutants is done effectively with AOPs; however, their adoption is hindered in developing regions due to the high cost and energy demand. Unknown byproducts that arise from the complex nature of the pollutants, along with inorganic ions, render oxidation efficiency useless, requiring unplanned pre-treatment adds to the issue. This increases energy and cost requirements, limiting the treatment of large wastewater volumes and hindering AOP adoption in wastewater management [55].

4.2 Adsorption

Adsorption is a vital method for water treatment, providing efficient and sustainable contaminant removal. The use of response surface methodology to optimize biomass, metal oxides, and nanomaterials as adsorbents heightens their performance. It is remarkable in harnessing emerging pollutants such as pharmaceuticals and pesticides [102].

Activated carbon continues to dominate demand due to its high surface area, while molecularly imprinted polymers (MIPs) accomplish selectivity. Newer carbon nanomaterials, including nanotubes and graphene oxide, outperform other materials, although their advanced capabilities come at a much steeper price. Sustainable materials, including clay, are cost-effective but typically allow lower adsorption and have a lower capacity. MIPs are revolutionary with customizable binding sites designed to cleave contaminants relevant under various operational conditions while enduring multiple regeneration cycles without significant performance degradation. Their versatile synthesis allows MIPs to be easily tailored for the removal of specific pollutants from the environment.

Molecular Imprinting Polymer (MIP) technology, developed by Wulff and Sarhan in the 1970s, has advanced significantly [103]. Their imprinting techniques allowed nucleophilic polymers to be tailored post-synthesis with specific features, ex-situ integration, selective, reconfigurable apertures. MIPS demonstrate unparalleled levels of phi binding along with superb selectivity, making them efficient "smart sponges" able to absorb the clean-up of pharmaceuticals and pesticides as well as endocrine disruptors, even in minute quantities. Their durability against harsh environments makes them ideal for various water treatment applications. Additionally, MIPS are reusable, cost-effective, and easily integrated into filtration and sensing systems, paving the way for eco-friendly solutions in water purification.

Molecularly Imprinted Polymers Technology

Molecularly Imprinted Polymers (MIPs) are cross-linked polymers with specific cavities for target analytes created through imprinting templates. MIPs operate similarly to antibodies in that they selectively bind to specific analytes [104, 105]. While recognition is essential for biomedical applications, factors such as instability and high prices pose significant impediments. These challenges are aided by MIPs created through molecular imprinting technology [106]. MIPs also efficiently eliminate environmental pollutants due to their high selectivity and reusability. As a result, modifying them to focus on specific EPs becomes straightforward. The synthesis procedure includes the polymerization of monomers around a contaminating pollutant and subsequent removal of the pollutant to create cavities designed for adsorption. MIPs function as adsorption materials, offering a more eco-friendly paradigm by utilizing natural solvents, which grants MIP sorbents ample reusability from the standpoint of economic efficiency, serving as environmental protectors [107, 108].

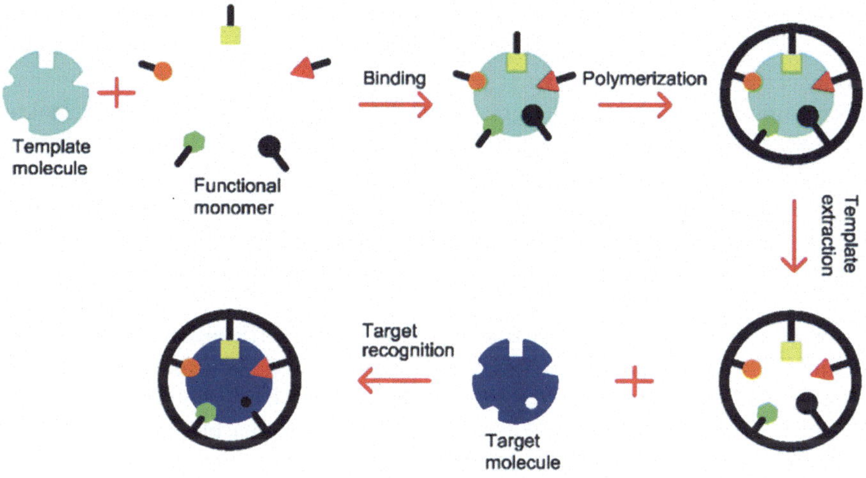

Fig. 6 Schematic illustration of the molecular imprinting process [113]

MIP composites improve sensitivity and selectivity while providing support at extreme pH is stagnant with modification, and stability with silica-based materials [109]. With a great surface-to-volume ratio, magnetic nanoparticles boost effectiveness, however, to reach optimal efficiency, they require functionalization for aggregation and to inhibit unwanted clustering [110]. Due to their strong adsorption ability, carbon-based materials are popular, but their high surface area and tendency for aggregation can affect performance [111]. Biopolymers are eco-friendly with high binding capacity, yet tend to be lower in selectivity [112]. The imprinting method entails polymerization of a monomer containing a cross-link with an initiator, while pore-forming solvents enhance solubility and refine pore structure based on monomer-template interactions (Fig. 6).

This technique strategically polymerizes a functional monomer with a crosslinker around a template molecule [114], crafting highly selective cavities that perfectly mimic the template's size and shape, enabling precise molecular recognition upon rebinding [115, 116]. procedure begins with the preparation of a pre-polymerization complex that subsequently undergoes polymerization into a three-dimensional network containing molecular recognition features [114]. The use of various templates, ranging from atoms and ions to entire microorganisms, to impart selectivity in synthetic polymers is one of the distinguishing characteristics of MIP technology [117].

Factors Influencing the MIP Process: The major driving forces of molecular imprinting include template molecule, polymer matrix, degree of crosslinking, solvent medium, and critical reaction conditions such as temperature and pH.

Functional Groups: MIPs undergo construction through the use of functional monomers that form specific cavities for the desired compounds, which allows recognition at a molecular level [118].

The effectiveness of MIPs greatly depends on what is chosen as a functional monomer to be added to the polymer scaffold, which in most cases is methacrylic acid as carboxylic or basic 4-vinyl pyridine or neutral styrene. Acidic monomers possess more attractive forces for greater affinity than their basic monomers, making it easier for them to bind to the template. The effectiveness of MIPs greatly depends on what is chosen as a functional monomer to be added to the polymer scaffold, which in most cases is methacrylic acid, as carboxylic or basic 4-vinyl pyridine or neutral styrene [119]. Acidic monomers possess more attractive forces for greater affinity than their basic monomers, making it easier for them to bind to the template [120].

The template molecule's functional groups impact MIP efficiency by shaping binding sites that recognise molecular size, shape, and chemical functionality [121]. Optimizing efficiency requires deep knowledge of the selective binding forces, such as hydrogen bonding, which are critical and underpin selective binding [122]. Methyl methacrylate, styrene, or acrylic acid are often employed to produce non-covalent MIPs, enjoying the benefits of sensitivity, flexibility, and design freedom [123]. Monomers such as 4-vinylbenzaldehyde are used in covalent approaches, while semi-covalent methods might use 3-isocyanato-propyltriethoxy [124].

Cross-linkers: Cross-linking agents, polymer-based materials, protective framework having structural integrity, distinct morphology, dimensions and value of mechanical strength. Concerning MIPs, their specificity is described in a more refined manner—they direct the functionalization of monomers spatially around the template during polymerization. In doing so, the construction of recognition cavities is further enhanced and, in addition, they are rendered sufficiently robust to resist deformation, monomer displacement, and template molecule removal, greatly improving the durability and selectivity of the imprinted polymer [125, 126]. For selection of cross-linkers, the choice should be made wisely as it impacts the polymer morphology, stability of imprinted sites and mechanical strength [127]. An insufficient number of cross-linkers can lead to nonfunctioning, poorly binding sites, while excessive amounts can make the polymer rigid, diminishing swelling capability and recognition [128]. Generally, possessing a greater quantity of cross-linker leads to greater stability of porous materials. However, surpassing 80% may lead to blocked sites hindering template removal [129, 130]. Among the most prevalent cross-linking agents in molecular imprinting are EGDMA, TRIM, and DVB. Among them, EGDMA stands out for its versatility and is frequently employed in the selective analysis of personal care products, owing to its excellent ability to form stable, high-fidelity polymer networks tailored for specific target molecules [131].

Porogens: Porogens are particularly important in the construction of MIPs since they assist in creating the porous framework during the polymerization stage [132]. As a consequence, the balance that has to be struck in choosing porogens in molecular imprinting is very intricate because they cannot be reactive with the monomer-template complex, as otherwise they would defeat the purpose of imprinting [133]. The solvents used for MIP synthesis are characterized by their polarity. Polar solvents are preferred in covalent imprinting because they provide controllable interactions, while non-polar solvents prevail in non-covalent approaches where binding is reversible and dynamic [118]. The choice of porogenic solvents changes the morphology of the polymer, its adsorption properties, and the sensitivity it exhibits towards template molecules [134]. In molecularly imprinted polymers (MIPs), porogens perform the dual function of enhancing the surface area and adsorption capacity as well as serving as scaffolds for structural frameworks. Assessing the performance of a monomer, cross-linker and template selection should be done carefully because they all impact synergistically. Solid porogen salt particles, which are soluble, serve as templates forming scaffolds and yielding more porous structures, allowing greater chemical coverage area. The cross-linker choice determines the polymer's stiffness, shape, and surface features, which in turn impact MIP performance [126].

Acetonitrile, chloroform, methanol, and tetrahydrofuran are some of the common solvents used, while ionic liquids are becoming more accepted because they do not catch fire and disperse easily [135]. However, after undergoing toxicological evaluations, a new solvent search is mandated [136]. The use of organic solvents can also be reduced by employing supercritical carbon dioxide for reactions [137] (Fig. 7).

***Monomers*:** Choosing the right monomers is pivotal in MIP synthesis, as they forge the essential interactions with the template that anchor a stable and selective pre-polymerisation complex [139, 140]. It's vital to synchronize the activities of monomers and templates for MIP stability. Preferred monomers exhibit cage effects, strong solvent interactions, and bonding stability [125]. Functional monomers, which include a recognition unit and a polymerizable unit, are crucial in determining the binding characteristics and cavity formation of molecularly imprinted polymers [141]. Methacrylic acid is notable for its strong hydrogen bonding and larger porosity [122, 142]. Examples of methyl methacrylate, styrene, and acrylic acid exemplify non-covalent MIPs, methyl methacrylate, styrene, and acrylic acid, capable of providing enhanced sensitivity and versatility. Monomers such as 4-vinylbenzaldehyde are used in covalent methods, whereas semi-covalent methods

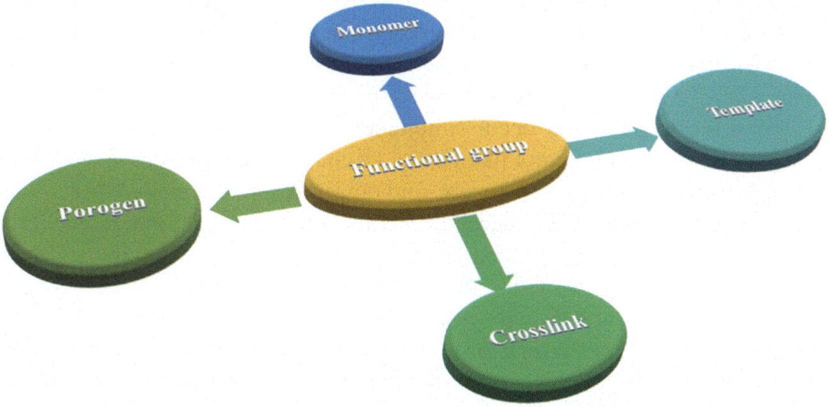

Fig. 7 The influence of functional groups on enhancing MIP efficiency [138]

may use 3-isocyanato-propyltriethoxysilane. These functional monomers aid in various preparatory techniques [124].

Currently, the precision extraction industry is undergoing an evolution with MIT's innovation on molecularly imprinted polymers (MIPs). These polymers, which are reusable and durable, can outperform traditional purification methods because they can withstand harsh conditions while selectively filtering out harmful substances. However, commercial scaling efforts run into issues like having guarded, sensitive, non-toxic MIPs and missing rapid methods for green synthesis.

4.3 Biodegradation

Biodegradation is an eco-friendly remedy for treating emerging contaminants ECs present in wastewater. It uses living microorganisms to change complex organic pollutants like pesticides and pharmaceuticals into harmless compounds. To complete these bioprocesses, the microbes need to transform the contaminants into non-toxic products like carbon dioxide and water. Its effectiveness is influenced by microbial community composition, contaminant structure, environmental conditions, and nutrient availability [143]. The combination of pollutant degradation with bioelectricity generation turns microbial fuel cells into revolutionaries within wastewater treatment. These bioreactors reinvent sustainability by merging two opposing functions into a single platform that redefines wastewater management.

Microbial Fuel Cells Process

The Microbial Fuel Cells (MFCs) restore the quality of wastewater by "treating" it with electricity, all while generating eco-friendly fuel. Organic pollutants in the wastewater are normally consumed by electroactive microbes that yield electrons and protons. For MFCs this means that electricity can be generated when the external circuit is powered, which enables the flow of electrons. Protons move to the cathode chamber where they undergo reduction by oxygen to clean water, thus, energy recovery and waste treatment occur simultaneously. This technique is capable of degrading difficult contaminants like pesticides and pharmaceuticals [144]. MFCs and biohydrogen production, biosensing, and power generation are used in conjunction with MFCs because of the merits they offer, such as direct energy conversion from substrates, controlled activated sludge deletion, no gas treatment or aeration energy needs, and no air disturbance during operation [145]. The idea was pioneered by M. C. Potter in 1911, however, it wasn't until the 1970s that researchers discovered that bacteria can transfer electrons through the anode [146]. Nowadays, the majority of MFCS focus on their eco-friendly means of electric power generation and waste pollutant treatment [147].

Important aspects for the functioning of a microbial fuel cell (MFC) system include its internal resistance, the electrode materials used, as well as the concentration of ions [148]. Considerable dependencies include the design of the cell, the type of electrodes used, as well as the proton exchange membrane (PEM), the mediators, substrates, microorganisms, and external factors [149]. They also require nutrients from wastes, for instance, swine or dairy [150]. Electrotrophic bacteria aid in detoxifying the environment by donating electrons from the electrodes [151]. It is assumed that the most used electrodes are made of carbon, metals and metal oxides, or conducting polymers. These electrodes should possess high conductivity and stability [152]. Microbial fuel cells have numerous advantages, like low production of sludge, reduced energy consumption, and the ability to treat complicated wastewater, which shows promise in utilizing them for environmental remediation [144].

MFCs mechanism: The below figure illustrates how exoelectrogens work to reduce toxic pollutants and aid in the bioremediation of oil spills. The expression for the electrotroph genes is distinct from that for the exoelectrogens [153]. At the anode, biological processes transfer electrons for the reductive precipitation of metal ions, aiding wastewater remediation and generating electricity. Anode biofilms oxidize organic substrates to enable electrochemical reductions at the cathode. The fate of reduced metals depends on solution chemistry, with successful reduction requiring the cathode's redox potential to be equal to or higher than the anode's [154]. A microbial fuel cell has an anode chamber for microbial substrate consumption and a cathode chamber with fresh water, separated by a proton exchange membrane. Microbes metabolize carbon sources, producing electrons, protons, and CO_2. Electrons flow through an external circuit while protons traverse the PEM. In the cathode chamber, electrons, protons, and oxygen react seamlessly, forming water and completing the cycle. The anode operates anaerobically, while the cathode is aerobic [145] (Fig. 8).

Wastewater from domestic, agricultural, and industrial sources contains materials that can be converted into renewable fuel for MFCs, which can function independently and treat wastewater without external energy. MFCs offer economic benefits, including improved energy efficiency, lower sludge production, fewer toxic byproducts, and recovery of valuable nutrients and electricity [155]. The performance of MFC is affected by biological activities like microorganism metabolism, as well as by physicochemical parameters such as the area of the electrodes and the rates of proton transfer [156]. Exogenic factors shaping performance are ohmic resistance, internal impedance, diffusion resistance, and charge transfer limitations [155].

5 Implications

Pollution Prevention Theory, the Precautionary Principle, and Systems Theory underlie the approaches to the challenges of emerging contaminants in water. The Pollution Prevention Theory provides support for waste and residual pollutants at their creation. It enables proactive measures to reduce pharmaceuticals, personal care products, and endocrine-disrupting chemicals before they enter wastewater at home.

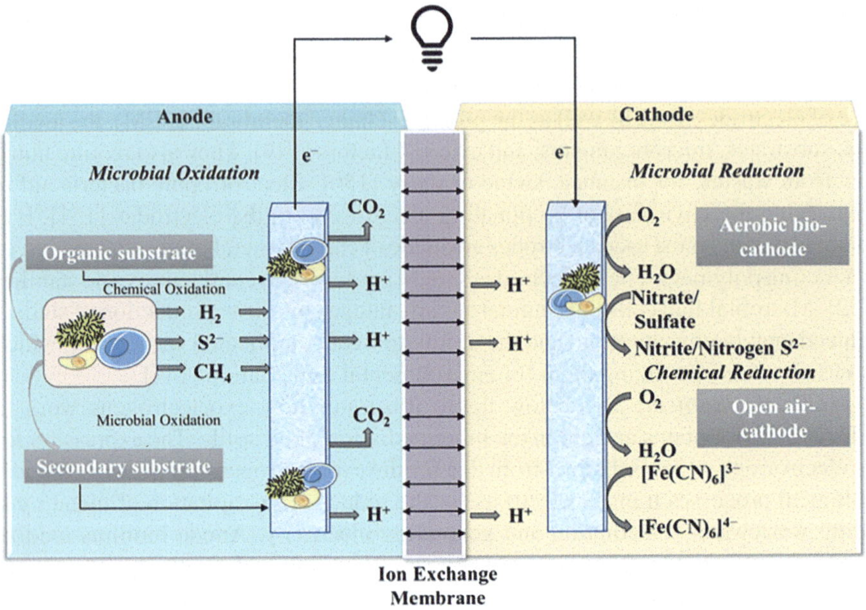

Fig. 8 Working principle of MFCS [155]

It supports the need to design sustainable products or approaches and the need to manage disposal methods. meta-playful recycling.

The Precautionary Principle calls for early action for environmental decision making, especially where there is inadequate scientific understanding. Many emerging contaminants are either not studied or poorly studied, calling for the application of the precautionary principle to address and inform regulatory structures and take proactive measures to reduce the overall potential for risk. Systems Theory emphasizes the relationships between wastewater treatment and environmental health, as well as the multifaceted challenges associated with emerging contaminants and the need for collaboration among scientists, engineers, policymakers, and policy makers to develop wastewater management systems. The theories require collaborative research concerning: the sources of emerging contaminants; behaviour, the targeting of risks, and facilitating strategies for managing emerging contaminants.

6 Scope for Future Work

Research on future work must address future and cost-effective technologies that can detect and remove emerging contaminants, particularly pharmaceuticals and personal care products. In addition, real-time monitoring and smart sensors of contaminants with new enhanced approaches, including bioremediation, nanotechnology, and artificial intelligence that require optimization, need reinforcement. Doing ecotoxicological research broadly and developing new regulations that we can similarly incorporate and regulate new contaminants or non-traditional pollutants are needed. Furthermore, increasing public awareness and promoting cooperation among scientists, technologists, and public policymakers will be essential for developing sustainable and effective wastewater management systems.

7 Conclusion

Emerging contaminants present difficult problems that need to be tackled with a thorough understanding of their origins, how they are used in industry, and how efficient the available treatment technologies are. The pharmaceuticals, textiles, petrochemicals, and electronics industries have emerged as primary sources of emerging contaminants, posing serious environmental risks. This research focuses on the sources and industrial use of emerging contaminants while studying novel treatment technologies. These include Advanced Oxidation Processes (AOPs), specially designed adsorbents, and even microbial biodegradation, which collectively offer hope to combat these pollutants.

While adsorption methods allow precise and highly selective removal of pollutants, AOPs allow rapid and complete breakdown of complex contaminants. Moreover, biodegradation utilizes natural microbial mechanisms for eco-friendly cleanup. The combination of sophisticated treatment techniques, continuous monitoring, and enforcement of laws is critical to addressing new pollutants and achieving enduring ecological protection. Striking a balance between emerging pollutant management requires intelligent monitoring technology, eco-friendly industrial processes, strong law enforcement, and systems thinking. An integrated approach that encompasses diverse fields will be necessary to defend against the growing risk these pollutants pose to the environment and human health.

Acknowledgements The author declares no conflict of interest.

References

1. Gavrilescu M (2010) Environmental biotechnology: achievements, opportunities and challenges. Dyn Biochem Process Biotechnol Mol Biol 4(1):1–36
2. Khan S, Naushad M, Govarthanan M, Iqbal J, Alfadul SM (2021) Emerging contaminants of high concern for the environment: current trends and future research. Environ Res 207:112609. https://doi.org/10.1016/j.envres.2021.112609
3. Deblonde T, Cossu-Leguille C, Hartemann P (2011) Emerging pollutants in wastewater: a review of the literature. Int J Hyg Environ Health 214(6):442–448. https://doi.org/10.1016/j.ijheh.2011.08.002

4. Koundouri P, Roseta-Palma C, Englezos N (2017) Out of sight, not out of mind: developments in economic models of groundwater management. Int Rev Environ Resour Econ 11(1):55–96. https://doi.org/10.1561/101.00000091
5. Wang X, Wang Z, Jiang S, Jordan RW, Gu Y (2023) Bioenrichment preference and human risk assessment of arsenic and metals in wild marine organisms from Dapeng (Mirs) Bay, South China Sea. Mar Pollut Bull 194:115305. https://doi.org/10.1016/j.marpolbul.2023.115305
6. Chen Z, Kahn ME, Liu Y, Wang Z (2018) The consequences of spatially differentiated water pollution regulation in China. J Environ Econ Manag 88:468–485. https://doi.org/10.1016/j.jeem.2018.01.010
7. Keerthanan S, Jayasinghe C, Biswas JK, Vithanage M (2020) Pharmaceutical and Personal Care Products (PPCPs) in the environment: plant uptake, translocation, bioaccumulation, and human health risks. Crit Rev Environ Sci Technol 51(12):1221–1258. https://doi.org/10.1080/10643389.2020.1753634
8. Kumar R, Qureshi M, Vishwakarma DK, Al-Ansari N, Kuriqi A, Elbeltagi A, Saraswat A (2022) A review on emerging water contaminants and the application of sustainable removal technologies. Case Stud Chem Environ Eng 6:100219. https://doi.org/10.1016/j.cscee.2022.100219
9. Thomaidis N, Asimakopoulos A (2013) Emerging contaminants: a tutorial mini-review. Glob NEST J 14(1):72–79. https://doi.org/10.30955/gnj.000823
10. Bolong N, Ismail A, Salim, Matsuura T (2009) A review of the effects of emerging contaminants in wastewater and options for their removal. Desalination 239(1–3):229–246.https://doi.org/10.1016/j.desal.2008.03.020
11. Morin-Crini N, Lichtfouse E, Liu G, Balaram V, Ribeiro ARL, Lu Z, Stock F, Carmona E, Teixeira MR, Picos-Corrales LA, Moreno-Pirajan JC, Giraldo L, Li C, Pandey A, Hocquet D, Torri G, Crini G (2022) Worldwide cases of water pollution by emerging contaminants: a review. Environ Chem Lett 20(4):2311–2338. https://doi.org/10.1007/s10311-022-01447-4
12. Boxall ABA, Brooks BW (2024) Pharmaceuticals and personal care products in the environment: what progress has been made in addressing the big research questions? Environ Toxicol Chem 43(3):481–487. https://doi.org/10.1002/etc.5827
13. Datta AR, Kang Q, Chen B, Ye X (2018) Fate and transport modelling of emerging pollutants from watersheds to oceans: a review. Adv Mar Biol. https://doi.org/10.1016/bs.amb.2018.09.002
14. Routti H, Atwood TC, Bechshoft T, Boltunov A, Ciesielski TM, Desforges J, Dietz R, Gabrielsen GW, Jenssen BM, Letcher RJ, McKinney MA, Morris AD, Riget FF, Sonne C, Styrishave B, Tartu S (2019) State of knowledge on current exposure, fate and potential health effects of contaminants in polar bears from the circumpolar Arctic. Sci Total Environ 664:1063–1083. https://doi.org/10.1016/j.scitotenv.2019.02.030
15. Previsic A, Rozman M, Mor J, Acuna V, Serra-Compte A, Petrović M, Sabater S (2019) Aquatic macroinvertebrates under stress: bioaccumulation of emerging contaminants and metabolomics implications. Sci Total Environ 704:135333. https://doi.org/10.1016/j.scitotenv.2019.135333
16. Pico Y, Belenguer V, Corcellas C, Diaz-Cruz, Eljarrat E, Farre M, Gago-Ferrero P, Huerta B, Navarro-Ortega A, Petrovic M, Rodríguez-Mozaz S, Sabater L, Santín G, Barcelo D (2018) Contaminants of emerging concern in freshwater fish from four Spanish Rivers. Sci Total Environ 659:1186–1198. https://doi.org/10.1016/j.scitotenv.2018.12.366
17. Wang F, Xiang L, Leung KS, Elsner M, Zhang Y, Guo Y, Pan B, Sun H, An T, Ying G, Brooks BW, Hou D, Helbling DE, Sun J, Qiu H, Vogel TM, Zhang W, Gao Y, Simpson MJ, Tiedje JM (2024) Emerging contaminants: a One Health perspective. Innovation 5(4):100612. https://doi.org/10.1016/j.xinn.2024.100612
18. Stefanakis AI, Becker JA (2019) A review of emerging contaminants in water. IGI Global eBooks. https://doi.org/10.4018/978-1-7998-1210-4.ch008
19. Iloms E, Ololade OO, Ogola HJO, Selvarajan R (2020) Investigating industrial effluent impact on municipal wastewater treatment plant in Vaal, South Africa. Int J Environ Res Public Health 17(3):1096. https://doi.org/10.3390/ijerph17031096

20. Zielinski W, Korzeniewska E, Harnisz M, Drzymała J, Felis E, Bajkacz S (2021) Wastewater treatment plants as a reservoir of integrase and antibiotic resistance genes—an epidemiological threat to workers and environment. Environ Int 156:106641. https://doi.org/10.1016/j.envint.2021.106641
21. Lapworth D, Baran N, Stuart M, Ward R (2012) Emerging organic contaminants in groundwater: a review of sources, fate and occurrence. Environ Pollut 163:287–303. https://doi.org/10.1016/j.envpol.2011.12.034
22. Buszka PM, Yeskis DJ, Kolpin DW, Furlong ET, Zaugg SD, Meyer MT (2009) Waste-indicator and pharmaceutical compounds in landfill-leachate-affected ground water near Elkhart, Indiana, 2000–2002. Bull Environ Contam Toxicol 82(6):653–659. https://doi.org/10.1007/s00128-009-9702-z
23. Fontes MK, Maranho LA, Pereira CDS (2020) Review on the occurrence and biological effects of illicit drugs in aquatic ecosystems. Environ Sci Pollut Res 27(25):30998–31034. https://doi.org/10.1007/s11356-020-08375-2
24. Selwe KP, Thorn JPR, Desrousseaux AOS, Dessent CEH, Sallach JB (2022) Emerging contaminant exposure to aquatic systems in the Southern African Development Community. Environ Toxicol Chem 41(2):382–395. https://doi.org/10.1002/etc.5284
25. Mahapatra S, Samal K, Dash RR (2022) Waste Stabilization Pond (WSP) for wastewater treatment: a review on factors, modelling and cost analysis. J Environ Manag 308:114668. https://doi.org/10.1016/j.jenvman.2022.114668
26. Samal K, Trivedi S (2020) A statistical and kinetic approach to develop a Floating Bed for the treatment of wastewater. J Environ Chem Eng 8(5):104102. https://doi.org/10.1016/j.jece.2020.104102
27. Samal K, Dash RR (2021) Modelling of pollutants removal in Integrated Vermifilter (IVmF) using response surface methodology. Clean Eng Technol 2:100060. https://doi.org/10.1016/j.clet.2021.100060
28. Mamta N, Bhushan S, Rana MS, Raychaudhuri S, Simsek H, Prajapati SK (2020) Algae-and bacteria-driven technologies for pharmaceutical remediation in wastewater. In: Elsevier eBooks, pp 373–408. https://doi.org/10.1016/b978-0-12-821014-7.00015-0
29. Feier B, Gui A, Cristea C, Săndulescu R (2017) Electrochemical determination of cephalosporins using a bare boron-doped diamond electrode. Anal Chim Acta 976:25–34. https://doi.org/10.1016/j.aca.2017.04.050
30. Hollman J, Dominic JA, Achari G, Langford CH, Tay J (2018) Effect of UV dose on degradation of venlafaxine using UV/H_2O_2: perspective of augmenting UV units in wastewater treatment. Environ Technol 41(9):1107–1116. https://doi.org/10.1080/09593330.2018.1521475
31. Garcia VSG, Tominaga FK, Rosa JM, Borrely SI (2024) Emerging pollutants in textile wastewater: an ecotoxicological assessment focusing on surfactants. Environ Sci Pollut Res 31(19):27817–27828. https://doi.org/10.1007/s11356-024-32963-1
32. Liu J, Liang J, Ding J, Zhang G, Zeng X, Yang Q, Zhu B, Gao W (2021) Microfiber pollution: an ongoing major environmental issue related to the sustainable development of textile and clothing industry. Environ Dev Sustain 23(8):11240–11256. https://doi.org/10.1007/s10668-020-01173-3
33. Konwar M (2020) Textile industry and its environmental impacts: a review. Indian J Pure Appl Biosci 8(3):134–139. https://doi.org/10.18782/2582-2845.7848
34. Norarmi NFB, Tajudin NHB, Hafizo NSHB, Luzi WNSBWM, Amin NaBM, Aziz SNFB, Musa ABH (2022) A review on textile and clothing industry impacts on the environment. Int J Acad Res Bus Soc Sci 12(10). https://doi.org/10.6007/ijarbss/v12-i10/11090
35. De Carvalho JGR, Augusto HC, Ferraz R, Delerue-Matos C, Fernandes VC (2024) Micro(nano)plastic and related chemicals: emerging contaminants in environment, food and health impacts. Toxics 12(10):762. https://doi.org/10.3390/toxics12100762
36. Abbas M, Abbas S, Hussain N, Javeed MT, Ghaffar A, Nadeem M, Khaliq M, Ullah S, Parveen Z, Khan KA, Ghramh HA (2024) Assessment of residues from common pesticides and associated risks in Pakistan. Environ Monit Assess 196(11):1061. https://doi.org/10.1007/s10661-024-13220-x

37. Houessionon MGK, Ouendo ED, Bouland C, Takyi SA, Kedote NM, Fayomi B, Fobil JN, Basu N (2021) Environmental heavy metal contamination from Electronic Waste (E-Waste) recycling activities worldwide: a systematic review from 2005 to 2017. Int J Environ Res Public Health 18(7):3517. https://doi.org/10.3390/ijerph18073517
38. Wang Z, Cousins IT, Scheringer M, Hungerbuhler K (2022) Global contamination of perfluorinated compounds from the electronics industry. Environ Pollut 2022(294):118659
39. Jowitt S, Mudd G, Thompson J (2020) The rare earth elements in modern technology: environmental and economic challenges. Resour Policy 2020(67):101722
40. Robinson BH (2009) E-waste: an assessment of global production and environmental impacts. Sci Total Environ 408(2):183–191. https://doi.org/10.1016/j.scitotenv.2009.09.044
41. Sharma 1D (2019) Impact of petrochemical industry on the environment. Acad J Earth Sci Geol Stud 01(01):10–14
42. Singh S, Sivaram N, Dhanjal DS, Assefa H, Singh J, Ramamurthy PC (2024) Navigating the complexity of emerging contaminants: sources, impacts, and remediation strategies. J Indian Inst Sci. https://doi.org/10.1007/s41745-024-00446-4
43. Huang H (2021) Integrated assessment of key mitigation technologies in petroleum and petrochemical industry under CO_2 emissions peaking and neutrality targets. In: Hai HH (ed) International conference on applied energy 2021, vol 18, p 113
44. Pal D, Sen S (2023) Emerging petroleum pollutants and their adverse effects on the environment. In: Environmental science and engineering, pp 103–137. https://doi.org/10.1007/978-3-031-48220-5_5
45. Krishnakumar S, Singh DSH, Godson PS, Thanga SG (2022) Emerging pollutants: impact on environment, management, and challenges. Environ Sci Pollut Res 29(48):72309–72311. https://doi.org/10.1007/s11356-022-22859-3
46. Sundt RC, Baussant T, Beyer J (2008) Uptake and tissue distribution of C4–C7 alkylphenols in Atlantic cod (*Gadus morhua*): relevance for biomonitoring of produced water discharges from oil production. Mar Pollut Bull 58(1):72–79. https://doi.org/10.1016/j.marpolbul.2008.09.012
47. Wang Y, Yu G, Deng S, Huang J, Wang B (2018) The electro-peroxone process for the abatement of emerging contaminants: mechanisms, recent advances, and prospects. Chemosphere 208:640–654. https://doi.org/10.1016/j.chemosphere.2018.05.095
48. Oller I, Malato S, Sánchez-Perez J (2010) Combination of advanced oxidation processes and biological treatments for wastewater decontamination—a review. Sci Total Environ 409(20):4141–4166. https://doi.org/10.1016/j.scitotenv.2010.08.061
49. Ikai H, Nakamura K, Shirato M, Kanno T, Iwasawa A, Sasaki K, Niwano Y, Kohno M (2010) Photolysis of hydrogen peroxide, an effective disinfection system via hydroxyl radical formation. Antimicrob Agents Chemother 54(12):5086–5091. https://doi.org/10.1128/aac.00751-10
50. Cooper WJ, Cramer CJ, Martin NH, Mezyk SP, O'Shea KE, Von Sonntag C (2009) Free radical mechanisms for the treatment of methyl Tert-Butyl ether (MTBE) via advanced oxidation/reductive processes in aqueous solutions. Chem Rev 109(3):1302–1345. https://doi.org/10.1021/cr078024c
51. Huang C, Dong C, Tang Z (1993) Advanced chemical oxidation: its present role and potential future in hazardous waste treatment. Waste Manag 13(5–7):361–377. https://doi.org/10.1016/0956-053x(93)90070-d
52. Ma D, Yi H, Lai C, Liu X, Huo X, An Z, Li L, Fu Y, Li B, Zhang M, Qin L, Liu S, Yang L (2021) Critical review of advanced oxidation processes in organic wastewater treatment. Chemosphere 275:130104. https://doi.org/10.1016/j.chemosphere.2021.130104
53. Mousazadeh M, Alizadeh S, Frontistis Z, Kabdaşlı I, Niaragh EK, Qodah ZA, Naghdali Z, Mahmoud A, Sandoval M, Butler E, Emamjomeh M (2021) Electrocoagulation as a promising defluoridation technology from water: a review of state of the art of removal mechanisms and performance trends. Water 13(5):656. https://doi.org/10.3390/w13050656
54. Titchou FE, Zazou H, Afanga H, Gaayda JE, Akbour RA, Nidheesh PV, Hamdani M (2021) Removal of organic pollutants from wastewater by advanced oxidation processes and its

combination with membrane processes. Chem Eng Process—Process Intensif 169:108631. https://doi.org/10.1016/j.cep.2021.108631
55. Tijani JO, Fatoba OO, Madzivire G, Petrik LF (2014) A review of combined advanced oxidation technologies for the removal of organic pollutants from water. Water Air Soil Pollut 225(9):2102. https://doi.org/10.1007/s11270-014-2102-y
56. Wang Z, Chen C, Ma W, Zhao J (2012) Photochemical coupling of iron redox reactions and transformation of Low-Molecular-Weight organic matter. J Phys Chem Lett 3(15):2044–2051. https://doi.org/10.1021/jz3005333
57. Sun C, Chen C, Ma W, Zhao J (2010) Photodegradation of organic pollutants catalyzed by iron species under visible light irradiation. Phys Chem Chem Phys 13(6):1957–1969. https://doi.org/10.1039/C0CP01203C
58. Parulekar PJ, Sahasrabudhe I, Hegiste M, Gupte S (2024) Review of advanced oxidation processes for wastewater treatment. Int J Eng Res Technol 13(02)
59. Litter MI, Slodowicz M (2017) An overview on heterogeneous Fenton and photo Fenton reactions using zerovalent iron materials. J Adv Oxid Technol 20(1):20160164. https://doi.org/10.1515/jaots-2016-0164
60. Lin P, Yeh C, Jiang J (2021) Theoretical insight into hydroxyl production via H2O2 decomposition over the $Fe_3O_4(311)$ surface. RSC Adv 11(57):36257–36264. https://doi.org/10.1039/d1ra06943h
61. Babuponnusami A, Muthukumar K (2013) A review on Fenton and improvements to the Fenton process for wastewater treatment. J Environ Chem Eng 2(1):557–572. https://doi.org/10.1016/j.jece.2013.10.011
62. Masood AS, Ali MS, Manzar MS, Khan NA, Khan AH (2023) Current situation of pharmaceutical wastewater around the globe. In: Elsevier eBooks, pp 19–52. https://doi.org/10.1016/b978-0-323-99160-5.00013-8.
63. Kulistakova A (2023) Removal of pharmaceutical micropollutants from real wastewater matrices by means of photochemical advanced oxidation processes—a review. J Water Process Eng 53:103727. https://doi.org/10.1016/j.jwpe.2023.103727
64. Homem V, Santos L (2011) Degradation and removal methods of antibiotics from aqueous matrices—a review. J Environ Manag 92(10):2304–2347. https://doi.org/10.1016/j.jenvman.2011.05.023
65. Xiong R, Lu Z, Tang Q, Huang X, Ruan H, Jiang W, Chen Y, Liu Z, Kang J, Liu D (2020) UV-LED/chlorine degradation of propranolol in water: degradation pathway and product toxicity. Chemosphere 248:125957. https://doi.org/10.1016/j.chemosphere.2020.125957
66. Song K, Mohseni M, Taghipour F (2016) Application of ultraviolet light-emitting diodes (UV-LEDs) for water disinfection: a review. Water Res 94:341–349. https://doi.org/10.1016/j.watres.2016.03.003
67. Jarvis P, Autin O, Goslan EH, Hassard F (2019) Application of Ultraviolet Light-Emitting Diodes (UV-LED) to Full-Scale Drinking-Water disinfection. Water 11(9):1894. https://doi.org/10.3390/w11091894
68. Wang W, Wu Q, Li Z, Lu Y, Du Y, Wang T, Huang N, Hu H (2016) Light-emitting diodes as an emerging UV source for UV/chlorine oxidation: carbamazepine degradation and toxicity changes. Chem Eng J 310:148–156. https://doi.org/10.1016/j.cej.2016.10.097
69. Verma S, Sillanpaa M (2015) Degradation of anatoxin-a by UV-C LED and UV-C LED/H_2O_2 advanced oxidation processes. Chem Eng J 274:274–281. https://doi.org/10.1016/j.cej.2015.03.128
70. Phugare SS, Kalyani DC, Surwase SN, Jadhav JP (2011) Ecofriendly degradation, decolorization and detoxification of textile effluent by a developed bacterial consortium. Ecotoxicol Environ Saf 74(5):1288–1296. https://doi.org/10.1016/j.ecoenv.2011.03.003
71. Rao Y, Chu W (2010) Degradation of linuron by UV, ozonation, and UV/O3 processes—effect of anions and reaction mechanism. J Hazard Mater 180(1–3):514–523. https://doi.org/10.1016/j.jhazmat.2010.04.063
72. Gogate PR, Pandit AB (2003) A review of imperative technologies for wastewater treatment I: oxidation technologies at ambient conditions. Adv Environ Res 8(3–4):501–551. https://doi.org/10.1016/s1093-0191(03)00032-7

73. Miklos DB, Remy C, Jekel M, Linden KG, Drewes JE, Hübner U (2018) Evaluation of advanced oxidation processes for water and wastewater treatment—a critical review. Water Res 139:118–131. https://doi.org/10.1016/j.watres.2018.03.042
74. Johnson G, Bullen L, Akil M (2017) Review on emerging pollutants and advanced oxidation processes. Int J Adv Res 5(2):2315–2324. https://doi.org/10.21474/ijar01/3412
75. Deng Y, Zhao R (2015) Advanced Oxidation Processes (AOPs) in wastewater treatment. Curr Pollut Rep 1(3):167–176. https://doi.org/10.1007/s40726-015-0015-z
76. Kishimoto N (2019) State of the art of UV/Chlorine advanced oxidation processes: their mechanism, byproducts formation, process variation, and applications. J Water Environ Technol 17(5):302–335. https://doi.org/10.2965/jwet.19-021
77. Cerreta G, Roccamante MA, Oller I, Malato S, Rizzo L (2019) Contaminants of emerging concern removal from real wastewater by UV/free chlorine process: a comparison with solar/free chlorine and UV/H_2O_2 at pilot scale. Chemosphere 236:124354. https://doi.org/10.1016/j.chemosphere.2019.124354
78. Carra I, Lozano JF, Autin O, Bolton JR, Jarvis P (2020) Disinfection by-product formation during UV/Chlorine treatment of pesticides in a novel UV-LED reactor at 285 nm and the mitigation impact of GAC treatment. Sci Total Environ 712:136413. https://doi.org/10.1016/j.scitotenv.2019.136413
79. Ghanbari F, Yaghoot-Nezhad A, Wacławek S, Lin KA, Rodríguez-Chueca J, Mehdipour F (2021) Comparative investigation of acetaminophen degradation in aqueous solution by UV/Chlorine and UV/H_2O_2 processes: kinetics and toxicity assessment, process feasibility and products identification. Chemosphere 285:131455. https://doi.org/10.1016/j.chemosphere.2021.131455
80. Khader EH, Abbood NS, Albyati TN, Mohammed TJ (2022) Performance evaluation of integral process for treatment of oilfield wastewater. AIP Conf Proc 2473:030035. https://doi.org/10.1063/5.0091914
81. Coleman H, Vimonses V, Leslie G, Amal R (2007) Removal of contaminants of concern in water using advanced oxidation techniques. Water Sci Technol 55(12):301–306. https://doi.org/10.2166/wst.2007.421
82. Kong CPY, Suhaimi NAA, Shahri NNM, Lim J, Nur M, Hobley J, Usman A (2022) Auramine o UV photocatalytic degradation on TIO_2 nanoparticles in a heterogeneous aqueous solution. Catalysts 12(9):975. https://doi.org/10.3390/catal12090975
83. Suhaimi NAA, Kong CPY, Shahri NNM, Nur M, Hobley J, Usman A (2022) Dynamics of diffusion- and immobilization-limited photocatalytic degradation of dyes by metal oxide nanoparticles in binary or ternary solutions. Catalysts 12(10):1254. https://doi.org/10.3390/catal12101254
84. Elmolla ES, Chaudhuri M (2009) Photocatalytic degradation of amoxicillin, ampicillin and cloxacillin antibiotics in aqueous solution using UV/TiO_2 and UV/H_2O_2/TiO_2 photocatalysis. Desalination 252(1–3):46–52. https://doi.org/10.1016/j.desal.2009.11.003
85. Vaitsis C, Mechili M, Argirusis N, Kanellou E, Pandis PK, Sourkouni G, Zorpas A, Argirusis C (2020) Ultrasound-assisted preparation methods of nanoparticles for energy-related applications. IntechOpen eBooks. https://doi.org/10.5772/intechopen.92802
86. Khader EH, Muslim SA, Saady NMC, Ali NS, Salih IK, Mohammed TJ, Albayati TM, Zendehboudi S (2024) Recent advances in photocatalytic advanced oxidation processes for organic compound degradation: A review. Desalin Water Treat 318:100384. https://doi.org/10.1016/j.dwt.2024.100384
87. Baruah A, Chaudhary V, Malik R, Tomer VK (2018) Nanotechnology-based solutions for wastewater treatment. In: Elsevier eBooks, pp 337–368. https://doi.org/10.1016/b978-0-12-813902-8.00017-4
88. Sun J, Jiang C, Wu Z, Liu Y, Sun S (2022) A review on the progress of the photocatalytic removal of refractory pollutants from water by BiOBr-based nanocomposites. Chemosphere 308:136107. https://doi.org/10.1016/j.chemosphere.2022.136107
89. Madi M, Tahir M, Tasleem S (2021) Advances in structural modification of perovskite semiconductors for visible light-assisted photocatalytic CO_2 reduction to renewable solar fuels: a review. J Environ Chem Eng 9(5):106264. https://doi.org/10.1016/j.jece.2021.106264

90. Ikhlaq A, Javed F, Niaz A, Munir HMS, Qi F (2020) Combined UV catalytic ozonation process on iron-loaded peanut shell ash for the removal of methylene blue from aqueous solution. Desalin Water Treat 200:231–240. https://doi.org/10.5004/dwt.2020.26135
91. Albayati TM, Sabri AA, Abed DB (2020) Functionalized SBA-15 by amine group for removal of Ni(II) heavy metal ion in the batch adsorption system. Desalin Water Treat 174:301–310. https://doi.org/10.5004/dwt.2020.24845
92. Hassaan MA, Nemr AE, Madkour FF (2016) Advanced oxidation processes of Mordant Violet 40 dye in freshwater and seawater. Egypt J Aquat Res 43(1):1–9. https://doi.org/10.1016/j.ejar.2016.09.004
93. Qahtan TF, Owolabi TO, Olubi OE, Hezam A (2023) State-of-the-art, challenges and prospects of heterogeneous tandem photocatalysis. Coord Chem Rev 492:215276. https://doi.org/10.1016/j.ccr.2023.215276
94. Liu Y, Ma Q, Yang M, Dong X, Yang Y, Wang J, Yu W, Liu G (2015) Flexible hollow nanofibers: novel one-pot electrospinning construction, structure and tunable luminescence–electricity–magnetism trifunctionality. Chem Eng J 284:831–840. https://doi.org/10.1016/j.cej.2015.09.030
95. Chan SHS, Wu TY, Juan JC, Teh CY (2011) Recent developments of metal oxide semiconductors as photocatalysts in advanced oxidation processes (AOPs) for treatment of dye waste-water. J Chem Technol Biotechnol 86(9):1130–1158. https://doi.org/10.1002/jctb.2636
96. Wang H, Zhang L, Chen Z, Hu J, Li S, Wang Z, Liu J, Wang X (2014) Semiconductor heterojunction photocatalysts: design, construction, and photocatalytic performances. Chem Soc Rev 43(15):5234. https://doi.org/10.1039/c4cs00126e
97. Lofranoa G, Pedrazzanib R, Libralatoc G, Carotenuto M (2017) Advanced oxidation processes for antibiotics removal: a review. Curr Org Chem 21:1–14
98. Legrini O, Oliveros E, Braun AM (1993) Photochemical processes for water treatment. Chem Rev 93(2):671–698. https://doi.org/10.1021/cr00018a003
99. O'Shea KE, Dionysiou DD (2012) Advanced oxidation processes for water treatment. J Phys Chem Lett 3(15):2112–2113. https://doi.org/10.1021/jz300929x
100. Almomani FA, Shawaqfah M, Bhosale RR, Kumar A (2016) Removal of emerging pharmaceuticals from wastewater by ozone-based advanced oxidation processes. Environ Prog Sustain Energy 35(4):982–995. https://doi.org/10.1002/ep.12306
101. Elmolla ES, Chaudhuri M (2010) Comparison of different advanced oxidation processes for treatment of antibiotic aqueous solution. Desalination 256(1–3):43–47. https://doi.org/10.1016/j.desal.2010.02.019
102. Baloo L, Isa MH, Sapari NB, Jagaba AH, Wei LJ, Yavari S, Razali R, Vasu R (2021) Adsorptive removal of methylene blue and acid orange 10 dyes from aqueous solutions using oil palm wastes-derived activated carbons. Alex Eng J 60(6):5611–5629. https://doi.org/10.1016/j.aej.2021.04.044
103. Wulff G, Sarhan A (1972) The use of polymers with enzyme-analogous structures for the resolution of racemates. Angew Chem Int Ed 11:341–344
104. Masqu N (2001) Molecularly imprinted polymers: new tailor-made materials for selective solid-phase extraction. Trends Anal Chem 20(9):477–486. https://doi.org/10.1016/s0165-9936(01)00062-0
105. Alexander C, Andersson HS, Andersson LI, Ansell RJ, Kirsch N, Nicholls IA, O'Mahony J, Whitcombe MJ (2006) Molecular imprinting science and technology: a survey of the literature for the years up to and including 2003. J Mol Recognit 19(2):106–180. https://doi.org/10.1002/jmr.760
106. Han S, Song Y, Liu S, Zhao L, Sun R (2022) Dual responsive molecularly imprinted polymers based on UiO-66-DOX for selective targeting tumor cells and controlled drug release. Eur Polym J 171:111219. https://doi.org/10.1016/j.eurpolymj.2022.111219
107. Zhao M, Shao H, Ma J, Li H, He Y, Wang M, Jin F, Wang J, El-Aty AA, Hacımftuoglu A, Yan F, Wang Y, She Y (2019) Preparation of core-shell magnetic molecularly imprinted polymers for extraction of patulin from juice samples. J Chromatogr A 1615:460751. https://doi.org/10.1016/j.chroma.2019.460751

108. Banan K, Hatamabadi D, Afsharara H, Mostafiz B, Sadeghi H, Rashidi S, Beirami AD, Shahbazi M, Kecili R, Hussain CM, Ghorbani-Bidkorbeh F (2021) MIP-based extraction techniques for the determination of antibiotic residues in edible meat samples: design, performance & recent developments. Trends Food Sci Technol 119:164–178. https://doi.org/10.1016/j.tifs.2021.11.022
109. Roshan S, Mujahid A, Afzal A, Nisar I, Ahmad MN, Hussain T, Bajwa SZ (2019) Molecularly imprinted Polymer-Silica hybrid particles for biomimetic recognition of target drugs. Adv Polym Technol 2019:1–7. https://doi.org/10.1155/2019/9432412
110. Ariani MD, Zuhrotun A, Manesiotis P, Hasanah AN (2022) Magnetic molecularly imprinted polymers: an update on their use in the separation of active compounds from natural products. Polymers 14(7):1389. https://doi.org/10.3390/polym14071389
111. Ding Q, Guo Z, Chen W, Yu H, Zhu X, Liu Q, Fu M (2021) Biomass activated carbon–derived imprinted polymer with multi-boronic acid sites for selective capture of glycoprotein. J Colloid Interface Sci 596:225–232. https://doi.org/10.1016/j.jcis.2021.03.151
112. Bagheri AR, Aramesh N, Lee HK (2022) Chitosan- and/or cellulose-based materials in analytical extraction processes: a review. Trends Anal Chem 157:116770. https://doi.org/10.1016/j.trac.2022.116770
113. Kadhem AJ, Gentile GJ, De Cortalezzi MMF (2021) Molecularly Imprinted Polymers (MIPs) in sensors for environmental and biomedical applications: a review. Molecules 26(20):6233. https://doi.org/10.3390/molecules26206233
114. Cimen D, Bereli N, Denizli A (2022) Molecularly imprinted surface plasmon resonance sensor-based devices for clinical applications. Deleted J 1(1):202–212. https://doi.org/10.1007/s44174-022-00022-3
115. Janczura M, Lulinski P, Sobiech M (2021) Imprinting technology for effective sorbent fabrication: current state-of-art and future prospects. Materials 14(8):1850. https://doi.org/10.3390/ma14081850
116. Turiel E, Martín-Esteban A (2019) Molecularly imprinted polymers-based microextraction techniques. Trends Anal Chem 118:574–586. https://doi.org/10.1016/j.trac.2019.06.016
117. Araz A (2021) Bovine serum albumin detection by using molecularly imprinted surface plasmon resonance sensors. Turk J Chem. https://doi.org/10.3906/kim-2109-6
118. Sajini T, Mathew B (2021) A brief overview of molecularly imprinted polymers: Highlighting computational design, nano and photo-responsive imprinting. Talanta Open 4:100072. https://doi.org/10.1016/j.talo.2021.100072
119. Marc M, Wieczorek PP (2019) Introduction to MIP synthesis, characteristics and analytical application. In: Comprehensive analytical chemistry, pp 1–15. https://doi.org/10.1016/bs.coac.2019.05.010
120. Nasraoui C, Jaoued-Grayaa N, Vanoye L, Chevalier Y, Hbaieb S (2022) Development of molecularly imprinted polymer for the selective recognition of the weakly interacting fenamiphos molecule. Eur Polym J 177:111441. https://doi.org/10.1016/j.eurpolymj.2022.111441
121. Xie X, Bu Y, Wang S (2016) Molecularly imprinting: a tool of modern chemistry for analysis and monitoring of phenolic environmental estrogens. Rev Anal Chem 35(2):87–97. https://doi.org/10.1515/revac-2016-0005
122. Hasanah AN, Maelaningsih FS, Apriliandi F, Sabarudin A (2020) Synthesis and characterisation of a monolithic imprinted column using a methacrylic acid monomer with porogen propanol for atenolol analysis. J Anal Methods Chem 2020:1–15. https://doi.org/10.1155/2020/3027618
123. Guo L, Deng Q, Fang G, Gao W, Wang S (2011) Preparation and evaluation of molecularly imprinted ionic liquids polymer as sorbent for online solid-phase extraction of chlorsulfuron in environmental water samples. J Chromatogr A 1218(37):6271–6277. https://doi.org/10.1016/j.chroma.2011.07.016
124. Chen L, Wang X, Lu W, Wu X, Li J (2016) Molecular imprinting: perspectives and applications. Chem Soc Rev 45(8):2137–2211. https://doi.org/10.1039/c6cs00061d

125. Vasapollo G, Del Sole R, Mergola L, Lazzoi MR, Scardino A, Scorrano S, Mele G (2011) Molecularly imprinted polymers: present and future prospective. Int J Mol Sci 12(9):5908–5945. https://doi.org/10.3390/ijms12095908
126. Mueller A (2021) A note about crosslinking density in imprinting polymerization. Molecules 26(17):5139. https://doi.org/10.3390/molecules26175139
127. Singh M, Singh S, Singh SP, Patel SS (2020) Recent advancement of carbon nanomaterials engrained molecular imprinted polymer for environmental matrix. Trends Environ Anal Chem 27:e00092. https://doi.org/10.1016/j.teac.2020.e00092
128. Zhang J, Wang M, Liu P, Zhang X, Huo T, Chen Z, Liu C, Zhang B, Zhao S, Chen Z (2021) Study on synthesis and adsorption properties of glutathione surface molecular imprinting polymer. Pigm Resin Technol 50(6):585–594. https://doi.org/10.1108/prt-08-2020-0091
129. Włoch M, Datta J (2019) Synthesis and polymerisation techniques of molecularly imprinted polymers. In: Comprehensive analytical chemistry, pp 17–40. https://doi.org/10.1016/bs.coac.2019.05.011
130. Song X, Li J, Wang J, Chen L (2009) Quercetin molecularly imprinted polymers: preparation, recognition characteristics and properties as sorbent for solid-phase extraction. Talanta 80(2):694–702. https://doi.org/10.1016/j.talanta.2009.07.051
131. Liang R, Kou L, Chen Z, Qin W (2013) Molecularly imprinted nanoparticles based potentiometric sensor with a nanomolar detection limit. Sens Actuators B 188:972–977. https://doi.org/10.1016/j.snb.2013.07.110
132. Azizi A, Bottaro CS (2019) A critical review of molecularly imprinted polymers for the analysis of organic pollutants in environmental water samples. J Chromatogr A 1614:460603. https://doi.org/10.1016/j.chroma.2019.460603
133. Lah NFC, Ahmad AL, Low SC, Zaulkiflee ND (2021) Isotherm and electrochemical properties of atrazine sensing using PVC/MIP: effect of porogenic solvent concentration ratio. Membranes 11(9):657. https://doi.org/10.3390/membranes11090657
134. Cajamarca FA, Tarley CRT (2022) Influence of synthesis parameters and polymerization methods on the selective and adsorptive performance of Bio-Inspired ion imprinted polymers. Separations 9(10):266. https://doi.org/10.3390/separations9100266
135. Liu H, Jin P, Zhu F, Nie L, Qiu H (2020) A review on the use of ionic liquids in preparation of molecularly imprinted polymers for applications in solid-phase extraction. Trends Anal Chem 134:116132. https://doi.org/10.1016/j.trac.2020.116132
136. Booker K, Holdsworth CI, Doherty CM, Hill AJ, Bowyer MC, McCluskey A (2014) Ionic liquids as porogens for molecularly imprinted polymers: propranolol, a model study. Org Biomol Chem 12(37):7201–7210. https://doi.org/10.1039/c4ob00547c
137. Viveiros R, Rebocho S, Casimiro T (2018) Green strategies for molecularly imprinted polymer development. Polymers 10(3):306. https://doi.org/10.3390/polym10030306
138. Gkika DA, Tolkou AK, Lambropoulou DA, Bikiaris DN, Kokkinos P, Kalavrouziotis IK, Kyzas GZ (2023) Application of molecularly imprinted polymers (MIPs) as environmental separation tools. RSC Appl Polym 2(2):127–148. https://doi.org/10.1039/d3lp00203a
139. Pratiwi R, Megantara S, Rahayu D, Pitaloka I, Hasanah AN (2018) Comparison of bulk and precipitation polymerization method of synthesis molecular imprinted solid phase extraction for atenolol using methacrylic acid. J Young Pharm 11(1):12–16. https://doi.org/10.5530/jyp.2019.11.3
140. Haupt K, Rangel PXM, Bui BTS (2020) Molecularly imprinted polymers: antibody mimics for bioimaging and therapy. Chem Rev 120(17):9554–9582. https://doi.org/10.1021/acs.chemrev.0c00428
141. Sharma PS, Iskierko Z, Pietrzyk-Le A, D'Souza F, Kutner W (2014) Bioinspired intelligent molecularly imprinted polymers for chemosensing: a mini review. Electrochem Commun 50:81–87. https://doi.org/10.1016/j.elecom.2014.11.019
142. Farooq S, Wu H, Nie J, Ahmad S, Muhammad I, Zeeshan M, Khan R, Asim M (2021) Application, advancement and green aspects of magnetic molecularly imprinted polymers in pesticide residue detection. Sci Total Environ 804:150293. https://doi.org/10.1016/j.scitotenv.2021.150293

143. Sharma B, Dangi AK, Shukla P (2018) Contemporary enzyme-based technologies for bioremediation: a review. J Environ Manag 210:10–22. https://doi.org/10.1016/j.jenvman.2017.12.075
144. Santoro C, Arbizzani C, Erable B, Ieropoulos I (2017) Microbial fuel cells: from fundamentals to applications. A review. J Power Sources 356:225–244. https://doi.org/10.1016/j.jpowsour.2017.03.109
145. Gajendra Prasad J, Panda S (2018) Microbial fuel cells: types of MFC and different source of substrate. Int J Latest Technol Eng Manag Appl Sci 7(5):158–165
146. Drendel G, Mathews ER, Semenec L, Franks AE (2018) Microbial fuel cells, related technologies, and their applications. Appl Sci 8(12):2384. https://doi.org/10.3390/app8122384
147. Kim K, Yang W, Logan BE (2015) Impact of electrode configurations on retention time and domestic wastewater treatment efficiency using microbial fuel cells. Water Res 80:41–46. https://doi.org/10.1016/j.watres.2015.05.021
148. Araneda I, Tapia NF, Allende KL, Vargas IT (2018) Constructed Wetland-Microbial fuel cells for sustainable greywater treatment. Water 10(7):940. https://doi.org/10.3390/w10070940
149. Chaudhuri SK, Lovley DR (2003) Electricity generation by direct oxidation of glucose in mediatorless microbial fuel cells. Nat Biotechnol 21(10):1229–1232. https://doi.org/10.1038/nbt867
150. Tremouli A, Antonopoulou G, Bebelis S, Lyberatos G (2013) Operation and characterization of a microbial fuel cell fed with pretreated cheese whey at different organic loads. Bioresour Technol 131:380–389. https://doi.org/10.1016/j.biortech.2012.12.173
151. Palanisamy G, Jung H, Sadhasivam T, Kurkuri MD, Kim SC, Roh S (2019) A comprehensive review on microbial fuel cell technologies: processes, utilization, and advanced developments in electrodes and membranes. J Clean Prod 221:598–621. https://doi.org/10.1016/j.jclepro.2019.02.172
152. Liu Y, Liu Z, Liu H, Liao M (2019) Novel porous nitrogen doped graphene/carbon black composites as efficient oxygen reduction reaction electrocatalyst for power generation in microbial fuel cell. Nanomaterials 9(6):836. https://doi.org/10.3390/nano9060836
153. Krzeminski P, Tomei MC, Karaolia P, Langenhoff A, Almeida CMR, Felis E, Gritten F, Andersen HR, Fernandes T, Manaia CM, Rizzo L, Fatta-Kassinos D (2018) Performance of secondary wastewater treatment methods for the removal of contaminants of emerging concern implicated in crop uptake and antibiotic resistance spread: a review. Sci Total Environ 648:1052–1081. https://doi.org/10.1016/j.scitotenv.2018.08.130
154. Xu Z, Xu X, Tao X, Yao C, Tsang DC, Cao X (2019) Interaction with low molecular weight organic acids affects the electron shuttling of biochar for Cr (VI) reduction. J Hazard Mater 378:120705. https://doi.org/10.1016/j.jhazmat.2019.05.098
155. Ye Y, Ngo HH, Guo W, Chang SW, Nguyen DD, Liu Y, Nghiem LD, Zhang X, Wang J (2019) Effect of organic loading rate on the recovery of nutrients and energy in a dual-chamber microbial fuel cell. Bioresour Technol 281:367–373. https://doi.org/10.1016/j.biortech.2019.02.108
156. Jatoi AS, Akhter F, Mazari SA, Sabzoi N, Aziz S, Soomro SA, Mubarak NM, Baloch H, Memon AQ, Ahmed S (2020) Advanced microbial fuel cell for wastewater treatment—a review. Environ Sci Pollut Res 28(5):5005–5019. https://doi.org/10.1007/s11356-020-11691-2

Bibliometric Analysis: Role of Artificial Intelligence and Blockchain in Boosting the Blue Economy

Ashwini Wasnik, Dinesh Chandra Pandey, Ashulekha Gupta, and Rajesh Tiwari

Abstract This chapter highlights the latest developments that have occurred over the years in the marine economy, with the use of artificial intelligence and blockchain as key enablers, driving its growth. Techniques like deep learning, the Internet of Underwater Things (IoUT), and blockchain have played crucial roles in boosting the marine economy. The themes that have evolved focus more on sustainability and innovation in the blue economy, disaster handling and waste management, the use of automation in supply chain management (SCM), the digitalisation of fisheries, and human resource development in the blue economy, to encourage more private and public sector participation. With the help of predictive modelling techniques and decision support systems, many marine economy issues have been identified, like coral bleaching, oceanic disruption and sea level temperature. A comprehensive study of the current literature for gauging the outcome of artificial intelligence and blockchain on the blue economy, which is both qualitative and quantitative. This body of work identifies the themes that have evolved and focuses on the challenges and opportunities in the blue economy by conducting a rigorous content analysis of the literature. The bibliometric technique is employed for the quantitative analysis to understand the study's quantitative aspects, including citation analysis, network analysis, etc. Databases like Google Scholar and Scopus have been used along with Vos Viewer and the Bibliometrics R package for analysis.

Keywords Artificial intelligence · Blockchain · Blue or marine economy · Bibliometric analysis · Content analysis

A. Wasnik (✉) · D. C. Pandey · A. Gupta · R. Tiwari
Department of Management Studies, Graphic Era (Deemed to Be University), Dehradun, India
e-mail: Ashuwasnik1992@gmail.com

D. C. Pandey
e-mail: dineshchandra@geu.ac.in

A. Gupta
e-mail: ashulekhagupta.mgt@geu.ac.in

V. Kandpal and T. Choudhury (eds.), *Blue Revolution: Sustainable Water Use and Circular Economy Innovations*, Urban Sustainability,
https://doi.org/10.1007/978-981-95-1916-3_13

1 Introduction

Development is an integral component of a healthy economy, and so are its resources. Oceans and marine life form a major chunk of the economy, providing bread and butter to a large section of society and facilitating commercial ventures like aquaculture, renewable energy generation, trade, commerce-related applications, and environmental conservation. The issues like marine pollution, climate change, unsustainable exploitation of marine resources, and so on are of major concern. Artificial intelligence and blockchain play major roles in combating these critical scenarios, as their application can be applied to all the existing scenarios.

Oceans and marine life are the crucial cornerstones of a developing economy, as they are one of the avenues for people to earn their livelihoods. Many economic sectors have witnessed the repercussions of the pace at which nations develop and resource overuse. In response to underlying concerns, state officials are directing their efforts to technology and sustainability as a panacea for all. As per the definition available on the United Nations (UN) website, "Marine Economy consists of various economic sectors and policies, inclusively defining the sustainable usage of oceans and their resources." Before creating awareness regarding sustainable oceans, it becomes crucial to identify the indicators that cause unsustainability in oceans. Additionally, it is observed that technological breakthroughs can be viewed as a solution for understanding environmental deterioration, a deeper understanding of oceans, and fostering sustainable interaction with oceans for developing a robust marine economy [1]. The three cornerstones of the marine economy are environmental sustainability, economic viability, social equity, and inclusion. Technology has been identified as one of the driving factors for the sustainable handling of oceans, with artificial intelligence (AI) and blockchain seeming to be a major de facto. Artificial intelligence (AI) has had a noteworthy impact on various industries over the years, and marine aquaculture is a crucial component of the blue economy, experiencing an exceptional surge in production. The AI-powered systems have helped increase aquaculture farms' overall efficiency in operations, production, and management,innovative applications for forecasting marine ecosystems,and waste management. The adoption of artificial intelligence in the marine economy will prove to be a long-term, sustainable approach for the marine industry and the well-being of the oceans [2]. The industry 4.0 revolution, promoting connecting devices with the Internet of Things and artificial intelligence (IoT), has made a paradigm shift in the way things are operating from labour-intensive to computerised systems, heavily relying on embedded innovation. The digitisation of every industry, covering fish farms and cultivation, has been the first step towards technological advancement with IoT and big data. Several changes have been observed after the digitisation of fisheries in terms of time, labour productivity, and money. This paradigm shift can be witnessed in stages of fish farming, such as the initial stage of locating cultivation sites using GIS, the intermediate stage of making use of resources like feeders and water testing equipment, and post-production, making use of web marketing

[3]. The Internet of Underwater Things (IoUt), integrated with blockchain technology, has become significant due to its applicability in areas like environmental screening, disaster handling, localisation, defence, underwater exploration, and so on. Blockchain is regarded as a futuristic component of the internet with the prospective to change the way systems operate, and with many benefits, it plays an important part in the internet of underwater things, too [4]. At the global level, certain observations can be witnessed for a thriving marine economy; it is important to integrate technologies like artificial intelligence (AI), big data analysis, and machine learning in aquafarming and cultivation. Shrimp cultivation takes place in regions like Khulna, Cox Bazar, Chittagong, and Bhola in Bangladesh. Issues like shrimp pond management, harvesting, grading, and production have been addressed by innovative solutions, leading to a potential increase in the production of shrimp [5]. The study conducted in Tasmania assessed a set of 36 substitutes located at the aquafarms and windfarms across it. The other options were tested against three parameters, such as scientific aspects, territory aspects, and location options, by employing a Bayesian model, and the uncertainties of each possible outcome were calculated. The evaluated options were positioned based on the maximum utility rule to select the optimum one [6]. The study conducted in Tasmania assessed a set of 36 alternatives located at the aquafarms and windfarms across it. The other options were tested against three parameters, such as technological aspects, environmental aspects, and site selection, by employing a Bayesian model, and the uncertainties of each possible outcome were calculated. The evaluated options were ranked based on the maximum utility rule to select the optimum one [7]. Climate change constitutes an alarming threat to the marine ecosystem in Kenya, including coral bleaching, ocean acidification, sea level rise, etc. Artificial intelligence technologies offer a wide range of solutions to these problems with the help of predictive modelling techniques, resource utilisation, and decision systems [8]. Over the years, it has been observed that technological breakthroughs like artificial intelligence and blockchain turned out to be a ray of hope for the sustainable management of marine ecosystems due its possible applications Forecasting sea surface temperature has multiple benefits in predicting climatic conditions around oceans, identifying fishing zones, and over-exploiting resources of marine life. In the past, many dynamic and statistical models were proposed for forecasting sea-surface temperature, but with artificial intelligence deep learning techniques, predicting sea surface temperature has become more reliable as it is built on a series of basic models that make use of a large set of datasets to give better predictions in comparison to their counterparts [9]. Customary maritime practices have provided many gains to humans, but at the same time, harmed the marine ecosystem. Artificial intelligence (AI) applications have the capability to innovate the capabilities of customary ranch businesses. China is making use of underwater robots and the Internet of Things to create a sustainable ranching industry [10]. The Internet of Underwater Things (IoUT) refers to technology that is used for monitoring underwater operations using underwater objects. This technology has made a novel paradigm shift for the blue economy. as it can communicate with autonomous underwater vehicles, sensing, actuating, and transferring data, and it can work in unpredictable ocean conditions [11]. The underwater image classification technique

is part of deep learning, which is used for classifying aquatic flora and fauna. This technique is significant for gauging the health and quality, and protecting the aquatic flora and fauna. Its applications can be observed in oceanography, marine economy, environmental protection, underwater mineral extraction, etc. [12]. Smart destinations make use of technology to deliver value and services to tourists. The research shows that those islands that have issues with accessibility and connectivity should include blockchain technology for developing the blue economy more efficiently and effectively, with good digital literacy for achieving sustainability [13]. Technological advancement solutions in financial and marketing sectors with the help of blockchain and IoT can be used to increase the production capacity of trawlers, operational control, commerce, and a reliable marketing approach for the expansion of business, empowering local communities and governance of the coastal areas [14] (Table 1).

The target of this work is to understand the influence of artificial intelligence (AI) and blockchain innovation on the blue economy by identifying the relevant themes and scenarios taking place in the sector, and the possible challenges and opportunities for future study. This body of work can be utilised by academicians, future researchers, organisations, and policymakers to gain insights into this topic. The body of work seeks to find answers to the following explorations.

Q1. What are the key insights after performing a bibliometric analysis?
Q2. What are the current themes that have evolved over the period on this topic?
Q3. What could be possible challenges and opportunities that can come while integrating technologies in the marine economy?

Table 1 Application of AI and blockchain in blue economy

	Applications	Advantages
Deep learning	Underwater Image Classification	For Classifying the Aquatic Flora and Fauna
	Predicting the Sea Surface Temperature	Predicting climatic conditions and identifying the fishing zones and mineral extraction sites
	Water Quality Assessment	Identifying the location and quality of water
Internet of Things and Internet of Underwater Things	Underwater Robots	Aquatic Ranching Industry
	Sensor-Based Models	For communicating with autonomous underwater vehicles
Blockchain	Smart Island	Good Digital Literacy, promoting tourism and local governance

Source Created by authors

The body of work is sectioned into four parts. The first part starts with an introduction built around the topic, the second part is a narrative review, the third part deals with quantitative and qualitative analysis, and the fourth part deals with the core of the research.

2 Thematic Literature Review

The thematic analysis is appropriate for those studies where the researcher is trying to establish a relationship between a set of variables sand a set of evidence for comparing the different scenarios in a body of work [15].

2.1 Promoting Sustainable Practices by Leveraging Artificial Intelligence Applications

It defines the broader perspective of the blue economy with a prime focus on environment-friendly practices in fisheries, aquaculture, and fish farming by incorporating technologies for fish management and aquaculture methods, by exploring the role of artificial intelligence in fisheries [16]. The study aims to collect streaming data on ocean conditions, with major focus areas being marine environmental monitoring, integration of marine infrastructure data, and coastal management for promoting sustainability for developing a robust coastal network [17].

2.2 Application of Artificial Intelligence in Scoping up Disaster Management and Facilitating Smooth Logistics Network

To create a platform by leveraging applications like big data and the Internet of Things (Iot) for better disaster management, particularly related to tsunamis and hurricanes, by incorporating remote sensing and image segmentation, big data for predicting ocean conditions, and IoT sensors for collecting real-time data for disaster management [18]. Identifying the potential use of artificial intelligence (AI) and automation in Caribbean logistics sectors with a special emphasis on ports, automated vehicles, artificial intelligence-based inventory management, and smart logistics in improving capacities and reducing costs [19].

2.3 Use of Artificial Intelligence (AI) and Blockchain Techniques in Monitoring Pollution Level in Blue Ecosystem

The study makes use of deep learning algorithms for assessing the quality of the water by predicting and monitoring the quality and nutrient load, checking pollution levels, and ecosystem health. Also, identify the role of big data for data-driven decision-making and understanding climate change and the blue ecosystem [20]. It explores areas like technological innovations, sustainable practices, and policy frameworks required for marine ecosystem management, with a special focus on sustainable development goals concerning a blue economy [20]. It focuses on using blockchain technology for waste management in marine ecosystems by employing a circular economy approach from collecting waste from source to recycling, from tracking down plastic waste to ensuring full transparency and accountability [21]. The study discusses the evolution of the marine economy's practices, policies, and frameworks over the last 30 years and critically analyses the pros and cons to gain insights for better governance of the marine economy in the future [22]. Technological innovations like satellite imagery, sensors, etc., can be used for monitoring and managing marine life. Predictive modelling will prove to be fruitful in aquaculture farming and cultivation while focusing on sustainability and the prevention of the ecosystem [23]. Technological innovations like satellite imagery, sensors, etc., can be used for monitoring and managing marine life. Predictive modelling will prove to be fruitful in aquaculture farming and cultivation while focusing on sustainability and the prevention of the ecosystem [24]. The application of artificial intelligence, blockchain, and predictive modelling can be used for the traceability of fishery and aquaculture commerce. Which in turn will help in achieving sustainability of fisheries by reducing costs, waste management, and well-established systems. At the same time, it will help in logistics and supply chain decision-making processes too [25].

2.4 Role of Artificial Intelligence (AI) and Blockchain in Increasing the Profitability in Blue Economy Business, Biodiversity Preservation and Initiatives by States in Promoting Blue Economy

Blockchain can help in meeting compliance standards and regulations by ensuring traceability and transparency through the value chain, detecting fraud and lack of operational efficiency in the global market, and having the potential to build customers' trust by ensuring credit rating systems on seafood items [26]. The implication of artificial intelligence (AI) and blockchain can be seen in economic performance, too, like profitability and growth in aquafarms at local and global levels [27].

Researchers have highlighted the significance of blue carbon contributing to biodiversity preservation, carbon sequestration, and sustainable management of marine ecosystems in hot climatic regions. The ability of marine ecosystems to sequester and store carbon serves as a natural antidote for climate change. Focusing on how blue carbon can be used and maximised in hot climatic regions while maintaining a balance between climatic goals and ecological preservation. At the same time, it highlights blue carbon's multifaceted benefits, including the protection of coastlines, providing habitat for marine life, and encouraging economies for sustainability [28]. To encourage investments in the marine economy and promote sustainable development, the study has investigated how the taxation process can be modified or implemented. By investigating the possibility of implementing new or altered tax laws to encourage marine economy sectors (fishing, aquaculture, shipping, maritime tourism, etc.). How the Nigerian government might impose tax laws to encourage marine economy expansion while maintaining environmental sustainability. It is the state official's responsibility to create an efficient legislative and regulatory framework that includes the marine economy in a country's overall economic expansion. By figuring out how the marine economy may contribute significantly to income through investments, taxes, and sustainable business practices [29]. Researchers have highlighted the importance of human assets in building the marine economy's long-term viability. With an emphasis on human resources development, education, and skill-building initiatives, the objective is to determine their abilities and provide necessary training to support the blue economy its allied sectors such as coastal tourism, aquaculture, and fisheries, to determine the kind of training programs, certifications, and educational activities required to empower the marine economy labour force to promote businesses. The marine economy can achieve long-term, sustainable growth through significant investments in human capacity development, which will lower unemployment and enhance resource management, ultimately leading to successful cooperation in developing capacity-building initiatives among governments, educational institutions, and private ventures [30] (Table 2).

2.5 Theoretical Background Supporting the Study

The study has focused on the following theories of sustainability:

Stakeholder Theory

Stakeholder theory describes the different stakeholder that forms a part of an ecosystem, and it can be any individual, community, employees, government, or whole society, and they can benefit mutually from the situation [32]. In this study, the stakeholders in the blue economy are the state, local communities, the private sector and people in fisheries, aquaculture, maritime tourism, shipping and ports,all

Table 2 Themes were identified after doing content analysis over the years

Authors	Objective	Research Gaps	Themes
CMFRI [16]	To emphasise the importance of environmentally friendly practices and technologies in the marine economy for fishery management and aquaculture methods	Fails to address the impact of climate on blue ecosystems and fails to explain the role of AI and automation in fish processing technologies	Sustainability and Innovation in Marine Resource Utilisation
Farcy et al. [17]	To develop a robust coastal observing network that addresses the scientific and societal challenges in the marine ecosystem	Lack of evidence on data standardisation, interoperability, and socio-economic integration	Monitoring environmental conditions along with integrating marine data infrastructure for coastal management
Zhang and Lu [18]	To create a platform for disaster management for the marine economy by leveraging the use of AI and IoT for better decision-making		Use of IoT and big data for disaster management for predicting tsunamis, hurricanes, and sea level rise
Foster et al. [19]	To identify the potential of AI and automation in logistics scenarios, especially ports, for enhancing operational efficiency and reducing costs	Fail to address the interoperability of present infrastructure with automated systems and impact on employment, especially in those regions where logistics is a source of livelihood for the major workforce	Automation and AI in logistics management and optimisation of the supply chain
Gambín et al. [20]	To identify how deep learning algorithms can be used for assessing the quality of water	Continuous monitoring of real-time data for prediction as well as quality is also a major concern	Deep learning in marine science and marine ecosystem health
Samad and Abbasi [31]	To study the impact of technological innovation, policy framework, and sustainable practices in harnessing blue ecosystems	Lack of focus on the marine circular economy	Technological innovations in the blue economy and sustainable goals

(continued)

Table 2 (continued)

Authors	Objective	Research Gaps	Themes
Gong et al. [21]	To study how blockchain can be used to address issues like waste management in the blue economy	Scalability and Implementation challenge of the blockchain solution	Blockchain for environmental sustainability and waste management in the circular blue economy
Leape et al. [23]	To study how data analytics, new management models, and technology can contribute to ocean resource management	Cost of technology adoption and data accessibility, and integration challenge	Technology in ocean resource management and data-driven decision making
Rowan [25]	To assess the current status of digital technologies in the fishery sector	Technological adoption barriers need to be studied in small and medium-sized aquaculture farms	Digital transformation in fisheries
Jiang et al. [27]	To assess the role of blue carbon in carbon sequestration, biodiversity conservation, and coastal ecosystem management in hot climatic regions	Further study needs to be done on blue carbon stock, particularly in hot climatic regions	Blue Carbon and climate change mitigation
Cinar et al. [28]	To understand how taxation mechanisms can encourage investment in a blue economy, followed by sustainable resource management	To design a proper taxation regime for a blue economy, the needs and characteristics of the blue economy have to be specified first, and how it will impact economically on small and medium businesses needs to be studied	Blue economy and taxation, and revenue generation
Anikwe et al. [30]	Identifying skill sets requires developing allied sectors of the blue economy	Needs to understand skill gap analysis as per the blue economy	Human Resource Development in the blue economy, encouraging public and private investments

Source Created by authors

of them together help in building the productivity of the blue economy by being the part of circular economy facilitating business avenue and growth for one another.

System Theory

It views an entity as a complex, open system composed of interconnected and interdependent subsystems. In the study, researchers have tried to gauge the impact of AI and Blockchain in boosting the blue economy. With technological advancement touching every corner of the economy, people have started taking this opportunity to promote, facilitate and integrate technology in their business models. The study highlights the application of AI and blockchain techniques that not only helped in preserving flora and fauna of the blue ecosystem but also formed an integral part in waste and disaster management, facilitating sustainability.

3 Research Methodology and Data Analysis

A systematic way to accomplish the research goal is described as a research methodology. A researcher employs several procedures in research methodology to achieve the study's purpose, and the rationale behind each step is explained. It becomes vital for researchers to understand methods and techniques and to employ appropriate methodology per the study's purpose [33]. The study looks into the prevailing body of work on the outcomes of artificial intelligence on the marine economy using a methodical literature review approach. The action plan includes examining pertinent papers and articles from databases like Google Scholar and Scopus. A timeline from 2018 to 2024 has been considered for analysis. Logical operators like "and" and "or" were used to input keywords like "sustainability," "blue economy," "marine economy," "blue economy," "artificial intelligence," and "blockchain.". The criteria for including papers and articles were based on subject area, document type, and linguistic style. Forty-six documents were obtained for the final assessment, and tools such as Vos Viewer and the Bibliometrics R package were utilised. The body of work seeks to find answers to the following explorations.

3.1 Data Analysis

Descriptive analysis is used to analyse the data's quantitative and qualitative aspects. To understand the key elements of the data technique, bibliometric analysis and content analysis have been performed, and tools like the Bibliometrix R package and Vos Viewer have been used for interpretation. Bibliometric analysis will help in understanding the quantitative growth of artificial intelligence and blockchain over the years in the marine economy, while content analysis will help in understanding the thematic evolution that has shaped the marine economy over the years with the integration of technology into it. For studying the quantitative aspects, parameters like annual scientific productions, average citations per year, most relevant source, most cited authors and their impact factor, most cited countries, and keywords. For content analysis, "clustering" will be used for tracing the themes that have evolved.

Average Scientific Production

This aspect describes a number of articles that have been published over the considered timeline related to the footprints of artificial intelligence and blockchain on the marine economy over the years, and the rate at which publications have increased (Fig. 1 and Table 3).

From the above table and Fig. 2, it can be observed that there is a constant growth in publications with an annual growth rate of 90% from 1 article published in the year 2018 to 10 articles in the year 2024.

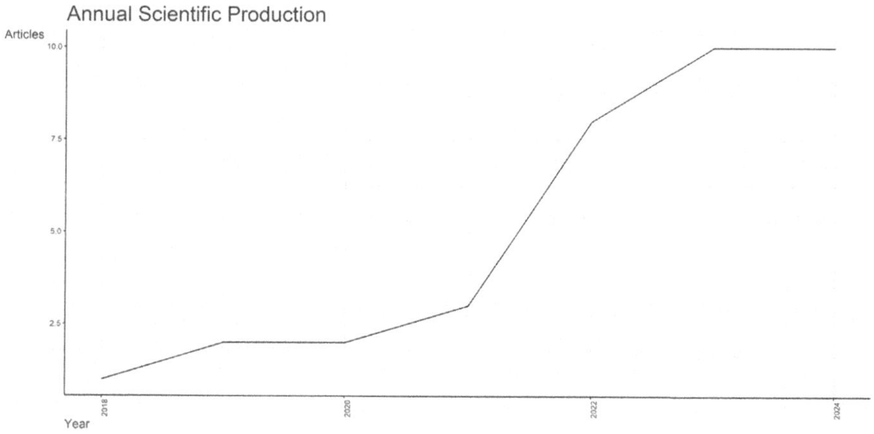

Fig. 1 Annual scientific productions over the considered timeline. *Source* Created by authors using the Biblioshiny R package

Table 3 Annual scientific production from 2018 to 2020

Year	Articles
2018	1
2019	2
2020	2
2021	3
2022	8
2023	10
2024	10

Source Created by authors using the Biblioshiny R package

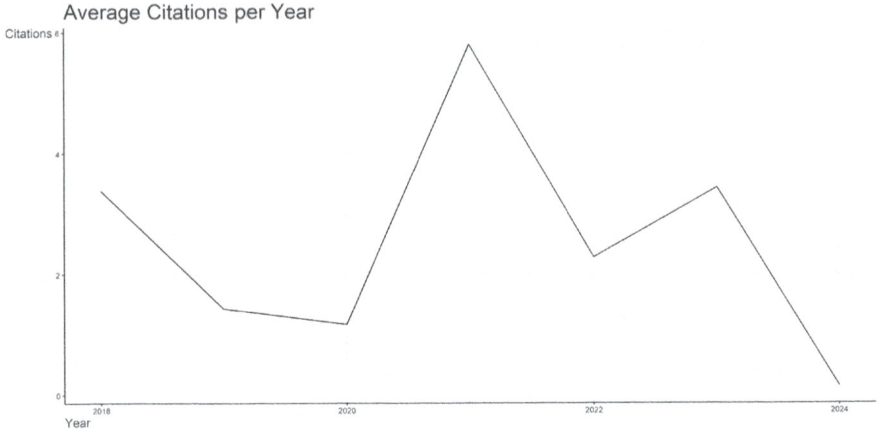

Fig. 2 Average citations per year. *Source* Created by authors using the Biblioshiny R package

Average Citations Per year

This refers to the sum of the total number of quotations received by associated articles divided by the sum of the total number of articles. It computes the impact of a publication over a specific timeline or a particular area of interest or field.

Where MeanTCperArt is the common indicator for gauging the impact of the research article. MeanTCperYear—it shows the average total citation received by a group of research papers cited by other authors/publications. From Fig. 3 and Table 4, it is evident that the impact factor of the research articles increased in the years 2018, 2021, and 2023, with a maximum average total citation in 2021, followed by 2023 and 2018.

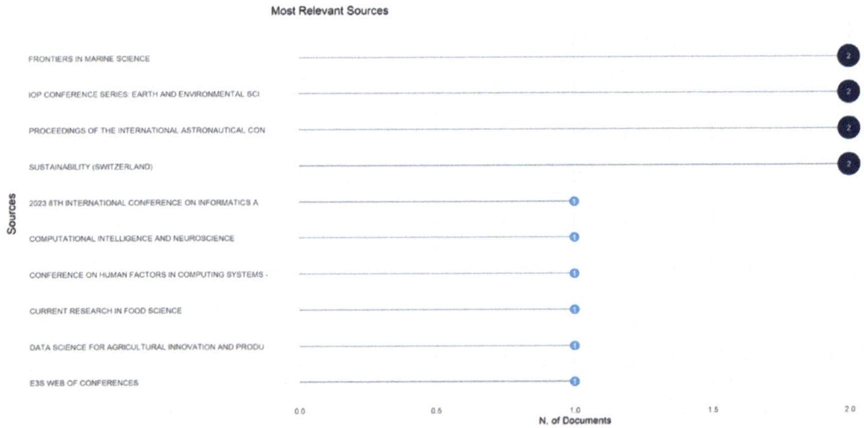

Fig. 3 Most relevant sources. *Source* Created by authors using the Biblioshiny R package

Table 4 Average citations per year from 2018–2024

Year	MeanTCperArt	N	MeanTCperYear	CitableYears
2018	27.00	1	3.38	8
2019	10.00	2	1.43	7
2020	7.00	2	1.17	6
2021	29.00	3	5.80	5
2022	9.12	8	2.28	4
2023	10.30	10	3.43	3
2024	0.30	10	0.15	2

Source Created by authors using the Biblioshiny R package

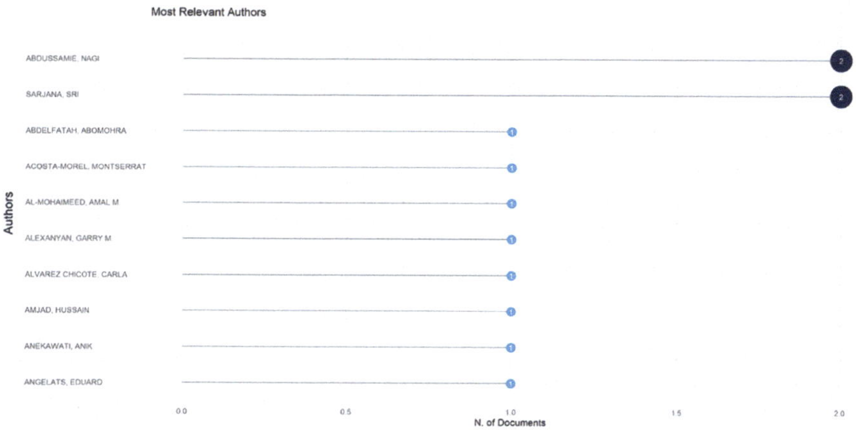

Fig. 4 Most relevant authors. *Source* Created by authors using the Biblioshiny R package

Most Relevant Sources

Figure 4 and Table 5 indicate that frontiers in marine science emerge as the most preferred source, followed by IOP Conference Series and Proceedings of the International Astronautical Congress, which produce, on average, 2 articles, along with the E3S Web of Conferences, which produces, on average, 1 article.

Most Cited Authors and Authors' Local Impact

Most cited authors are those whose work is referred to by most researchers because of the impact their work has created through their research findings and recognition, whereas the author's local impact parameter measures how their research papers are referred to at the local level by using statistical metrics.

Table 5 The most relevant sources for this topic

Sources	Articles
Frontiers In Marine Science	2
IOP Conference Series: Earth and Environmental Science	2
Proceedings Of the International Astronautical Congress, IAC	2
Sustainability (Switzerland)	2
2023 8th International Conference on Informatics and Computing, ICIC 2023	1
Computational Intelligence and Neuroscience	1
Conference On Human Factors in Computing Systems – Proceedings	1
Current Research in Food Science	1
Data Science for Agricultural Innovation and Productivity	1
E3S Web of Conferences	1

Source Created by authors using the Biblioshiny R package

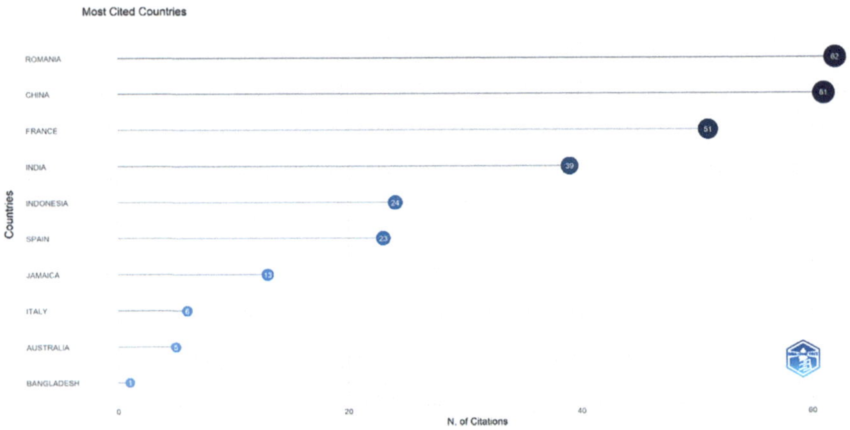

Fig. 5 Most cited countries and their contributions. *Source*: Created by authors using the Biblioshiny R package

From Fig. 5 and Table 6, it is indicated that authors Abdussamie N, Sarjana Sri and Acosta-Morel M are the most cited authors with prominent indices h_index, g_ index and m_index.

Most Cited Countries

These are those countries whose research works are mostly referred to by researchers in their work, with major contributions in a particular field of research. From the table below, it has been observed that Romania has some contribution in studying the impact of artificial intelligence and blockchain in marine ecosystems, with a total

Table 6 Most relevant authors and their local impact factor

Author	h_index	g_index	m_index	TC	NP	PY_start
Abdussamie N	1	2	0.25	5	2	2022
Acosta-Morel M	1	1	0.333	6	1	2023
Al-Mohaimeed Am	1	1	0.5	1	1	2024
Alexanyan Gm	1	1	0.25	1	1	2022
Alvarez Chicote C	1	1	0.125	27	1	2018
Amjad H	1	1	0.25	56	1	2022
Angelats E	1	1	0.2	23	1	2021
Anthony V	1	1	0.333	6	1	2023
Aryai V	1	1	0.25	5	1	2022
Aziz A	1	1	0.25	6	1	2022

Source Created by authors using the Biblioshiny R package

citation of 62, whereas Bangladesh has the least contribution with a total citation of 1.

From Fig. 6 and Table 7, which describe the most cited countries and their contributions from this, it is observed that countries like Romania, China, and France contribute more significantly, whereas countries like Italy, Australia, and Bangladesh are at the bottom position in terms of contributions.

Keywords Analysis

This describes the most frequent keywords used by researchers while studying the footprints of artificial intelligence and blockchain on the marine economy. The following Table 8 shows the most frequent keywords, where artificial intelligence and sustainable development are the most frequently used keywords and carbon least used an occurring 11 and 5 times.

Bibliographic Coupling

Figure 7 helps in understanding that Bibliographic coupling is a parameter that pins down conceptual similarity shared by two or more documents that are cited together. For study purposes, 36 documents were considered out of a total of 46 documents. The unit of analysis was a document, and for every 36 documents, the total association strength was calculated, and 2 clusters were formed, representing the bibliographic association. Cluster 1 is represented by the green colour, containing a total of 3 items, and cluster 2 is represented by red, containing 3 items.

i. Cluster 1 (green): The underlying association of items in the cluster emphasises the incorporation of artificial intelligence (AI) for screening and handling blue

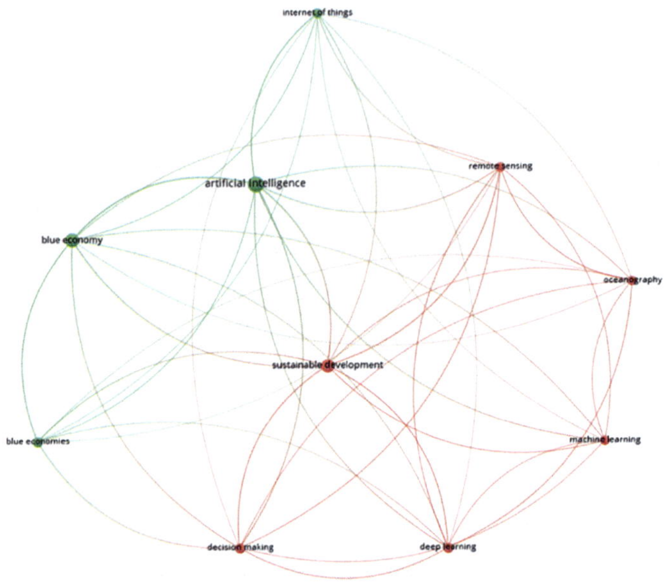

Fig. 6 Cluster diagram for keyword analysis. *Source* Created by authors using the Vos Viewer

Table 7 Most cited countries and their contributions

Country	TC	Average Article Citations
Romania	62	62.00
China	61	12.20
France	51	25.50
India	39	19.50
Indonesia	24	12.00
Spain	23	11.50
Jamaica	13	13.00
Italy	6	6.00
Australia	5	5.00
Bangladesh	1	1.00

Source Created by authors using the Biblioshiny R package

ecosystems while addressing marine problems like coral bleaching, ocean acidification, sea level rise, and oceanic disruption by using predictive modelling, resource optimisation, and decision support. To encourage the allied sectors of the marine economy, technologies like big data, AI, and machine learning should be encouraged in aquafarms.

Table 8 Keyword analysis

Words	Occurrences
Artificial intelligence	11
Sustainable development	11
Deep learning	9
Remote sensing	8
Decision making	7
Blue economies	6
Internet of Things	5
Oceanography	5
Carbon	4
Human	4

Source Created by authors using the Biblioshiny R package

Fig. 7 Cluster diagram for bibliographic coupling. *Source* Created by authors using the Vos Viewer

ii. Cluster 2 (Red): Focuses on encouraging the application of deep learning techniques in forecasting sea surface temperature and water quality estimation and forecasting, along with making use of multiple technologies like remote sensing, IoT, big data, and AI, etc.

The following themes can be identified from the study of the above clusters:

i. The application of artificial intelligence can be utilised for addressing the issues related to the marine ecosystem and its allied sectors.
ii. Deep learning techniques can prove to be a boon in the marine ecosystem.

4 Implications

The structural pillars of the blue economy are the Ocean economy, Sustainability and Environment [34]. Implementing strategies to promote marine finance for economic development and foreign investment should be encouraged for thriving technological advancement [35]. Prudent use of funds in important areas of the blue economy for achieving the short-term returns and attracting long-term investment, stability and sustainability. Financial aid, free-market capitalism, and technological research and development should focus on judicious circulation of funds for leveraging territorial advantage and building new avenues for the blue economy [36]. Targeted areas should also include low-carbon initiatives, digitalisation, and optimise resource allocation to mitigate potential substitution effects [37]. Support community empowerment and the role of the state in coastal areas for improving the quality of life of people, and support sustainable tourism and fishing practices [14]. Artificial intelligence techniques like remote sensing, IoTs, big data, and cloud computing should be advocated for sustainable management of the marine environment, and focus on decision-making policies should be on transfer learning, knowledge fusion, reinforcement learning, and edge computing [20]. Mandating the integration of AI techniques into the blue economy allied sector for better sustainability and environmental monitoring [2].

5 Future Scope

After exploring the current literature, the footprints of artificial intelligence and blockchain have a vital role to play in boosting the marine economy, as they open avenues of technology gateways to strengthen it. The expansion of real-time monitoring networks for better environmental prediction using technology like IoT sensors, the introduction of green technologies for reducing carbon footprints, the usage of blockchain for tracing seafood waste management, smart aqua farm management using AI-based sustainable practices, machine learning models for oceanographic study, and blockchain adoption for the secure and transparent logistics sector can be explored. The incorporation of AI into ocean conservation and the advancement of digital technology in marine farming ought to be its main topics. To maintain aquaculture management for record provenance, growth, water quality, and carbon footprint, as well as to build a robust decision support system for evaluating the salt quality, financial backing is essential for businesses creating AI technologies in marine aquaculture, which can be explored further. We need to look for cross-country sustainability practices which other countries can further adopt, also tourism opportunities that are present on marine islands need to be harnessed, but at the same time proper strategy must be developed for handling the problem of overcrowding, waste management and issues related to supply chain management in coastal areas.

Since, for studying purposes limited number of research papers are referred is the biggest roadblock, but still, it has identified some of the major concerns affecting the blue economy. Difficulties have been identified in acquiring global maritime data for training AI models, the high cost of underwater IoT connectivity, and limited awareness among market players of the marine economy. Addressing security challenges such as privacy, traceability, and transparency is still a major concern in IoUT systems using blockchain.

The IoUT's numerous uses present substantial commercial opportunities. AI's ability to optimise processes and lessen environmental effects in marine aquaculture is transforming sustainable production systems. AI is crucial for maintaining the health of the oceans since it can be used to manage diseases, optimise production, monitor health, and improve spatial planning for sustainable aquaculture methods. Effective management of the blue economy and strong digital literacy have a big impact on blockchain. The management of marine plastic garbage now faces difficulties that can be resolved using blockchain technology. Public awareness of marine plastic garbage governance can be raised by a blockchain-based digital token system and identity recognition system.

The integration of technology in the marine economy opens paths for new business models for the circular economy practices in allied sectors of the marine economy, with the help of AI conservation and identification of marine species would become easy using image recognition techniques, use of deep learning for early detection of pollution and disruptions in marine economy and encouraging private players for involvement in data analytics and ocean observations tools.

6 Conclusion

With technological interventions touching every pillar of the economy, it can no longer be said that the blue economy is untouched. The footprints of AI and blockchain seem to play a crucial role in uplifting the marine economy, as it can be observed that countries like Romania, China, and France are heavily investing in technological aspects to boost their marine economy. Quantitative studies focus on annual scientific production observed with an annual growth rate of 90%, average citation per year, keyword analysis, list of sources and bibliographic coupling identifying the mutual association between the shared work. Themes that are identified throughout the study are leveraging the AI and blockchain in promoting sustainable practices, disaster management using AI and blockchain, monitoring the health of the blue economy, business models and biodiversity preservation. Players of the blue economy are heavily relying on technology for addressing marine issues like conservation of marine life, climate forecasting, effective fishery management, and more focus on waste management. A technique like deep learning seems to be a boon for marine ecosystems. There should be more focus on policy initiatives from the government side to encourage technology, not only to boost the economy but also to conserve it, more involvement of private players for funding, and more projects need to be undertaken by the government, which focuses on circular economy aspects like business models and practices that boost blue economy and its allied sectors. Along with a strong security challenge must be addressed while implementing blockchain technology, big data, AI and machine learning. Though this body of work has identified a limited number of themes purveying the blue economy but there are allied sectors which need to be explored, benefiting both people and the economy, and also focus areas like waste management, factors that promote tourism on islands, sustainable practices for protecting flora and fauna must be preached to the people with a good awareness drive.

References

1. Agarwala N (2024) Role of technology in a sustainable blue economy. Springer Nature, pp 1–29. https://doi.org/10.1007/978-3-031-32671-4_38-1
2. Er-Rousse O, Qafas A (2024) Artificial intelligence for the optimisation of marine aquaculture. E3S Web of Conferences 477:00102. https://doi.org/10.1051/e3sconf/202447700102
3. Mohammed Sanusi S, Invinder Paul S, Muhammad Makarfi A (2024) Internet of fisheries things (IOFT) for blue economy & ecosystem. In Data science for agricultural innovation and productivity. Bentham Science Publishers, pp 127–162
4. Bhattacharya S, Victor N, Chengoden R, Ramalingam M, Selvi G, Maddikunta P, Donta P, Dustdar S, Jhaveri R, Gadekallu T (2022) Blockchain for internet of underwater things: state-of-the-art, applications, challenges, and future directions. Sustainability 14(23):15659. https://doi.org/10.3390/su142315659
5. Bir J, Golder MR, Zobayer MF, Das KK, Zaman S, Chowdhury LMD, Paul PC (2020) A review on blue economy in Bangladesh: prospects and challenges. Int J Natural Soc Sci 7(4):21–29
6. Villalba J, Abdussamie N, Aryai V, Nikolova N, Tenekedjiev K, Wang C-M, Penesis I (2022) Assessment of uncertain alternatives for co-located aquaculture and offshore wind farm in Tasmania. Ocean Eng 249:110949. https://doi.org/10.1016/j.oceaneng.2022.110949
7. Efendy M, Darmawan AK, Umam BA, Anekawati A, Komarudin A (2023) AHP-TOPSIS-based decision support system (DSS) for assessing the salt quality for Indonesian salt e-farming. In 2023 Eighth International Conference on Informatics and Computing (ICIC). IEEE, pp 1–7
8. Gesami BK, Nunoo J (2024) Artificial intelligence in marine ecosystem management: addressing climate threats to Kenya's blue economy. Front Mar Sci 11. https://doi.org/10.3389/fmars.2024.1404104
9. Marndi A, Patra GK (2022) Attention-based ensemble deep learning technique for prediction of sea surface temperature. In Lecture Notes on Data Engineering and Communications Technologies. Springer Nature, Singapore, pp 425–438
10. Liu T, Wu F (2022) A sensor-based IoT data collection and marine economy collaborative innovation method. Comput Intell Neurosci. https://doi.org/10.1155/2022/3421999
11. Mohsan SAH, Mazinani A, Othman NQH, Amjad H (2022) Towards the internet of underwater things: a comprehensive survey. Earth Sci Inf 15(2):735–764. https://doi.org/10.1007/s12145-021-00762-8
12. Mittal S, Srivastava S, Jayanth JP (2023) A survey of deep learning techniques for underwater image classification. IEEE Trans Neural Netw Learn Syst 34(10):6968–6982. https://doi.org/10.1109/TNNLS.2022.3143887
13. Pranita D, Sarjana S, Musthofa BM, Kusumastuti H, Rasul MS (2023) Blockchain technology to enhance integrated blue economy: a case study in strengthening sustainable tourism on smart islands. Sustainability 15(6):5342
14. Sarjana S, Claudia SA, Ramadhina AT, Suyanti L (2024, April) A sustainable blue economy: blockchain and internet of things integration in tourism villages. In IOP Conf Ser Earth Environ Sci 1324(1):012020
15. Alhojailan MI (2012, October) Thematic analysis: a critical review ofits process and evaluation. In WEI international European academic conference proceedings, Zagreb, Croatia.
16. CMFRI K (2018) Report of blue economy working group-3 on fisheries, aquaculture & fish processing
17. Farcy P, Durand D, Charria G, Painting SJ, Tamminen T, Collingridge K, Puillat I (2019) Toward a European coastal observing network to provide better answers to science and to societal challenges: the JERICO research infrastructure. Front Marine Sci 6:529
18. Zhang Y, Lu K (2020) Research on disaster prevention technology of remote sensing image segmentation of ocean big data platform based on internet of things. IOP Conf Seri Earth Environ Sci 558(3):032026
19. Foster MN, Rhoden SL, Foster MN, Rhoden SL (2020) The integration of automation and artificial intelligence into the logistics sector: a Caribbean perspective. Worldwide Hospitality and Tourism Themes 12:56–68

20. Gambín ÁF, Angelats E, González JS, Miozzo M, Dini P (2021) Sustainable marine ecosystems: deep learning for water quality assessment and forecasting. IEEE Access 9:121344–121365
21. Gong Y, Wang Y, Frei R, Wang B, Zhao C (2022) Blockchain application in circular marine plastic debris management. Ind Mark Manage 102:164–176
22. Chen J, Chen H, Smith TF, Rangel-Buitrago N (2023) Analyzing the impact and evolution of ocean & coastal management: 30 years in retrospect. Ocean Coastal Manage 242:106697
23. Leape J, Abbott M, Sakaguchi H, Brett A, Cao L, Chand K, Myksvoll MS (2023) Technology, data and new models for sustainably managing ocean resources. The blue compendium: from knowledge to action for a sustainable ocean economy. Springer International Publishing, Cham, pp 185–211
24. Hanine S, Dinar B, Meftah S (2023) The blue economy 3.0: innovation for a sustainable ocean economy in Morocco. Int J Econ Stud Manage 3(6):2043–2064
25. Rowan NJ (2023) The role of digital technologies in supporting and improving fishery and aquaculture across the supply chain- Quo Vadis? Aquaculture Fish 8(8):365–374
26. Tolentino-Zondervan F, Ngoc PTA, Roskam JL (2023) Use cases and future prospects of blockchain applications in global fishery and aquaculture value chains. Aquaculture 565:739158
27. Jiang Y, Huang L, Liu Y, Wang S (2024) Impact of digital development and technology innovation on the marine fishery economy quality. Fishes 9(7):266
28. Cinar M, Hilmi N, Arruda G, Elsler L, Safa A, van de Water JA (2024) Blue carbon as a nature-based mitigation solution in temperate zones. Sustainability 16(17):7446
29. Umenweke M (2024) The introduction of the blue economy and its implications on the Nigerian tax regime. Int J Comp Law Legal Philosophy 1
30. Anikwe SO, Unachukwu LC, Onah FN (2024) Human capacity development and sustainable growth in the blue economy: opportunities, challenges, and strategies for Nigeria. African Bank Finance Rev J 15(15)
31. Samad G, Abbasi J (2022) Advancement of science and technology: future prospect of blue economy. Springer International Publishing, Cham
32. Freeman RE (2010) Strategic management: a stakeholder approach. Cambridge University Press, Cambridge
33. Kothari CR (n.d.) Research methodology: methods and techniques. New Age International.
34. Attri VN (2018) The blue economy and the theory of paradigm shifts. The blue economy handbook of the Indian ocean region. Africa Institute of South Africa, pp 15–37
35. Wu X, Wang P, Wu Y (2020) Marine finance in China: implications, development, and innovation. J Coastal Res 106(sp1):38. https://doi.org/10.2112/si106-010.1
36. Liang Y, Zhang S, Li J, Guo L (2022) Research on financial support, technological improvement and marine economic development for China's coastal regions. Water 14(17):2740. https://doi.org/10.3390/w14172740
37. Sun S, Zhang Z, Tan M (2024) Financial mechanism for sustainable development of the marine economy with respect to technology, digitalization, and low carbonization. Water 16(19):2841. https://doi.org/10.3390/w16192841

Blue Bonds: Financing Ocean Conservation and Sustainable Development

Nupur Gupta, Nandini Saraf, and Aakash Vijayvergiya

Abstract The extensive biodiversity crisis and ocean conservation have garnered significant attention recently, particularly regarding the marine ecosystem and water sustainability. Approximately 71% of the Earth's surface is composed of water bodies. In contrast, the Sustainable Development Goal (SDG) linked to water conservation, SDG 14, which addresses life below water, is the most underfunded amongst the 17 SDGs. Blue bonds play a pivotal role in addressing water scarcity, conserving the marine ecosystem, and ensuring the availability of clean water for consumption. This chapter examines how blue bonds revolutionise the sustainability of ocean economies, particularly addressing the issue of water scarcity in developing economies located near coastlines. The ocean economy has enormous potential in terms of returns generated from investments made under climate financing. In terms of yield benefits, investing USD 2–4 trillion in such projects is expected to generate over USD 8 to 23 trillion in the next 30 years. This chapter critically discusses the case studies of Blue Bond-issuing countries, namely the Seychelles, the Nordic countries, China, Indonesia, and Belize. It discusses how AI and Fintech are being utilised to measure the impact of issuing Blue Bonds. Investors interested in sustainable projects tend to prefer small-scale bonds, unlike the Seychelles blue bond, which is small-scale but has a high transactional cost. The challenges associated with Blue Bonds typically stem from the use of proceeds, which often require repaying investors regardless of the project's success. The environmental impact of blue bonds is evaluated, and a future scope of research is suggested to improve transparency, standardized reporting formats, and enhance global collaboration.

N. Gupta (✉)
Department of Accounting & Finance, SK Somaiya School of Commerce & Business Studies, Somaiya Vidyavihar University, Mumbai, Maharashtra, India
e-mail: nupur.g.bhattacharya@gmail.com

N. Saraf · A. Vijayvergiya
Jaipuria Institute of Management Indore, Indore, India
e-mail: nandini.saraf.26i@jaipuria.ac.in

A. Vijayvergiya
e-mail: aakash.vijayvergiya.26i@jaipuria.ac.in

V. Kandpal and T. Choudhury (eds.), *Blue Revolution: Sustainable Water Use and Circular Economy Innovations*, Urban Sustainability,
https://doi.org/10.1007/978-981-95-1916-3_14

Keywords Blue Bonds · SDG 14 · Water sustainability · Ocean economy · Marine conservation · Climate financing

1 Introduction

1.1 Need for Conservation and Ocean Sustainability

The extensive biodiversity crisis and ocean conservation have garnered significant attention in recent years, particularly concerning the marine ecosystem and water sustainability. Throughout the years, extensive research and data collection have been conducted using multiple methods for sustainable management. Overexploitation of natural resources has significantly impacted marine biodiversity, leading to alarming situations worldwide. Furthermore, numerous other social and economic changes, such as overpopulation, pollution, and urbanisation, have created significant tensions related to climate change. There is an urgent need to find sustainable and economically viable ways to conserve the marine ecosystem and initiate a concrete blue revolution in terms of both water sustainability and marine conservation.

There are four types of sustainable debt instruments: green bonds, social bonds, sustainability bonds, and sustainability-linked bonds. Based on different objectives, these bonds are divided into sustainability-linked bonds and use-of-proceeds bonds. These sustainable debt instruments adhere to the framework outlined by the International Capital Market Association [1]. Like green bonds, one of the latest additions in climate financing is the concept of blue bonds, which is currently categorised as a subset of green bonds. According to the OECD [2] report, it is estimated that the ocean economy will surpass US$3 trillion in terms of gross value added by 2030. This expansive growth is made possible with the help of a financial instrument- Blue Bonds, a debt security issued to raise funds for supporting the execution of the SDGs associated with oceans and marine biodiversity. Businesses, financial institutions, private sector companies and local governments can issue blue bonds to promote ocean-related projects. This thereby boosts the implementation of various sectors covered under the ocean economy umbrella, known as the 'Blue Economy'. It encompasses conventional marine development initiatives, new technological adoption projects, ocean-focused research projects, mangrove and coral preservation activities, as well as tourism [3].

To finance initiatives specifically aligned with the conservation and sustainable development of oceans and marine ecosystems, the World Bank has defined debt financing instruments such as Blue Bonds, issued by Governments, development banks, and private companies, which adhere to the Sustainable Development Goals, including SDG 14. It is designed in a way which provides a channel to fund such projects and focus on the protection of marine environments [4].

Historically, in 2018, the Republic of Seychelles seized this opportunity and issued the first-ever sovereign blue bond, thereby becoming a pioneer in this segment. As a result, the world was exposed to the benefits and advantages of such financing methods, and it also shed light on the fact that both sustainability and financing can be achieved simultaneously, leading to good returns. With massive traction for this innovative idea in just four years, from 2018 to 2022, the total collection reached 5 billion USD, derived from 26 blue transactions, with a CAGR of 92% [5]. Ironically, ocean financing remains critically underfunded, and by 2019, only USD 2.3 billion had been allocated to SDG 14. This highlights the lack of priority given to private sector commitments for SDG 14 [6]. Moreover, the importance of multilateral development banks lies in providing a well-structured risk-sharing framework that can attract the attention of the private sector [7].

In the Blue Economy of the Caribbean, blue bond funding is utilised to support sustainability and development in key sectors. As presented by March et al. [8] in their report, these funds are used to preserve ecosystem services, advance aquaculture and mariculture, enhance fish farm monitoring, and reduce pollution in water bodies. Upholding investments in habitat helps restore mangroves, coral reefs and seagrass beds, which leads to an increase in biodiversity and fisheries productivity by as much as 60%. In the field of aquaculture, blue bonds are utilised to enhance technologies and infrastructure, primarily by focusing on sustainability and value chain integration. Additionally, the funds are crucial for enhancing waste management and promoting the adoption of circular economy principles to address ocean-bound plastic and untreated wastewater. Blue bond money can be invested in marine protected areas as they serve conservation, help manage fishing and open doors for eco-tourism. Investing in these areas is crucial to achieving SDGs 14, 8, and 17 in regional practices.

The research paper by March et al. [9] showcases how Barbados, Grenada, St. Vincent and the Grenadines shared how blue financing helped fisheries and aquaculture by supporting the environment and adding value to seafood. To support Barbados's tuna industry, blue financing is utilised for ecosystem-based strategies and to modernise the processing lines, which could earn twenty times more by selling processed fish instead of whole fish. The country utilises funds from blue funds to restore damaged habitats to health and promotes the use of sustainable management practices for fisheries, thereby enhancing the resilience of fish populations. In St. Vincent and the Grenadines, efforts to restore coral reefs utilise blue financing, and the conversion of fish waste into a valuable resource is being explored. Investing in blue bonds may help save species, support sustainability, and benefit communities near the oceans by utilising innovative market strategies.

1.2 Evolution of Blue Bonds and Blue Bond Taxonomy

The period from 2020 to 2024 introduced a vital financial solution known as Blue Bonds, which directs investment toward ocean and freshwater conservation and sustainable use. Blue Bonds operate as debt instruments that mirror traditional debt instruments yet dedicate their specific funding towards achieving sustainability goals related to water resources. These financial instruments serve as a central bridging mechanism between the United Nations Sustainable Development Goals and global capital markets, enabling investors to incorporate quantified environmental impact and sustainability objectives into their portfolio performance.

The concept of blue bonds originated in the small archipelagic nation of Seychelles in East Africa. The World Bank supported the Seychelles, as its economy relies heavily on tourism and fisheries. It became a pioneer in issuing the concept of blue financing. On October 29 2018, the Government of Seychelles, in collaboration with the World Bank, announced the issuance of the first Sovereign Blue Bond. Being one of the planet's biodiversity hotspots, Seychelles had to find a way to reconcile its ambition for economic growth with its requirements. Following the Seychelles' issuance of its first blue bond in 2019, the Nordic Investment Bank also issued a SEK 2 billion (Swedish currency) Blue Bond. The Seychelles is thus regarded as a prime case study to investigate how these financing mechanisms operate to promote a sustainable blue economy initiative, a nascent concept [9].

Bosman and de Mariz [5] identifies governments, multilateral development banks (MDBs), private corporations, financial institutions, and conservation partnerships as the primary issuers of blue bonds between 2018 and 2022 and it can be inferred that the most initiative funding comes from the public sector and MDBs, followed by corporations and interest groups.

The growth of the blue bond market is being limited by major issues, including the lack of agreed-upon definitions and metrics for measuring its effects. According to Bosmans and de Mariz [5], having multiple frameworks for blue investments creates challenges in identifying qualified projects, which negatively impacts investor confidence. If the "blue" criteria are not defined precisely, issuers often apply them differently, which makes it harder to compare these instruments. This lack of certainty makes it more likely for issuers to use vague terms and label routine projects as environmentally friendly. They erode trust in the market and make it more challenging to produce numerous impactful and legitimate projects. A clear taxonomy and more standardisation in blue bonds definition, with the removal of entry barriers and infusing more private capital in ocean financing [7].

Moreover, only half of issuers now disclose their impact. For those who disclose the details, the rigour of the disclosure is often insufficient, as there is a significant discrepancy in what the metrics mean. Lacking standardised instruments for measurement hinders the evaluation of a firm's environment and finances. To be as transparent as possible, reduce reputation damage, and maximise blue bond business, these gaps need to be filled through proper global standards and improved reporting impacts.

1.3 Economic Potential

The ocean economy has enormous potential in terms of returns generated from investments made under climate financing. According to studies on this topic, investing

USD 2–4 trillion in such projects is expected to yield benefits of over USD 8 to 23 trillion in the next 30 years [10]. Statistically, it means that for every one dollar spent on such projects, the returns generated are approximately between 3 to 12 dollars, making the cost-to-benefits ratio of 1:3 to 1:12. Thus, unlike other traditional bonds, the main agenda of blue bonds is both creating a sustainable ocean economy and focusing on issuers' needs and financial returns. These results have led 72% of asset managers and asset owners to believe in and invest in Blue Bonds, ultimately contributing to a sustainable blue economy Responsible Investor Credit Suisse [11].

With the increasing popularity and returns generated by such initiatives, investors' willingness to invest in these bonds is growing, reflecting their growing awareness of environmental and marine ecosystem concerns. The popularity of blue bonds indicates that a larger audience is interested in methods that address such social issues and target aquatic problems on a broader scale. Various factors influence the psychology of investors related to sustainable bonds, including knowledge and awareness, risk involved, return on investment, and a preference for promoting social change [12].

Similarly, the framework is designed for issuers to provide all information transparently, with a significant focus on the benefits of sustainable development. All objectives should be clearly stated, and important financial metrics should be mentioned. Studies have shown that younger investors are more receptive to taking risks and investing in ESG-related bonds, which demand higher returns [13]. Transparency increases investors' trust, and with increasing willingness, the blue sustainability market is expanding [14].

2 Sustainable Development Goal (SDG) Linkages with Blue Bonds

SDG-embedded Blue Bonds demonstrate that stakeholders recognise the need for climate-informed infrastructure development to incorporate sustainable resource distribution and eco-focused management methods. Such bonds trigger wide-ranging systemic changes across multiple SDG targets, as they effectively leverage financial resources to support water conservation, climate adaptation activities, and marine biodiversity initiatives. These programs demonstrate considerable impact through their advancement of SDG 6 and other targets by creating and implementing sustainable management systems.

When water stewardship strategies aimed at achieving sustainable and equitable water management are specifically funded, Blue Bonds themselves automatically meet the goals of Sustainable Development Goal 6, which is to ensure the availability and sustainable management of water and sanitation for all. These investment tools are utilised in critical fields, including wastewater treatment, rainwater collection, groundwater recharge, and decentralised sanitation, to serve vulnerable and climate-affected populations. By encouraging a cycle-focused approach to water use, Blue Bonds will facilitate the development of innovative, integrated systems that minimise waste and promote water reuse and optimal water management at all levels of operation.

Additionally, these instruments facilitate the implementation of Integrated Water Resources Management (IWRM) and the development of governance frameworks that integrate land use planning with water management policies across various sectors and communities. Through this approach, the mechanisms strengthen institutional capacity and local capabilities, thereby aligning total efforts to secure water supplies.

Through investment from Blue Bonds, the rollout of advanced water solutions, including AI-powered monitoring systems and state-of-the-art purification technologies, has been made possible. In the Segura Hydrographic Basin in Spain, it has been demonstrated that combining artificial intelligence with hydrological modelling and remote sensing enhances water allocation, enabling the distribution of over 642 million cubic meters of water within six months and significantly reducing CO_2 emissions and deficits [15].

Further analysis in northern and southern Italy confirmed that large-scale constructed wetlands possess appreciable ecological and fiscal benefits, with many projects claiming a benefit-cost ratio of more than one. Sometimes the benefit is as high as 10 times the cost [16]. Funding these nature-based solutions using Blue Bonds supports important decentralised sanitation and water cleaning efforts, making them more effective towards achieving SDG 6.

A practitioner's guide was adopted by the International Finance Corporation (IFC) in collaboration with the International Capital Market Association (ICMA), the United Nations Global Compact, the United Nations Environment Program Finance Initiative (UNEPFI), and the Asian Development Bank (ADB), to increase the standardization and effectiveness of Blue Bonds in 2023. The guidance outlines explicit criteria, procedures, and models for issuing Blue Bonds to facilitate financing for sustainable blue economy projects, including water and sanitation initiatives.

Additionally, a thorough analysis of the Blue Bond Sector, covering the period from 2018 to 2022, identified 26 transactions valued at approximately USD 5 billion [5], with a focus on wastewater treatment, biodiversity preservation, and sustainable fisheries. The report emphasises the need for standard definitions and measurement rules to ensure the success and transparency of blue bonds in managing water resources effectively. The overall results highlight the critical functions that Blue Bonds have in promoting sustainable water management, inducing innovative thinking, and facilitating infrastructure development. Through the capitalisation of quasi-projects for water quality improvement, accessibility enhancement, and governance improvement, Blue Bonds are essential tools in the global undertaking of SDG 6.

The aim is to develop reliable infrastructure networks, sustainable industrial procedures, and innovative projects. Linking sustainable finance and technological progress is essential to SDG 9-Industry, Innovation, and Infrastructure, because it seeks to update industrial processes, safeguard infrastructure from potential hazards, and innovate new leading hospitals to enhance sustainable and inclusive growth with stakeholder plans translated to action interventions for water infrastructure development and industrial development through blue bonds as key financing mechanisms.

Utilising Blue Bond resources for next-generation water technologies promotes resource enhancement, yielding maximum sustainability and enhanced environmental outcomes. Some key innovations include real-time data monitoring, water leak detection technology using AI-backed tools, interconnected water networks utilising IoT technology, and automatic wastewater treatment solutions, all integrated with blockchain. With the introduction of these tools, the effectiveness of water-intensive industries is improved, data-driven strategies are enhanced, preventive maintenance is made possible, and the construction of flexible infrastructure is made possible, which helps create lasting sustainability and resilience.

The study by Karkou et al. [17] reports that projects funded using Blue Bonds are modernising water infrastructure by utilising technology, such as digital systems and techniques from the circular economy. These innovations significantly impact sectors such as manufacturing, textiles, energy generation, and Agri-processing, which urgently need advanced systems of managing resources to address significant water use and pollution issues. Financial support from Blue Bonds enables industries to adopt closed-loop water management and onsite wastewater analysis, resulting in reduced freshwater use, associated costs, and ecological footprint savings.

An in-depth investigation of the Blue Bond market from 2018 to 2022 revealed 26 transactions totalling approximately USD 5 billion, with a notable focus on sustainable water and wastewater management. A significant fraction of these investments has been devoted to improving water and sanitation infrastructure in underserved areas, aligning with the primary goals of SDG 6. The investigation emphasises the significance of standardised definitions and performance reporting indicators to maximise the advantages and benefits of Blue Bonds for the management of global water resources.

Blue Bonds are critical to water infrastructure resiliency in areas facing climate-water imbalances, increasing cities, and degrading infrastructure. Blue bond-driven investments facilitate the update and decentralisation of the water system, ultimately establishing strong treatment facilities, efficient distribution networks, and secure storage mechanisms that strengthen services to vulnerable populations.

Accelerating institutional innovation, Blue Bonds creates partnerships among public agencies, private companies, research organisations, and civil society groups. These measures enhance cooperation among research-oriented innovation networks, facilitating knowledge sharing and accelerating the adoption of new technologies. Blue Bond projects are also designed to involve green certification protocols and performance evaluation systems, thus strengthening investors' trust and introducing ESG criteria to the industrial sector.

The alignment of infrastructure financing with the ambition to attain sustainable development outcomes is embedded in Blue Bonds. By aligning technological solutions with environmentally conscious industrial processes and reinforcing infrastructure, Blue Bonds plays a crucial role in achieving SDG 9, which facilitates the inclusive and sustainable development of industrial practices. The more countries adopt Blue Bonds, the more these tools become decisive in building industrial landscapes centred around ecological stewardship, carbon reduction, and resilience.

The existing representation of Blue Bonds enables the radical transformation of industrial resource utilisation, which is necessary to achieve the goals of Sustainable Development Goal 12 (SDG 12), focusing on responsible consumption and production. At its core, this goal promotes ecological sustainability by decoupling economic growth from environmental degradation through robust resource management, the development of eco-friendly infrastructure, and the stimulation of green technologies.

Blue Bonds are built on the foundation of the circular economy to create targeted schemes for financing that empower industrial organisations to adopt recycling systems for water, transforming linear business models into sustainable cycles. Important industry sectors include manufacturing products from chemistry, textile production, and agriculture, which consume a significant amount of water and are typically processed once, after which they generate substantial waste that is often discharged into the Blue Bonds. The availability of Blue Bonds provides industries with a mechanism for enhancing their water treatment processes to achieve zero-liquid discharge levels and implementing resource recovery systems, thereby significantly reducing the amount of water used and waste produced.

Blue bonds play a crucial role in stimulating technological innovation and fostering industry cooperation, ultimately leading to more sustainable production patterns. For example, underfinanced by blue bonds, the textile industry has adopted green dyeing technologies accompanied by biological wastewater treatment, resulting in decreased chemical discharge and freshwater needs. With the help of blue bonds, investment has increased in precision irrigation methods, hydroponic farming, and waste resource transformation, maximising crop and nutrient efficiency in the agricultural sector.

In aggregate, the previous research findings suggest that Blue Bonds have a significant impact in promoting radical changes in industrial attitudes toward resource consumption. Blue Bonds support the adoption of circular economy strategies, drive technological innovation, and promote responsible behaviour in all sectors, playing a crucial role in the success of SDG 12, which promotes responsible consumption and production.

Water security and climate resilience are crucial focus areas in the global battle against climate change. The financial instrument known as a Blue Bond plays a crucial role in achieving the objectives stated in SDG 13, specifically in Climate Action, by emphasising the need to develop measures for climate adaptation and mitigation promptly. The increasing number of climate-related disasters, including droughts, floods, cyclones, and altered precipitation patterns, necessitates immediate action towards developing climate-resilient water infrastructure.

The direct function of Blue Bonds is to fund the development of adaptive and resilient water systems. Adaptive systems that absorb climatic shocks include flood-resistant water networks, climate-smart irrigation technologies, green infrastructure, and early warning systems. The funding provided by Blue Bonds reduces the exposure of settlements, humans, agricultural lands, and ecological systems to extreme weather events and the impacts of climate change.

Blue Bonds promote circular water systems, substantially reducing the use of limited freshwater resources, such as groundwater aquifers and surface reservoirs, which are under stress due to increasing global temperatures. Blue bonds facilitate circular water economies, accessible through wastewater reclamation, rainwater harvesting, and stormwater reuse, thereby preserving and protecting resources and mitigating climate-induced water shortages for communities. The approach combines water management at the systemic level to strengthen adaptability, particularly in areas of the global south that face both water scarcity and climate-related vulnerabilities.

According to International Finance Corporation [18], blue bonds deliver observable climate benefits through their targeted investments, which achieve better adaptation practices as well as mitigation effects for projects looking to support national climate plans under the Paris Agreement through Nationally Declared Commitments (NDCs) National carbon neutrality goals receive support when Blue Bond-financed projects implement measures to decrease water-related emissions such as methane in untreated sewage and water distribution energy usage. These instruments fund infrastructure, strengthening communities during climatic pressure by maintaining water access, thus preventing community conflicts and displacement and sustaining livelihoods.

Blue Bonds are directly involved in achieving progress on SDG 14- Life Below Water. Building oceanic and marine resources sustainability and preservation as their key goal forms the basis of Blue Bonds' association with the Life Below Water initiative. As the specialised role of financial tools with water environmental sustainability as their target, Blue Bonds steer money towards actions that improve ocean water quality, reduce pollution, and protect marine ecosystems to achieve this goal.

One of the primary connections between Blue Bonds and SDG 14 is that Blue Bonds facilitate the implementation of measures to prevent the discharge of pollutants of terrestrial origin into marine environments. Blue bond investments drive improvements in sewage management, involve waste-free industries, and encourage the use of natural barriers, such as wetlands and mangroves, to prevent excessive farm and urban runoff from affecting water ecosystems. Intercepting sources minimises the chemical, biological, and plastic pollutants released into coastal and marine environments.

With the aid of Blue Bonds, organisations can construct watershed and river basin management systems that demonstrate the effect of upstream land use on marine environments. All elements of the water cycle are addressed here in an environmentally friendly manner, with protective programs that safeguard farming methods, urban sanitation, and industrial water usage from undermining marine biodiversity. The financial aid provided by Blue Bonds predominantly focuses on protecting ecologically vulnerable regions, such as coral reefs, estuaries, or deltaic regions, that are at risk from nutrient accumulation, sedimentation, and industrial pollution.

Organisations can achieve distinct accomplishments by applying Blue Bond frameworks, thereby enhancing the preservation of marine biodiversity and the sustainable use of water resources. For example, in 2018, the Seychelles Blue Bond was launched, marking the first sovereign bond primarily targeting ocean conservation. It provided \$15 million for expanding marine protected areas, enhancing fisheries management, and promoting climate-resilient communities in coastal zones. In 2021, Belize implemented a debt-for-nature swap, which required annual expenditures of \$4 million in Belizean dollars for marine conservation and aimed to protect 30% of its ocean by 2026.

Blue bonds are a key and scalable financial system that promotes SDG 14 by advancing pollution elimination, establishing marine reserves, and linking economic development with environmental sustainability. Today's adoption trends in developing and developed economies suggest that these financial instruments may become the primary funding source for the blue economy, aligning with national development goals. The International Finance Corporation (IFC), International Capital Market Association (ICMA), United Nations Global Compact, United Nations Environment Program Finance Initiative (UNEPFI), and the Asian Development Bank (ADB) have issued a global practitioner's guide to aid the development of blue bonds through defined criteria and procedures for issuers and investors.

Blue Bonds provide the central backbone for the success of SDG 17, as capital-raising strength and the development of coordination between several partners are critical for Partnerships for the Goals. This overarching sustainability goal underscores the crucial need for business collaborations and consistent operational strategies across various areas to promote international sustainability. Through their integrated solutions, Blue Bonds link national governments, development banks, international financial organisations, private investors, non-governmental organisations, and civil society organisations as purpose-oriented financial mechanisms. The collaborative actions of participants continue from the inception and funding of Blue Bond programs to their operation and assessment, aiming to achieve tangible sustainability and social impacts in water-based efforts.

Publications of collaborative efforts between stakeholders in Blue Bonds reflect the guidance provided by the International Capital Market Association (ICMA), which offers best practice guidelines for the sustainable issuance of bonds, including those related to Blue Bonds. These recommendations emphasise the vital role of transparency, accountability, and stakeholder engagement in achieving positive environmental and social outcomes.

Moreover, through Blue Bonds, public concessional funding is complemented by institutional support, enabling effective collaboration that attracts private sector investment. Implementing this approach is crucial for scaling projects within regions with significant investment barriers, particularly in developing countries. A case in point is the Bahamas, which is considering the introduction of blue bonds to finance sustainable marine ecosystem projects, illustrating the impact of strong, established institutional systems on success.

Underpinning the Blue Bond framework are the transparent and shared metrics that support shared accountability and the development of a governance model that drives well-executed and flexible project implementation. Through the shared development of impact assessment tools, sustainability benchmarks, and performance-oriented bond structures, stakeholders observe blue bond projects as they contribute to the Sustainable Development Goals for which they were established. Such mechanisms strengthen results and motivate partners to collaborate more closely, establishing trust and promoting joint efforts.

Additionally, blue bonds strengthen South-South and North-South partnerships by facilitating the exchange of water-wise technologies, policy reforms, and sustainable financing methods. Such programs in Seychelles, Latin America, and Southeast Asia are indicative of how other regions can also achieve financial and environmental objectives. By sharing these practices, the organisations create a global knowledge base that promotes the overall embedding of sustainability in important financial and infrastructure aspects.

Cooperative governance and global solidarity are essential in achieving the objective of Sustainable Development Goal 17, which is to benefit the attainment of Blue Bonds. By collaborating, Blue Bonds increase the breadth and intensity of water sustainability initiatives and provide greater global capabilities for implementing several integrated SDGS, particularly focusing on climate action, industrial growth, and environmental conservation. The growth of funding and partnerships enables Blue Bonds to significantly enhance the pace and effectiveness of sustainable development measures.

Blue bonds help avoid climate change by supporting the restoration of ecosystems, the designation of marine protected areas (MPAs) and the creation of sustainable infrastructure. Blue bonds are used, according to Cinar et al. [19], for financing blue carbon ecosystems, which include seagrass meadows, salt marshes and mangrove forests. Over an extended period, they prevent atmospheric CO_2 from entering the air by storing it in the ground or plants. Restoration projects help animals and plants coexist, improve the water system, mitigate erosion, and reduce carbon in the atmosphere. By funding MPAs with blue bonds, carbon-rich places that are important for the climate are protected. Likewise, blue bond investments support the development of environmentally friendly infrastructure, including biodegradable anchors and new habitats for estuary animals, thereby protecting the area's ecology. Blue bonds support the achievement of climate targets worldwide by promoting both ecological and carbon finance measures.

Blue bonds utilise innovative financing mechanisms, such as mezzanine debt, to support long-term growth in sustainable infrastructure. According to Gonzalez-Ruiz et al. [20], mezzanine debt, positioned between equity and senior debt, plays a crucial role in ensuring the financial viability of water-related sustainable infrastructure systems (SIS). When mezzanine debt is part of a blue bond, it can be converted into equity for lenders if the project's operations and maintenance fulfil water sustainability requirements (e.g., those defined by the Climate Bonds Initiative). Thanks to this organisation, the risk of funding drops, while private companies are enticed through better financial results to act in an eco-friendly way. Evaluating the investment model suggests that when debt is converted to equity, the sponsors gain Financial Captured Value (FCV). When blue bonds utilise mezzanine debt and public-private partnerships (P3S), this type of funding provides strong support for both environmental and financial security.

2.1 SDG Connection: Challenges

Finance in any field is always associated with two key factors, i.e. risk and return. Similarly, with probable potential, there are specific challenges. Blue bonds, which typically follow the use of proceeds, are designated to repay investors through the general budget of the issuing Government, irrespective of the project's success and environmental impact. If the bonds struggle to achieve their targets, sovereign bonds may face financial strain in repaying investors through public funds. Moreover, investors involved in such projects typically prefer small-scale bonds, unlike the Seychelles blue bond, which was small-scale but had high transactional costs. One of the significant challenges faced is the well-established concern of Greenwashing, wherein due to a lack of regulations and data handling, the projects are named to be sustainable without having any visible or proven environmental impact, eventually harming the investors' trust and reputation, like the case of 'clean coal' in Chinese green bonds. Greenwashing is a significant concern that translates into false or misleading claims and deceptive techniques related to investments and the proceeds generated from such bonds and debt instruments.

2.2 SDG Connection: Gaps and Lack of Funding

With multiple SDGS linked with this concept out of the 17 Sustainable Development Goals the United Nations gave in 2015, SDG 14 specifically aligns with "conserve and sustainably use the oceans, seas and marine resources for sustainable development". However, these goals and the financing resources should align to achieve the objectives. According to the research [6], SDG 14, with multiple subcategories, is the least funded of the SDGs. Without monetary participation, timely success and fulfilment are at risk. This will require an infusion of monetary funds to the tune of $174.52 billion per year, and hence, raising awareness and attracting more investors to trust the initiative is important.

3 Literature Review

The financial sector has been an integral part of many industries and is now equally recognised as an integral part of the sustainability sector. Different financial instruments have been utilised and researched across various industries for diverse purposes. One of the important instruments that has been widely researched and brought into use is sustainable bonds, which are debt securities that function similarly to other bonds issued. According to Smith [21] financing sustainable initiatives is primarily achieved with the help of bond securities, as bonds generally offer less expensive capital than bank loans. Companies often raise capital by issuing stocks or bonds, which have several advantages over other financing options. Generally, bonds are considered safe, low-risk financial instruments with fixed returns and defined maturity, which makes them even more attractive to investors, such as blue bonds.

According to Pratsch's [22] research, the European Investment Bank initiated the Green Bond initiative in 2007. Green bonds are widely established, and with significant research and development, this financial instrument utilises proceeds specifically for environmentally friendly projects and climate change mitigation. According to reports published by the Climate Bond Initiative in 2018, the growing market of green bonds is valued at $389 billion, which has become a pioneer in introducing this innovative concept of bond proceeds, essentially stemming from the concept of blue bonds for ocean conservation and the protection of aquatic life. Looking at the increasing appetite for green bonds over the years, with approximately 173 billion dollars in proceeds issuance, making it a 31% YoY increase in 2019, the trust and awareness of sustainable financing increased. According to the research conducted by Flammer [23], the issuance of green bonds by most corporate and private sector firms is a sign of Corporate Social Responsibility, a culmination of peer pressure, and an impactful business strategy. The eco-friendly projects align themselves with the four principles of green bonds and comply with the relevant regulations.

With the same initiative and considering the lack of preservation of marine ecosystems and oceans, Blue Finance emerged, addressing the risks and returns associated with financing sustainable initiatives in line with sustainable development. This emerged in 2018 with the Seychelles Blue Bond case study—the first sovereign blue bond studied by Benzaken et al. [24]. Aligning with the goals and objectives of the United Nations, many corporate houses and private sector firms issue these bonds to transform and support the marine ecosystem.

The financial segment of this initiative has a specific approach that combines traditional ocean development projects, marine information, past research data, science and technology, and conservation strategies with financial and capital instruments. The Seychelles blue bond was supported by the World Bank, providing the world with a new perspective on studying and implementing these financial instruments following a well-thought-out taxonomy. The essential step in issuing blue bonds is the framework established by the ICMA.

4 Case Studies on Blue Bonds

In this section, we analyse the need, function, methodology, and specific gaps in the blue bonds financing method, which are studied through different case studies.

4.1 Case Study 1: Seychelles: Sovereign Blue Bond—The First of Its Kind Initiative for Sustainable Fisheries

As part of the most crucial topics that came to the surface, the conservation of marine ecosystems and finding means to finance ocean development projects are key concerns. The concept of blue bonds emerged as a transformative initiative sparked by the issuance of the first-ever sovereign bond, the Seychelles Blue Bond. The following case study examines the key factors that drove the need for financing activities and projects related to the Ocean economy and its development. With the benefits associated, the World Bank also stepped in support of this initiative, making it an introductory gambit after Green Bonds.

Seychelles, a small nation with its prosperity directly linked to the marine and its surrounding aspects, had to actively seek ways to carry out projects and protect its natural resources. It is a collection of 115 islands, making it an exclusively marine economic zone, which means its economy relies heavily on marine resources, fisheries, and tourism. The Ministry of Fisheries and Blue Economy, with the support of The Nature Conservancy (TNC) and The World Bank, launched a blue financing method in the form of a Blue Bond in 2018, a means for conservation. Before issuing the Blue Bond, Seychelles had another unique way to finance its need to conserve the marine ecosystem with the help of a mechanism known as Debt for Nature Swap, which generated around $20 million in national debt, creating opportunities to fulfil the ongoing and upcoming projects, essentially focusing on the blue economy [24]. The same investment scheme, introduced later in 2018, was later extended into a blue bond for the next five years, making it a completely ethical, legally compliant, innovative, and first-of-its-kind financial tool that offered returns not just in the form of financial benefits but also in the form of sustainability. The pioneer of the blue bond was initially a debt-for-nature tool, which raised $15 million for international investors after the updated issuance. It was inherently managed by the Seychelles Conservation and Climate Adaptation Trust, popularly known as "Seyccat," and the Development Bank of Seychelles (DBS) [25]. This trust was established in 2015 to manage the bond's proceeds and diversify them into potentially high-impact projects. The interest rate at which this bond was issued was 4%.

The bond is currently divided, with only $3 million allocated to research and development. However, the maximum stake is allocated for fisheries, aquaculture preservation, and related projects. A significant portion of the Seychelles' blue economy is allocated to fisheries; hence, approximately $12 million is allocated for this purpose. Success of this bond initiative has its own merits and demerits from different standpoints, which have now been critically studied. Since this bond initiated the financial tool, the debt instrument cannot be a perfect resource for global application. However, Seyccat's formation was a significant step towards a non-biased implementation and an impartial allocation of funds, thereby gaining the trust and confidence of investors [9].

This reform opened doors for many countries to understand and implement financing methods, such as Blue Bonds, which focus particularly on funding the sustainable development of oceans and marine ecosystems.

4.2 *Case Study 2: Nordic Investment Bank's Blue Bond for Marine Protection*

Following the issuance of the first sovereign blue bond, a Nordic investment bank introduced its SEK 2 billion blue bond in January 2019 to protect and conserve the Baltic Sea. This initiative typically followed the footsteps of the Seychelles, and considering its objective and social benefits, the Nordic Investment Bank (NIB) issued this blue bond in the market. The Nordic Blue bond was the first debt financial tool used to protect and rehabilitate the marine ecosystem. According to the analysis done immediately after the blue bond's release, it successfully raised awareness. It helped create an avenue for investment houses to raise funds, which were used to protect the Baltic Sea. The Baltic Sea is heavily affected by eutrophication, which in turn harms the marine ecosystem. This bond played a crucial role in the treatment of the Baltic Sea, and due to its long history of green bonds since 2011, investors were optimistic about future returns.

This bond was issued under the NIB environmental bond framework and is intended for clean water projects, aquatic ecosystem conservation initiatives, and rehabilitation projects. In financial terms, NIB has an excellent credit rating, rated AAA, which ensures the financial stability of ongoing projects by channelling the use of proceeds. Following a successful oversubscription of investors in 2020, it launched its second SEK 1 billion bond, due in 2025. NIB has a structured and well-researched pricing approach, which, in the long term, increases the chances of marine sustainability projects being backed by investors and creates capital for other initiatives. On the other hand, Nordic blue bonds have attracted considerable attention to the SDG goals, especially SDG 6 and 14, which explore how private capital can be leveraged to create sustainable sources for protecting marine ecosystems and water resources [26].

4.3 *Case Study 3: Bank of China's Blue Bond for Marine Renewable Energy*

Financial tools predominantly gain popularity when risks and returns are analysed appropriately, and this turns out to be a win-win situation in terms of capital gains and, in this case, sustainability. China has been involved in sustainable bonds and financial tools for quite some time, which has also led to exploring the concept of Blue Bonds. As a country driven by numbers and data and characterized by a fast-paced environment, China issued Asia's first blue bond in December 2020 in association with the Bank of China. This was a significant move globally, as a commercial bank issued a bond. Blue bonds have started to gain considerable attention, and investors are increasingly embracing projects related to sustainability, particularly those focused on marine life and aquatic ecosystems.

This two-year blue bond was issued in the amount of CNY (Chinese Yuan) 3 billion, which was utilized for initiatives related to marine ecosystems and aquatic life. This highlighted their potential in finance and showcased their attentiveness towards handling the rising concerns related to the blue economy. A commercial bank has a lot more potential, and the reach is massive; thus, the Bank of China successfully brought many investors on board whose vision aligned with the initiative and had knowledge about the same. The bond attracted investors, with orders exercised up to CNY4.5 billion in Asia [27].

Apart from focusing on marine ecosystems and the sustainable blue economy, the eligible projects under the Bank of China (BOC) primarily included renewable energy projects, which aimed to foster marine renewable energy and wastewater treatment initiatives. This was a very alarming situation in the past. The process and methodology followed were like the framework used during the issuance of green bonds, as outlined by ICMR.

The primary analysis that can be drawn from this is that innovative financing tools have the power to fund projects that are not directly related to traditional formats but focus on ocean-based and environmentally friendly initiatives that will help preserve and protect natural resources.

Interestingly, an article published by the World Economic Forum [28] highlights some unique aspects, stating that oceans cannot be neglected. With increasing concerns worldwide, innovative concepts are attracting private investors, foreign institutional investors, commercial banks, asset managers, and insurance companies, who aim to increase awareness and support such initiatives. This success is expected to result in a 20% increase in ocean-based jobs and related livelihoods. Moreover, it is well established that the ocean economy is crucial in providing jobs and supporting countries that are heavily reliant on blue economy financing instruments. Taking advantage of this, China's blue potential is paying off, with the Gross Ocean Product reaching approximately CNY 9 trillion. The China Banking and Insurance Regulatory Commission is also optimistic about this trend in the future.

4.4 Case Study 4: Indonesia's Blue Bond Strategy for Coastal Resilience

Indonesia is a tropical country facing the tremendous effects of climate change. It comprises 17,504 islands with a vast coastline. One of the interesting aspects of its government is the concept of *Climate Budget Tagging*, which calculates and allocates funds required for sustainable projects and explores various ways to raise capital.

With an archipelagic geography, meaning the land is surrounded by sea and marine zones, the threat of various disasters, including floods, has a direct impact on its maritime economy. This posed a profound threat to the blue ecosystem, and thus, there was an urgent need to safeguard ocean resources and Sealife. One of Indonesia's highly successful initiatives regarding sustainable bonds was its Green Sukuk bond, which had a significant impact and served as an innovative financial mechanism in combating the hazardous effects of climate change that the country was experiencing. Indonesia's Ministry of Finance has always been proactive in discussions related to sustainable financing instruments. Thus, after realising the need to protect the marine ecosystem, the Government of Indonesia, in association with the Japanese debt capital market, issued a blue bond at JPY 20.7 bn in May 2023 with a maturity of 7 and 10 years [2]. It was the first time the bond had been issued publicly, and it perfectly aligned with the United Nations' Sustainable Development Goals. The Government's long-term mission was to innovate in the sustainable financing of oceans and coastal resources, with the primary aim of aligning with the SDGs and contributing to coastal protection. The country also experienced a rise in its SDG index ranking, reflecting advancements in developmental initiatives [29].

Although the Blue Bond has significantly contributed to funding initiatives to conserve marine biodiversity and strengthen its blue economy, it has faced numerous regulatory and legal challenges within the current framework. Analysing Indonesia's case study becomes important primarily through two aspects: firstly, the gap in current regulations, and secondly, the urgency for revised and well-applicable solutions to sustainable financing tools. Financial mechanisms are based on mandated legal compliance, and the Government of Indonesia has also suggested changes to the regulatory framework. It mentioned it as a legal vacuum with many gaps, while blue bonds are regulated or made public. Moreover, thematic bonds, such as sustainable bonds, raise funds for a particular cause, in this case, related to the environment. The basic convention of financial tools needs revision, which must be disclosed. Furthermore, the maritime economy encompasses a wide range of activities and managing them within a single framework is a technical challenge that necessitates a comprehensive legal framework. To address this issue, one recent initiative is the Blue Economy Act, which compares the regulatory frameworks of the Philippines and Indonesia, highlighting the legislative measures that should be considered for a revised and more comprehensive regulation.

4.5 Case Study 5: Blue Bonds Issued by Latin American Countries

Latin American sub-sovereigns and governments quickly adopted blue bonds—special debt instruments whose proceeds finance ocean- and water-themed climate adaptation costs—to raise private capital for coastal resilience, marine-ecosystem protection, and sanitation at a pace which has seen they account for close to half the global issuance since 2019; sovereign deals include Belize's USD 364 million Aa2 (political-risk insured with DFC), refinancing Superbond and paying for marine-protected areas comprising 30% of its EEZ, and Costa Rica's USD 500 million, 15-year Aa2/SDG-stepped bond that finances wastewater management and watershed improvement, both to ICMA/UN ocean standards and third-party verified by e.g., CICERO, EY; sub-sovereign and corporates—like Panama's USD 150 million IDAA-serviced, sewage-upgrades IDB Invest-guided issue and Puerto Rico's USD 200 million PRASA, backed by government grant and FEMA-granted, storm-resilience pipelines issue—are examples of good practice in the use-of-proceeds presentation, independent reportage of impacts, and credit enhancements that facilitate institutional investor coverage and oversubscription; mobilizing billions in the process, these instruments enhanced water access by hundreds of thousands, safeguarded millions of hectare of ocean habitat, improved SDG 6 and 14, created a replicable template—homogenized reportage, tight legal frameworks, multilateral engagement—for emerging market economies to upgrade blue finance scaling, with potentially up to a further USD 3 billion arriving in 2025 as more issuers emerge (e.g., Mexico, Ecuador) as well as venues [30].

4.6 Case Study 6: Fintech in Blue Bonds

Embedding AI and FinTech within blue-bond impact measurement squarely addresses the chronic transparency deficit, as only roughly two-thirds of blue-bond issuers currently disclose standardized impact metrics. By providing automated collection and analysis of data, this approach meets the increasingly investor-demanding requirements of ESG-conforming assets. Machine-learning models based on high-resolution satellite imagery can identify changes in coastal vegetation, water turbidity, and shoreline erosion, supporting near–retail–time monitoring of mangrove restoration with more than 90% accuracy and significantly reducing the need for expensive field surveys. Based on central-bank "Gaia" pilots, natural—language—processing systems robotically cull climate KPIS and use-of-proceeds reporting from unstructured local-language and annual reports, aligning data across jurisdictions and synchronising bond disclosure with ICMA's Green and Blue Bond Principles. Blockchain-powered FinTech registries create indelible records of every tranche of bond proceeds and vendor payments, eliminating the risk of misallocation and enabling investors to track funds directly to applications such as wastewater treatment improvements or reef rehabilitation. Concurrently, API-based environmental data feeds from IoT-capable water-quality sensors—publishing measures such as biological oxygen demand and contaminant concentrations—can be combined into investor dashboards to drive covenant compliance notifications and smart-contract disbursement triggers. Multilateral development banks, such as the IFC, ICMA, and ADB, now collaborate on consensus digital finance standards that propose interoperable APIs, AI model validation protocols, and on-chain verification to ensure SDG obligations in digital bond markets. Early pilots, such as development bank pilots integrating satellite AI, IoT sensors, and blockchain, have demonstrated up to 30% lower project delay rates and a 20% better reporting cadence than legacy monitoring approaches. By reducing monitoring expenses, improving data integrity, and increasing investor trust, these AI and FinTech technologies are expected to generate an additional USD 14 billion for ocean and water initiatives by 2030, thereby accelerating the achievement of SDG 6 (clean water) and SDG 14 (life below water).

Discussing the need for more specificity and clarity in the regulatory framework, Indonesia's case study examines the necessity of a framework. This will undoubtedly help to clear the legal vacuum and make the process more accessible and manageable for both issuers and investors. In May 2023, Indonesia issued the Sovereign Blue bond, yet it failed to adhere to specific critical legal mandates, especially for such instruments. Such cases bring considerable uncertainty regarding the authenticity and potential implementation of the bond [31].

Indonesia recognises this well, and it specifically highlights the urgency to organise the regulations and legalities involved in the process. This also includes the associated SDGs and the global standards. This case study revealed a significant issue of national interest that required prioritisation.

To enhance international cooperation, factors such as legal mandates, global standards requirements, SDG regulations, international agreements, national laws, and Government norms should all be considered to create more effectiveness and credibility for blue bonds.

Fintech has now become an integral part of the financial landscape. Hence, AI and blockchain help to revolutionise the financial sector, as well as promote innovative launches, such as blue bonds. Integrating such technologies into the system not only enhances awareness but also brings about significant efficiency and cleaner operations. All this together promotes the most significant factor of sustainability and global financial stability with reduced costs and interested investors on board.

Moreover, apart from conservation and sustainability, finance encompasses a multitude of elements that must be seamless and well-connected. This connection is demonstrated by the econometric models, which incorporate new cutting-edge technology, identify the risks involved, verify the model's correctness, and develop a framework for the volatility of blue bonds. Overall, such models help improve efficiency and give much more analytical and research-backed data, which is technology-driven and more feasible. Blockchain technology improves the collaboration of resources among all stakeholders, promoting the advancement of blue economies more effectively and efficiently while ensuring sustainability [32].

Furthermore, all analysis and quantitative data must be mapped appropriately to identify key gaps and challenges. Tools such as Cite Space can help visualise data and transform it into patterns and charts [33]. This highlights the unexplored areas, analyses the intricacies involved, and provides key data points in brief. The data becomes more manageable and easily accessible for researchers and policymakers, thereby facilitating and promoting sustainability initiatives. There is scope for bibliometric analysis on Blue Bonds, a powerful tool for research mapping and trend analysis [34].

5 Conclusion

Blue Bonds incorporates strong transparency and accountability mechanisms, requiring shared metrics and stakeholder collaboration to ensure effective implementation, adaptable governance, and continuous learning. Performance-linked instruments and sustainability assessment frameworks allow for precise tracking of progress toward SDG targets. These practices foster trust, deepen partnerships, and enhance stakeholder engagement.

Blue Bonds facilitate the exchange of water-smart technologies, policy innovations, and sustainable financing models, enabling South-South and North-South cooperation. Successful implementations in regions such as the Seychelles, Latin America, and Southeast Asia serve as replicable models for global adoption. This transfer of best practices builds a global knowledge ecosystem that integrates sustainability into the financial and infrastructure sectors.

The effectiveness of Blue Bonds hinges on cooperative governance and global solidarity, key pillars of SDG 17. Their focus on partnerships expands the reach of water sustainability initiatives while enhancing global capabilities in environmental conservation, industrial development, and climate resilience.

Beyond financing, blue bonds enable institutions to implement participatory governance systems and make data-driven policy decisions, thereby supporting environmental restoration and economic stability. As instruments of impact-based finance, they bridge the gap between ecological responsibility and economic progress. Blue bonds are a transformative tool in the global shift toward integrated, sustainable development.

Extending the dimensions of analysis, a future perspective about newer policies and comprehensive yet impactful assessments should also be evaluated. Blue finance is a relatively nuanced topic; hence, measured performances are significant for building the foundation for future studies and global cooperation [35]. A worldwide partnership promotes marine sustainability and measures the interdisciplinary collaboration between scholars and knowledge-based experts. This creates a meaningful transformation and development for which such blue finance initiatives are started. Effective and inclusive governance can promote long-term sustainability and safeguard marine ecosystems, serving as the cornerstone of successful blue finance efforts, according to Partelow et al. [36], provided that quantifiable results and interdisciplinary cooperation support it.

A blue bond is a relatively new financial tool, and as such, it presents numerous loopholes and challenges that need to be addressed. The above case studies highlight the potential environmental concerns and financial issues that arise when issuing sustainable bonds. One such issue is concerning the least funded SDG 14, which focuses on life underwater. Due to inadequate financial support, most projects and initiatives are halted, and these issues persist over time. Another major challenge is the lack of a proper framework and standardised guidelines, which increases the risk of greenwashing, and investors lack the trust to invest in such initiatives. Blue bonds alleviate the financial burden on the issuing bodies, provided the targets are met, and the returns are well-defined. Hence, the most significant challenge faced is regulating the system, where sustainability meets finance and targeted returns are generated with complete transparency, accountability, and growth.

References

1. ICMA Green Bond Principles (June 2021) https://www.icmagroup.org/assets/documents/sustainable-finance/2021-updates/green-bond-principles-june-2021-100621.pdf
2. OECD (2016) The ocean economy in 2030. OECD Publishing, Paris. https://doi.org/10.1787/9789264251724-en
3. Mathew J, Robertson C (2021) Shades of blue in financing: transforming the ocean economy with blue bonds. J Invest Compliance 22(3):243–247. https://doi.org/10.1108/JOIC-04-2021-0020
4. World Bank (2018, October 29) Sovereign blue bond issuance: frequently asked questions https://www.worldbank.org/en/news/feature/2018/10/29/sovereign-blue-bond-issuance-frequently-asked-questions
5. Bosmans P, de Mariz F (2023) The blue bond market: a catalyst for ocean and water financing. J Risk Financ Manage 16(3):184. https://doi.org/10.3390/jrfm16030184
6. Johansen DF, Vestvik RA (2020) The cost of saving our ocean: estimating the funding gap of sustainable development goal 14. Marine Policy 112:Article 103783. https://doi.org/10.1016/j.marpol.2019.103783
7. Shiiba N, Wu HH, Huang MC, Tanaka H (2022) How blue financing can sustain ocean conservation and development: a proposed conceptual framework for blue financing mechanism. Mar Policy 139:104575. https://doi.org/10.1016/j.marpol.2021.104575
8. March ALA, Failler P, Bennett M (2023a) Challenges when designing blue bond financing. ICES J Marine Sci. 80(8):2244–2251. https://doi.org/10.1093/icesjms/fsac238
9. March A, Failler P, Bennett M (2023b) Caribbean fishery and aquaculture financing needs in the blue economy: identifying opportunities & constraints in Barbados, Grenada, and St. Vincent and Grenadines. J Sustainability Res 5(1):1–37. https://doi.org/10.20900/jsr20230004
10. Konar M, Ding H (2023) A sustainable ocean economy for 2050: approximating its benefits and costs. In Lubchenco J, Haugan PM (eds), The blue compendium. Springer, pp 681–714. https://doi.org/10.1007/978-3-031-16277-0_18
11. Responsible Investor & Credit Suisse (2020, March) Investors and the Blue Economy. Responsible Investor Research, pp 26–27. https://www.scribd.com/document/464523877/BlueEconomy-report-Final
12. Aruga K (2024) Comparing the willingness to invest returns among green, sustainability, and blue bonds. J Environm Manage 364:121458. https://doi.org/10.1016/j.jenvman.2024.121458
13. Zheng Z (2024) The Impact of ESG Report Transparency on Investor Behavior. Advances in Economics, Management and Political Sciences 129:151–156 https://doi.org/10.54254/2754-1169/2024.18396
14. Thompson D, Ngoma A (2022) Blue bonds for marine conservation and a sustainable ocean economy. Mar Policy 144:105219. https://doi.org/10.1016/j.marpol.2022.105219
15. Otamendi U, Maiza M, Olaizola IG, Sierra B, Flores M, Quartulli M (2024) Integrated water resource management in the Segura hydrographic basin: an artificial intelligence approach. J Environ Manage 370:Article 122526. https://doi.org/10.1016/j.jenvman.2024.122526
16. García-Herrero L, Lavrnić S, Guerrieri V, Toscano A, Milani M, Cirelli GL, Vittuari M (2023) Cost–benefit of green infrastructures for water management: a sustainability assessment of full-scale constructed wetlands in Northern and Southern Italy. arXiv. https://doi.org/10.48550/arXiv.2305.06284
17. Karkou E, Papadopoulos A, Zafeirakopoulos I (2024) Industrial circular water uses practices. J Environ Manage 370:122596. https://doi.org/10.1016/j.jenvman.2024.122596

18. International Finance Corporation, International Capital Market Association, United Nations Global Compact, United Nations Environment Programme Finance Initiative, & Asian Development Bank (2023, September) Bonds to finance the sustainable blue economy: A practitioner's guide. https://doi.org/10.22617/TCS230328-2
19. Cinar M, Hilmi N, Arruda G, Elsler L, Safa A, van de Water JAJM (2024) Blue carbon as a nature-based mitigation solution in temperate zones. Sustainability 16(17):7446. https://doi.org/10.3390/su16177446
20. Gonzalez-Ruiz JD, Arboleda A, Botero S, Rojo J (2019) Investment valuation model for sustainable infrastructure systems. Eng Constr Architectural Manage 26(5):850–884. https://doi.org/10.1108/ECAM-03-2018-0095
21. Smith L (2023, October 31) Why companies issue bonds. Investopedia. Retrieved July 9, 2025, from https://www.investopedia.com/articles/investing/062813/why-companies-issue-bonds.asp
22. Pratsch M (2020, February 11) The 2020s—The decade of sustainable bonds. Environmental Finance. Retrieved from https://www.environmental-finance.com/content/the-green-bond-hub/the-2020s-the-decade-of-sustainable-bonds.html
23. Flammer C (2021) Corporate green bonds. J Fin Econ 142(3):499–516. https://doi.org/10.1016/j.jfineco.2021.01.010
24. Benzaken D, Adam JP, Virdin J, Voyer M (2024) From concept to practice: financing sustainable blue economy in small island developing states, lessons learnt from the Seychelles experience. Marine Policy 163:Article 106072. https://doi.org/10.1016/j.marpol.2024.106072
25. Silver JJ, Campbell LM (2018) Conservation, development and the blue frontier: The Republic of Seychelles' debt restructuring for marine conservation and climate adaptation program. Int Soc Sci J 68(229–230):241–256. https://doi.org/10.1111/issj.12156
26. Nordic Investment Bank (2019) NIB launches the inaugural 5-year SEK 2 billion Nordic–Baltic Blue Bond. https://www.nib.int/files/7bb314a4d17f64324262d6fb884d2f90398009ea/9207-2019-nib-seb-joint-press-release.pdf
27. HKEX (2020) Asia's first blue bonds to finance marine-related projects. https://www.hkex.com.hk/-/media/65E1863B880245918950C0A9EBBA3A83.ashx
28. Andersen TT, Drew M (2025, June 3) Why 2025 can be a breakthrough year for the blue economy. World Economic Forum. https://www.weforum.org/stories/2025/01/regenerative-blue-economy/
29. Sachs JD, Lafortune G, Fuller G (eds) (2025) The sustainable development report 2025. Sustainable Development Solutions Network. https://sdgtransformationcenter.org/reports/sustainable-development-report-2025
30. Latinex. (2023) Guide for issuers: thematic and sustainable bonds. Latin American Stock Exchange. https://www.latinexbolsa.com/documents/223/Guia_para_la_emision_de_instrumentos_financieros.pdf
31. Endarto B, Taufiqurrahman, Mardiana F (2024) The urgent need for blue bond regulation in Indonesia. Juridical Tribune-Rev Comp Int Law 14(4):530–546. https://www.tribunajuridica.eu/arhiva/y14v4/1.pdf
32. Yunas AB, Rumanto Y, Albart N (2025) Blockchain and Green Sukuk Integration for Coastal Community Empowerment in Sustainable Blue Economy. Research Horizon 5(2):415–426. https://doi.org/10.54518/rh.5.2.2025
33. Shi Y, Liu X (2019) Research on the literature of green building based on the web of science: a scientometric analysis in CiteSpace (2002–2018). Sustainability 11(13):3716. https://doi.org/10.3390/su11133716
34. Donthu N, Kumar S, Mukherjee D, Pandey N, Lim WM (2021) How to conduct a bibliometric analysis: an overview and guidelines. J Bus Res 133:285–296. https://doi.org/10.1016/j.jbusres.2021.05.016
35. Claudet J, Bopp L, Cheung WWL, Devillers R, Escobar-Briones E, Haugan P, Heymans JJ, Masson-Delmotte V, Matz-Lück N, Miloslavich P (2020) A roadmap for using the UN decade of ocean science for sustainable development in support of science, policy, and action. One Earth 2(1):34–42. https://doi.org/10.1016/j.oneear.2019.10.012

36. Partelow S, Gatto M, Chu E, Cotta B, Hattam C (2023) Aquaculture governance: five engagement arenas for sustainability transformation. Front Marine Sci 10:1178453. https://doi.org/10.3389/fmars.2023.1178453

Water: The Ethereal Potion of Existence and Spirituality

Suheel Rasool Mir and Arshad Bhat

Abstract The water has the quintessential significance in the cultural and faith driven systems thus forming the core of socio-cultural interactions through antiquity. From the Phoenicians to Greeks and all the way to current times, the role of water in having the cohesive and centrifugal effect on the meandering course of social patterns is vivid. Water carries an associative meaning of sacredness from the diverse religious perspectives. The divine and the sacred nature of water define the symbol of religious beliefs, sentiments and emotionality it brings forth. While using the concepts like sacred and profane borrowed from Durkheiman sociology this article looks; how faith can make ordinary things sacred and gives a sense of awe feeling and solidarity among different religious groups.

Keywords Water · Spirituality · Sacredness · Sociological imagination

1 Introduction

Water, so frequently believed to be an earthly need, has been held as a life-sustaining sacred force in many societies historically. Aside from its function as a sustainer of life, water was held to possess consciousness and memory in ancient societies. In modern science today, this ancient wisdom is brought alive in the research of Emoto [8], wherein water crystals expressed diverse forms based on words, music, and human emotion shared with them. These experiments show us that water isn't passive at all-it's responsive, a carrier of fine information, alive with the timeless Indigenous knowing that water contains in its being the memory of the Earth itself. Water as a threshold of worlds existed throughout religious rituals across many civilizations. In Celtic society, sacred wells and springs were thin places-liminal sites-where the

S. R. Mir
Government Degree College Tulail, Bandipora, Jammu and Kashmir, India

A. Bhat (✉)
Amity Institute of Liberal Arts, Amity University Mumbai, Maharashtra, India
e-mail: bhatarshad09@gmail.com

V. Kandpal and T. Choudhury (eds.), *Blue Revolution: Sustainable Water Use and Circular Economy Innovations*, Urban Sustainability,
https://doi.org/10.1007/978-981-95-1916-3_15

boundary between human and spiritual worlds was weak [17, 9]. Pilgrims made pilgrimages to these waters to gain visions, healing, and spirit contact. Likewise, in ancient Egyptian ritual, priests submerged deity images into sacred lakes to symbolize renewal and the cyclical death-rebirth theme so dominant in their cosmology [19]. Taoist philosophy made yet another profound observation, one seeing water as a spiritual symbol. Taoists observed water's yielding and unyielding nature exemplifying the greatest spiritual path-selfless, yielding, yet irresistibly powerful [16]. Taoist philosophy drew inspiration from such concepts as wu wei (effortless action), wherein it said that by being like water, one was harmonious with the Tao, or the general stream of existence. Also, in Himalayan traditions, sacred places like Mansarovar are not only pilgrims' places, but are stated to purify accumulated karmas of a few thousand lives. Pilgrims bathe in the water with a belief that it will destroy cycles of suffering, indicating the role of water as perceived by human beings to cleanse not only physical impurities, but impurities of the soul as well. Interestingly, contemporary scientists have studied "structured water" or "exclusion zone water," a condition of water with molecular arrangements different from regular H_2O. Naturally found within human cells and unspoiled nature, this water could have improved energetic qualities [20]. This indicates that ancient rituals involving sacred or "living" water may have tapped into actual, albeit subtle, physical processes yet unidentified today.

Water has a symbolic meaning of life and vitality; however, it is nuanced in its scope of meaning. Water has the quintessential significance in the cultural and faith-driven systems, thus forming the core of socio-cultural interactions throughout antiquity. From the Phoenicians to Greeks and all the way to current times, the role of water in having a cohesive and centrifugal effect on the meandering course of social patterns is vivid. In the mythological interpretation and religious scriptures, the cycle of life is supposed to start from water; simultaneously, it is concomitant with origin and fertility. In its functional approach, water is related to purification and cleansing with characteristics of transformation, renewal, and regeneration. From the classical four elements–earth, fire, air, and water- it creates the completeness of this universal consistency that was applied in various domains of knowledge. Water is considered as one of the essentialities of human existence with enormous social signifince. Morehouse highlighted the idea of water as a deep metaphor without which the human life cycle is impossible [9]. From the literary sense, water can be used to replicate the unconscious human state or the psychological emotional state of mind, the scene of stable water, as in a lake, can represent the emotional stability of mind and complacency. On the other side, the storm and torrential water convey an understanding of a disturbed and unsettled mind. However, the stable dirty water may mean an impediment or death, while the stirring water can mean the ongoingness of life. Water can designate satiation of thirst, desires, craving, and fulfilment of needs. Unquenched thirst on the other side can mean lacking fulfilment or having incompleteness in life, marking impermanence and uncertainty.

Water carries an associative meaning of sacredness from the diverse religious perspectives. Water in the Christian religion is an imperative thing used for baptizing a child. Furthermore, the event of Jesus baptism is correlated with water for the purpose of purification. In Buddhism, water is metaphorically allied to spiritual purification and cleansing. Buddhist pours water on Statues to signify the respect and adoration in their rituals. In Sikhism, the Amrit Sarovar (Pool of Nectar) water body located at the Harmandir Sahib called Golden Temple in India is believed to have healing properties. Apart from that, the Indigenous religion of Japan is Shintoism, the followers of Shintoism regard water bodies as having spiritual quality, Suijin- the god of water is revered by Japanese. The rivers, lakes, and springs in Japan are adored for being sacred sites. They are called to be “Shintai”- locations considered to be inhabited by Kami- spirits or deities [14]. In Jainism, water is used to carry various religious rituals, and they practise routinely ritual ablutions called “abhisheka”. Additionally, Muslim religion also revere water for multiple spiritual connotations, the daily practise of five times prayer -a day ‘Salah’ is only possible after ablution by water called ‘Wadu’. It also connects a deeper meaning of connection to God only after purifying yourself from sins or abstaining from sins. Apart from that many American native tribes venerate water and water bodies cherishing them for spiritual connections.

2 Water the Spiritual Potion for Diverse Religions

Water is of profound and universal significance in the world’s prevailing religious societies, symbolizing purity, life, renewal, and divine authority. In Christian tradition, water is an integral part of the sacrament of baptism, symbolizing purification of sin and birth into the new life in Christ [22]. Jesus’ baptism by John the Baptist in the River Jordan is employed again to stress the sanctity of water in relation to initiation by God and grace. Water is purifier and giver of life, a center of daily ceremonies and grand rituals, to Hinduism. Rivers like the Ganges are worshipped as goddesses, personifying the divine feminine and giving spiritual liberation (moksha) to pilgrims who bathe in their waters [4]. Holy water bodies, like the Ganges or the lake of Mansarovar at Mount Kailash, are believed to cleanse not only physical impurities but also karmic loans accumulated over centuries. Islam also places high spiritual importance on water. Purification by means of wudu (ablution) before prayer is necessary, emphasizing the need to come before God clean in body and soul [18]. The Zamzam well in Mecca is especially sacred, considered a miraculous well that was bestowed upon Hagar and her son Ishmael. Pilgrims employ water from this well during Hajj as a reminder and benediction of God’s providence [2]. Water is applied in rituals as an offering to Buddha images as a symbol of devotion and to cultivate virtues such as purity and detachment [12]. Buddhists go on pilgrimages to holy lakes, like Tibet’s Manasarovar, anticipating bathing or even to touch the water to provide higher spiritual merit.

Both native and animistic cultures similarly treat water with reverence as a living being. Lakes, rivers, and springs are sacred beings to be respected and protected by numerous Native American societies. Water ceremonies strive to respect the life-giving capability of water and to maintain nature in harmony [3]. The role of water in religious rites is deeper and more intricate than as a physical manifestation of purity,it is an overlap of the physical and the sacred. Water comes in Judaism as the central feature of the mikveh, a ritual bath for cleansing from the spiritual. Submersion in the mikveh is necessary for conversion, purification following some life events, and preparation for sacred events [10]. The action is a symbolic rebirth, meaning a new beginning or an enhanced state of spiritual readiness. Water is also holy in Shinto, Japan's indigenous faith. Misogi, a ritual purification in which the practitioners cleanse their bodies by bathing in natural bodies of water like rivers or waterfalls in a bid to purify themselves of spiritual filth and achieve inner balance, is part of Shinto ceremonies [15]. Waterfalls like Nachi Falls are sacred places that are considered sacred as being the sites of kami (sacred spirits) in nature.

Christianity's reverence for water goes far beyond baptism to its frequent symbolic manifestation in the Bible. Jesus calls Himself "the living water" (John 4:14), highlighting water's symbolic association with God's life and nourishment. Equally, in the Book of Revelation, there is a "river of the water of life" issuing from the throne of God, symbolizing eternal life and healing for nations (Revelation 22:1–2, [22]).

Hinduism's association with water is closely tied to cosmology. Holy rivers such as the Yamuna, Saraswati, and most significantly the Ganga are not geographical features but divine mothers who feed the body and the soul. Immersion in the Ganges during festivals such as Kumbh Mela is said to cleanse sins and result in liberation [4].

In Islam, after ritual purification (wudu), is ghusl, the purification of the body after certain events like marital relations or menstruation. Water is mentioned in the Qur'an as the origin of all living creatures: "We made from water every living thing" (Qur'an 21:30, [1], further indicating its divine source and necessity for life.

Buddhist reverence for water further enters sacred geography and temple architecture. The majority of temples have been constructed by watersides, illustrating the flow of the Dharma and the unbroken wheel of death and rebirth and existence [12]. Bows of pure water in bowls representing practice of developing perfection in generosity are common on Tibetan shrines, where offerings are given daily.

In Australian, American Indigenous, and African cultures, water gods or spirits are often worshipped. In the West African Yoruba tradition, Oshun, the goddess who rules rivers, love, and fertility, is worshipped. Holy lakes and rivers are usually thought of as being occupied by ancestral spirits, and ceremonies along these bodies of water are thought of as holding the community together and bringing fertility and wealth [6].

In *The Elementary Forms of the Religious Life* (1912), the French sociologist Emile Durkheim categorizes social facts into sacred and profane. Water being a basic element of human existence has two connotations sacred and profane here in connection with water bodies of diverse religious affiliations. The water which is commonly available comes under the domain of profane; the water having emotional and sacred symbolism attached comes under the domain of sacredness [7].

2.1 The Zam Zam Spring

In the same course of ideas, *Zam Zam* spring has been highly revered by Muslims around the world thus amplifying its scope as the sociological symbolism and iconography. The divine and the sacred nature of this spring define the symbol of religious beliefs, sentiments and emotionality water brings forth.

From the scriptural evidence and more historical references, the impact and relevance of this 'place' has never dimmed throughout the centuries [11]. Zamzam water comes from an underground spring in Makkah, Saudi Arabia. It is known to contain large amounts of calcium, fluoride, and magnesium, necessary nutrients for good health. The genesis of the *Zam Zam* water is given in the Quran [2].–the religious scripture of Muslims. Quran has mentioned the story of the Prophet Abraham (Ibrahim), his wife Hagar (Hajrah) and their baby son, Ishmael. According to Islamic sources, Hagar along with his son Ishmael were in barren valley of Mecca where they were left by Prophet Ibrahim to follow the commandments of Allah as a sacrifice to please his Almighty. There was no water and vegetation in the valley and Hagar was desperately searching for water to quench the thirst of his baby son, Ishmael [11].

Hagar ran overs the two hills of Safa and Marwa to search for water undauntedly, which was located close to Kaaba in Makkah. The desperate mother ran over Safa to see water and then went down and did the same on Marwa, she repeated her acts of running over hills for seven times. Her perseverance and hope that caused her to struggle pleased Almighty and consequently, miracle occurred, and water was seen gushing under the feets of baby boy-Ishmael. In the Quran, in Surah Al-Baqarah (2:158), mentions about Safa and Marwah, "Surely, al Safa and al-Marwah are the symbols of Allah. Hence, whoever performs Hajj (Full Pilgrimage) to the House (of Allah) or makes 'Umrah (Minor Pilgrimage), will find that it is no sin for him to ambulate between the two" (Islamic Studies). Today, Muslims from all over the world visit the sacred place, and while performing Hajj or Umrah- the Muslim pilgrims repeat the same act of running over Safa and Marwa, repeatedly in memory of Hagar and in pursuit to please Almighty. Nonetheless, the incident and water which is revered for its miraculous and spiritual existence also carry a loaded meaning of being hopeful and suggest the purpose for toil and labor amidst challenging circumstances of life. In contrary to existential crisis of questioning one's purpose and meaning, it provides an alternative ideology of God being omnipotent and all pervasive looking and listening. The profound message from divinity that embraces the perfect fairness over his creatures for which their deeds and perseverance counts and who would be rewarded by his unfathomable wisdom that would grant them impeccable rewards beyond the measure of this world.

This adds to socio-cultural and religious significance of this place among Muslims and its sacredness. Large number of Muslims visits this spring to drink the spring's sacred water having the spiritual healing effects. Since the origin of the spring and until now, its waters have never dried up. This water illuminate's faith among Muslims because the same water saved the son of Prophet Abraham and his wife Hagar, hence containing the sacredness and emotionality.

Muslims who perform Hajj pilgrimage always visit this spring to drink *zam zam* water which they believe will detoxify the impurities (both within and outside the body). After returning back to their native places, the pilgrims take some bottles filled from the spring as blessing from the pilgrimage itself. Every visitor is greeted by this *returning*home when people come out to. Then the person who drinks from the same water faces the direction of Kabba (Holy place among Muslims) and blesses himself with the same water with devotion and high reverence. Such a spiritual perspective on this water- not only maximizes the value and meaning but adds social value as well. Furthermore, the sacredness attached with this water improves the essence of Haj pilgrimage and the spring not only provides spiritual quench to believers but also becomes the symbol of unity and oneness where all the believers across the world unite to drink it and enhancing its value as water being the globalized element as well. There is a strange mix of reactions and surprise when Muslims come across the water from this spring making the symbols, emotions and sacred a unique triad of social and globalised existence. The feelings and affective spells that are accrued on consuming this water make its waters more sustained spiritual symbol.

Deep ecology is both a school of philosophical thought and movement that emerged during the 1970s and owes a great deal to Norwegian philosopher Arne Naess. Deep ecology simply demands an eccentric worldview in which the inherent value of all living organisms and ecosystems is valued irrespective of their utility to human beings. This philosophy stands contrary to the widely held anthropocentric perspective that places human beings at the forefront of environmental concerns. Nature holds value in deep ecology, and planetary health is the same as that of all its inhabitants. Deep ecology is a call for action and philosophy. It has become a movement that demands drastic transformations in human behaviour, looking towards a sustainable and respectable human-nature relationship. It demands a halt to the devastation of nature and rather working on the root causes of nature destruction, not just treating the symptoms. It includes working towards the preservation of biodiversity, reducing human consumption, and implementing policies that give higher importance to ecological health than economic growth.

This divine and spiritual uniqueness of this water always evinces the positive emotions with respect to religion and Ummah. In non-Arab Muslim societies this water is not used only for the common purposes but on certain occasions like illness and the sacred functions. This forms one of the important elements of the intangible identity of Muslims. Prior to drinking it, Muslims recites some sacred verses. The sacred and emotional bounding of Muslims with *zam zam* forms the base of our understanding of how faith and water shapes our everyday lives. Whenever the word Haji comes into the mind, the idea of this spring repeatedly comes along with the spiritual feelings and the emotional sensibilities. And the same spiritual and emotional feelings maintain the group cooperation and cohesion across a broad spectrum of Muslims.

The water we have been studying and understating only in terms of geomorphological and geographical parlance certainly has affective and cultural symbolism inherent and embedded in its character. The springs from Himalayas to the *zam zam* water itself enhances the scope and importance of water being an essential parameter when analysed and studied in socio-spatial configurations. The springs have rather marked sanctity and iconographic status attached to them when it is tallied with the place sensibilities.

2.2 Jordan River

The Jordan river that gets hundreds of mentions in Christian theological texts as in Testaments, and Biblical texts, narrates the story of John the Baptist baptised Jesus in the river of Jordan, "the prophets Elijah and Elisha also crossed the river dry-shod; and the Syrian general Naaman was healed of leprosy after washing in the Jordan at Elisha's direction" (Holy Land). Jordan River is esteemed by community of Jews significantly as they believe Joshua along with Israeli tribes crossed the river to enter the Promised Land after years of boondoggle roaming in deserts. The Jordan River is notable in Jewish history for having the ability to heal and cure in addition to purifying sins. General Naaman of Syria, who was expected to contract leprosy as retribution for his transgression:

Naaman arrives like Pharaoh "with his horses and chariots" yet halts at the threshold of Elisha's home like a supplicant (2 Kings5:9). Elisha does not go out to meet him, sending instead a messenger with the instruction, "Go and bathe in the Jordan seven times and your flesh will be restored, and you will be purified" (5:10). The servant girl, the king of Aram, and Naaman all use the phrase "to be cured of leprosy ...Elisha employs the language of purity: "wash... and you will be clean" [13].

For the community of Jews, the Jordan River is supposed to possess the power of healing as well as the power of spiritual salvation and purification of the soul:

> A convinced Naaman goes down to the Jordan, then immerses in the river seven times... As Naaman is not only healed and purified but also restored to a youthful state, the little stream in the vanquished land of Israel reveals an unsuspected potency. [13]

As an outcome, Naaman realises the power and potency of the Jews' God -Yahweh, and due to this reason, Jordon becomes holy for Jews. The consumption of Naaman's story of miraculous healing and rejuvenation in initial Christianity has influenced the concept of baptism effects on the body, "the connection between the Naaman story and the baptism ritual becomes apparent through the translation of the Hebrew verb "to dip"...into the Greek ...(baptise)" [13]. Christians adore the Jordan River as it symbolises to them after life at baptism that separates them from the present world, in addition to distinguishing them from Jews:

That Jesus goes down into the river rather than crossing it introduces the reader to the premises that the horizontal plane where political contest transpires is but another arena of the struggle between God and Satan that transpires along a vertical axis. This becomes all the clearer when the spirit that settles on Jesus at baptism drives him to the wilderness to contend with Satan...baptism influenced Christian baptism or the Christian rite required a story of Jesus immersion as a charter, the scene both mirrors and inverts the prophetic rites of passage at the Jordon [13].

The Jordan River has been a consistent source of attraction over the centuries for its historical and religious significance, which attracts a good number of Christian pilgrims from across the world. The water of Jordan is considered holy, and baptism is performed there even today by a large number of Christians. Baptism is the central ritual for a new-born Christian member, as it is this act that qualifies the man to be a follower of the Christian religion:

The Gospels establishes a transition for the baptized from a state of impurity to one of purity while transforming the meaning of "impure" and "pure". "Pure primarily indicates a state of moral righteousness that can be achieved through a combination of repentance and immersion. This radical recontextualization of priestly ablutions that dissolves states of physical impurity marks an important stage in the transformation of the concept of purity. The association of impurity with sin [13].

The immersion represents rebirth as well as purification, cleaning, acknowledgment, and rootedness to God.

2.3 Mansarovar Lake

Mansarovar Lake is known for astounding beauty for its colour of deep blue scene, embraced by Mount Kailash. The ashes of Mohandas Karam Chand Gandhi were sprinkled in the lake water, who is Father of Nation in India. Additionally, it comes as a hallowed space among different religions as for Buddhist, Hindus, Jains, and for Bon faith holders. Mansarovar is considered the holy lake, from ancient Vedas it is said to be inhibited by *Vishnu Bhagwan* and *Maa Lakshimi* calling the lake by a different name *Sheesh Sagar*. The Tibetan Buddhist use the nomenclature of *Mansarovar* and *Mapham Yumtso*, both mean the 'invisible lake' or 'victorious lake'. The devotees relate this lake with Anotatta Lake, the place supposed to be bathing place of Queen Maya-mother of Gautama Buddha. Tibetans call it with other names, additional to above mentioned ones: mTsho, Mapham, mTsho-Ma-dros-Pa and others to regard the place with reverence.

Mansarovar is the holy lake, as it was believed to be the bathing place for gods during Brahma Mahurta (the time period of worship, an hour before sunrise). According to Hindu mythology, the most celebrated myth is that the lake first developed in the mind of Brahma. Mansarovar could be broken into two words: 'Manas'- means 'mind' and 'intellect, and 'Sarovar' means 'lake'. The holy lake of Mansarovar is symbolic of the mind of the Lord, which is stable and calm. The Hindu religion maintains that immersing oneself in water and having some drops of holy water cleanse one from sins and purify one's soul. "According to the Hindu poet Kalidasa, drinking water of holy Mansarovar lake can clean the sins of hundreds of lifetimes" (Wonders of Tibet). Mythology says that Dev Parvati also used to take baths in this lake, and she was Lord Ganesha's creator. The lake of Mansarovar is known by another name, like the Lake of Compassion, because of that particular mythical connection. The Bon religion has its existence much before Buddhism religion in Tibet, they used to call Mansarovar as Matri Tso or Matri or Machui Co, and these names refer to Dragon king of Bon religion. The Bon religion claims its founder, Tonpa Sherab, took a bath in Mansarovar lake when he first arrived in Tibet. Jainism believes that the lake is connected to the first Tirthankara, Lord Rishabh Deva, who attained Nirvana (spiritual liberation). Buddhism considers Mansarovar to be a place where Gautam Buddha's mother conceived Buddha, and this place is related to Buddha as he learned the art of meditation. According to Sikhism, Mansarovar is the place where Guru Nanak Dev learned to meditate.

2.4 Ganga River

The river Ganga has stirred imagination in the minds of many poets and has a rich mythical history. The Ganga River has been the voracious inspiration and enchanted muse of interest for writers. From its origin from icy mountains of Himalayas to the Northern plains of Bay of Bengal, the river has multifaceted meaning and connotations. The mysterious Ganga signify divine mother bestowing salvation and also causing death, and destruction. The verse in sacred books quotes Ganga as:

> What need of expensive sacrifices,
> Or of difficult penances?
> Worship Ganga, asking for happiness.
> And good fortune,
> And she will bring you heaven.
> And salvation- Padma Purana V.60.39. (qtd. In Darian 1978, 18)

In another spiritual interpretation, Ganga River is given the feminine persona, the female goddess with triple facets that flows in three spaces: heaven, earth, and underworld. It is believed to flow from heaven where she is called by name of Mandakini, on the earth it is called Bhagirathi and below earth it is said to be Patala Ganga or the Bhogavati [4]. Ganga river water further expounds the meaning related to salvation and purgation apart from having dense historical and mythological spiritual connotations.

In the traditional beliefs of ancient India, the water of the Ganga had sanctity for over four thousand years from the beginning of civilizations in the Indus Valley. In Hinduism, Ganges water is used in various rituals for birth and death. In weddings, Ganga water serving has been an unchangeable custom. Ganga water is presented before deities in the morning and evening routines. In addition to that, in the Upanaya Ceremony (an equally meaningful ritual for Hindus as Baptism is for Christians), for some Hindus, the ritual of wearing sacred thread is followed by serving bread and Ganges water to that person. For Hindus, the sacred water of Ganga holds great significance even after death; the ashes of some cremated bodies are scattered in the sacred river of Ganga [4]. The most celebrated myth of the Ganga River is believed to be the celestial river that flowed from heaven, on the acceptance of years of austerity and sincere penance of king Sagara, whose 60,000 sons' ashes needed purification. The myth maintains that Lord Shiva controlled the movement of the Ganga River that flowed from heaven to earth, implying that the Ganga River flowed through the matted hair of Lord Shiva,otherwise, the Ganga River might have flooded the earth [4].

From the mythical and historical interpretations and connotations of Ganga water, it symbolises purification, rejuvenation, and salvation. Ganga is believed to possess the restorative and healing powers described in the religious scripture *Mahabharta.* In the great war of Vyasa, the slain armies from the depths of the river are called back by the author of the epic for rebirth, reconciliation, and regeneration [4]. Ganga is revered even today for its resurrection and regeneration power:

> After two thousand years, the goddess is still worshipped as a source of life and generation. Even today in Bihar, at the start of the ploughing season, before the seeds are sown, farmers put Ganga water in a pot and set it in a special place in the field to ensure good harvest. Among those who live along the river, a newly married woman unfolds her sari to Ganga and prays for children and the long life of her husband [4]

3 Conclusion

The sociological imagination attached to the character of water in the sacred sites represents a more pronounced ritualization of the process throughout. Water has long been portrayed as a representation of chastity, purity, dignity, and morality, both in traditional conceptions and literary folklore. In the German folktale known as *The Girl without Hands*, *The Helpless Maiden,* or *The Handless Maiden* (1812) tales collected by Brothers Grimm, which is about a young maiden, the young maiden's relationship to water serves as an allegory and a symbol to highlight her dedicated chastity. The folklore centres on the account of a young girl who belonged to a poor miller; according to the legend, the miller was inclined to give the devil custody of his chaste daughter. When the devil arrives to abduct the father's daughter, the young woman, who has been submissive, and pious, all along, washes herself with water. This causes the devil to grow enraged since he is unable to touch her, and he departs saying to the father, keep water away from your daughter. The devout maiden cries and her tears fall on her hands, making them clean, so the second time he tries to seize her, he is unsuccessful. On the third occasion, the devil comes to take her, but in fury, he orders the miller to chop off his daughter's hands. The young maiden again sobs bitterly and tears clean her stamps, which were initially her hands. Therefore, the devil fails to take her, as she becomes most pious and eventually is saved from getting seduced by him. Later, in the story, she loses hold of her baby son in water being handled; she cries and tries to catch hold of her baby in the stream, and her hands come back in water, implying the power of water to heal and regenerate [5]. Another German folktale- *The Water of Life* (1812), makes a similar point, in which a brave-hearted prince risks his life to obtain a cup of water from the water of life so that he might give it to his sick father, who heals after drinking it. The water of life becomes unattainable to his avaricious, crafty, and lustful brothers. As a result, he is successful in obtaining both the thrones of two Kingdoms and the hand of a lovely princess. In this way, water discourses have been portrayed as having ethically admirable qualities and spiritually enlightening traits: it is a spiritual purifier, associated with upright and pious people, sustains humanity and life, and is transparent, free from impurities like greed, sins, lust, jealousy, and inhumanity. The mainstream public consciousness and identity generally attributed to the water as more of a spatial and consumption entity largely shadows the more ontological and sociological character of the water. The present world which is full of greediness and gargantuan consumerism has resulted in commodification of environment and nature. However, the spring of *zam zam, river of Ganga, Jordan and Mansarovar Lake* in its essence and affectivity in the broad lexicons of religious spectrum as the holy 'spatial markers' inherits the more socio-religious basis of sacred. The sentiments and emotiveness in the wholeness of *'religious water'* then extends to the more complex interactions in formation of uniting 'whole'. While the use of religious water as a sacramental in profane spaces is infrequent, it attests to the beliefs of the significance of holy water. Water is ubiquitous, omnipotent, and restorative, and these meanings are derived from the water itself. From a sacred stance, this draws attention to the holy and holiness in the making, because these quintessential makings are not only what separates corporeal self from immortality in eternity, but they are also divine attributes and favours to humankind. It is noteworthy that water is abstracted and used as consecrated entity on similar planes within belief-systems transcending time space continuum. The cultural mores vary, but the practices often resemble analogous beliefs in demystifying the how and why of religious water. Water is a lively force and commonly accessible conduit for the disciples. For Winiwarter, *"the nexus of arrangements and practices linking humankind with Nature should be labelled 'socio-natural sites'"*-a term that considers the conceptualisation of water in its various conformations as *'active'* if not agent in daily human condition. Subject to beliefs, the divine powers might be narrow or all pervasive. The waters then may have particular and established influences of sin eradication (*like ablution or baptism*), or it may be believed to enable the religious duty. However, it can be noted the universal water available to all humans of diverse religions carry one mission of realisation that is purification or salvation of sins, inching closer mankind to universal values of *Truth, Beauty and Goodness*. Therefore, as these typologies of water are both characterized as 'holy,' they have customarily and devotedly definite ways of functioning, guiding the thoughtfulness to the constituting apparatuses at work.

From the same perspective, *zam zam, Ganga, Jordan,* and *Mansarovar* attains the topophiliac and sacred positionality from the place itself and then transporting the same into the different 'spaces' when the water is taken as spiritual souvenir form the water bodies. This augments the scope and essentiality of these water bodies as the locus of sacred and spatial embedding of water as more fundamental basis of study for understanding and deconstructing the unique social structure.

Moreover, the water conveniently available consumed in daily life can be a subject of profanity as it is supposed to become scarce. The scarcity of water in another way may mean that humans might lack spiritual potions and are likely to get swamped in animalistic behaviours. The apprehension of water scarcity exacerbates conflicts and tension between water sharers as undoubtedly trans-boundary water disputes occur due to unequal distribution of water. The persistent conflict over river Nile between upstream and downstream is one such example and the never-ending clash on Indus-river between India and Pakistan river sharing are some of authentic realities, where water consumption is more attached to profanic idea than to sacredness.

There is a need to revise and revisit the conventional, traditional sacred sources to regain the sanctity of water as being nectar of spirituality. Self-reflection on water is necessary for spiritual insights to fill the voids in our hearts and promote the flourishing growth of our moral principles. It is heartfelt prayer, more nuanced interpretations and adjustment befitting humanity and universal values will sprout out.

References

Ali AY (2004) The meaning of the Holy Qur'an (10th ed.). Amana Publications

Boshra A, Mariod A, Ali F, Abdalrhman E, Abbas S, Hassan A, Mahamedan M, Ela-Tta N, Kh H, Masaad A, Hamid A, Fadlalmola HA (2021) Composition, hydrology, and health benefits of Zamzam water, Archivo Historico Revista Bionatura 6(4):2326–2330. https://revistabionatura.com/files/2021.06.04.30.pdf

Cajete G (1994) Look to the mountain: AN ecology of Indigenous education. Kivaki Press

Darian SG (1978) The Ganges in myth and history. The University Press of Hawaii

Davidson J (2017) Dark night of the soul mythology: the Handless Maiden. Jessica Davidson. https://jessicadavidson.co.uk/2017/06/08/the-handless-maiden-complete-story/, 8 June 2017

Drewal HJ, Drewal MT (1983) Gelede: art and female power among the Yoruba. Indiana University Press

Durkheim É (1915) The elementary forms of the religious life: a study in religious sociology. G. Allen & Unwin; Macmillan

Emoto M (2004) The hidden messages in water. Atria Books

Euzen A, Morehouse B (2011) Special issue introduction: water: what values? Policy Soc 30(4):237–247. https://doi.org/10.1016/j.polsoc.2011.10.005

Fishbane M (2008) Sacred attunement: a Jewish theology. University of Chicago Press

Ford DB (1860) Scriptural evidence of the deity of Christ. Bibliotheca Sacra 17(67):67–82

Harvey P (2013) An introduction to Buddhism: teachings, history and practices (2nd ed.). Cambridge University Press

Havrelock R (2011) River Jordan: the mythology of a dividing line. The University of Chicago Press

Terry, Tokyo, (2023), Best Mythical Creatures of Japan: Water Edition. https://sakura.co/blog/best-mythical-creatures-of-japan-water-edition?srsltid=AfmBOopmGYnrXR5GRA4sBSu_gmK HTZKuWw2chMElKYz_mqS3xAdDu6mO

Kuroda T (1981) Shinto in the history of Japanese religion. J Jpn Stud 7(1):1–21. https://doi.org/10.2307/132163

Lao T (1988) Tao te ching (trans: Mitchell S). HarperPerennial

Matthews C (1994) The Celtic spirit: daily meditations for the turning year. HarperSanFrancisco

Nasr SH (2003) Islam, science, and the challenge of history. Oxford University Press

Pinch G (2002) Handbook of Egyptian mythology. ABC-CLIO

Pollack GH (2013) The fourth phase of water: beyond solid, liquid, and vapor. Ebner and Sons Publishers
Tafheem ul Quran: Surah 2 Al-Baqarah, Ayat 158–158 (2024) Islamic studies. Retrieved from https://islamicstudies.info/reference/reference.php?sura=2&verse=60-60
The Holy Bible, New International Version (2011) Zondervan

The manufacturer's authorised representative in the EU is Springer
Nature Customer Service Centre GmbH, Europaplatz 3, 69115 Heidelberg,
Germany. If you have any concerns regarding our products, please
contact ProductSafety@springernature.com

Printed and bound by CPI Group (UK) Ltd, Croydon, CR0 4YY
07/10/2026
02229349-0007